Studies in Fuzziness and Soft Computing

Volume 397

The series "Studies in Fuzziness and Soft Computing" contains publications on various topics in the area of soft computing, which include fuzzy sets, rough sets, neural networks, evolutionary computation, probabilistic and evidential reasoning, multi-valued logic, and related fields. The publications within "Studies in Fuzziness and Soft Computing" are primarily monographs and edited volumes. They cover significant recent developments in the field, both of a foundational and applicable character. An important feature of the series is its short publication time and world-wide distribution. This permits a rapid and broad dissemination of research results.

Indexed by ISI, DBLP and Ulrichs, SCOPUS, Zentralblatt Math, GeoRef, Current Mathematical Publications, IngentaConnect, MetaPress and Springerlink. The books of the series are submitted for indexing to Web of Science.

More information about this series at http://www.springer.com/series/2941

Tofigh Allahviranloo

Fuzzy Fractional Differential Operators and Equations

Fuzzy Fractional Differential Equations

Springer

Tofigh Allahviranloo
Faculty of Engineering and Natural Sciences
Bahcesehir University
Istanbul, Turkey

ISSN 1434-9922 ISSN 1860-0808 (electronic)
Studies in Fuzziness and Soft Computing
ISBN 978-3-030-51271-2 ISBN 978-3-030-51272-9 (eBook)
https://doi.org/10.1007/978-3-030-51272-9

This Springer imprint is published by the registered company Springer Nature Switzerland AG
The registered company address is: Gewerbestrasse 11, 6330 Cham, Switzerland

A thousand hours of my loneliness and suffering in the quarantine period is dedicated to Laala who took the responsibility of Sahand for success and progress and also made my academic life prosperous.

I Thank you.

Tofigh Allahviranloo

Preface

I tried to use my all knowledge and experiences during 22 years on the fuzzy subjects, to write this book. To this end, I studied many sources that all are in the sources section but the writings and contributions come from myself independently. The present book contains useful information for identifying fuzzy fractional differential equations. By studying this book, you can get acquainted with different types of vague information and learn how to use it efficiently in fractional systems. A detailed study of fuzzy fractional differential equations has been performed from a theoretical and numerical point of view. For this purpose, numerical and semi-analytical methods for solving these differential equations have been investigated. Besides, an analysis of the complete error of the solutions is also provided. Interesting applications of these systems in engineering and biology have also been reported.

Istanbul, Turkey — Tofigh Allahviranloo

Contents

Chapter 1
Introduction to Fuzzy Fractional Operators and Equations

1.1 Introduction

Uncertainty and hesitation always live with us and this is the reality. All human beings are accustomed to doubting of everything around them and asking themselves or others why? Because they are not sure and their information is not incomplete or accurate. Now imagine that with all this incomplete information, we are in an environment where there is confusion.

It is a fact that we do not know how to answer many of our real questions. This attitude of mind-this attitude of uncertainty-is very important for scientists. Instead of trying to combat uncertainty, our job is to figure out how to embrace it and work with it. Because there is uncertainty about the growth, facilities, and life you want. We need to make sure that uncertainty creates an environment for our further growth and learning.

Truth cannot be explained or evaluated simply by right or wrong, true or false. Sometimes it's almost right, or sometimes it's probably wrong. Most of the time, the answer is given by personal feelings, for example, "I hope it's true" or "I think". This is why mathematicians have to consider linguistic propositions in mathematical logic, and this is an entrance for uncertainty in mathematics. In fact, uncertainty has a history of human civilization, and humanity has long been concerned with controlling and exploiting this type of information.

For example, probability theory is one of the ambiguities. Tass gambling was the beginning of what is now called probability theory. In the sixteenth century, there was no way to quantify chance. If someone spun two and a half dice during the game, people would think it's just luck. This means that we can measure an event and know how lucky we are to be working.

One of the oldest and most obscure concepts has been the word "lucky". Gambling and dice have played an important role in the development of probability theory. In the fifteenth century, Gerolamo Cardano was one of the most well-known figures in a formal algebraic activity. In "Game of Chance," he presented his first

T. Allahviranloo, *Fuzzy Fractional Differential Operators and Equations*, Studies in Fuzziness and Soft Computing 397, https://doi.org/10.1007/978-3-030-51272-9_1

analysis of the rules of chance. In this century, he solved such numerical problems. In 1657, Christine Huygens wrote the first book on probability, "In Calculating Chance." This book was the real birth of probability. Probability theory was developed mathematically by Blaise Pascal and Pierre in the seventeenth century, which sought to solve mathematical problems in some gambling problems.

The concept of "predictive value" is now an essential part of economics and finance. By calculating the expected value of an investment, the value of each party can be understood. Since the seventeenth century, probability theory has been consistently developed and applied in various disciplines. Today it is very important in most fields of engineering and instrument management and even in medicine, ethics, law, and so on.

In classical logic, the values of "true" and "false" or "zero" and "one" are the values of a decision in binary logic. But with multi-valued logic, it will not work well with the hierarchy of truth. Another aspect of uncertainty appears in multi-value logic. Multi-value logic is a non-classical logic that is a special case of classical logic because it is the embodiment of the principle of the performance of truth. Multi-value logic uses the degree of its truth. It is very difficult to discuss the nature of "degrees of truth" or "values of truth".

As a practical example, in the literature of artificial intelligence and uncertain modeling, there is a long misconception about the role of fuzzy set theory and multi-value logic. The frequent question is that for the performance of a compound calculation and the validity of the rules there is an exception between mathematical meanings. This confusion, despite some early philosophical warnings, involves early developments in probable logic. Given this fact, three main points can be realized. First, it shows that the root of the disagreement lies in the unpleasant confusion between the hierarchy of beliefs and what logicians call "degrees of truth." The latter is usually complicity, while the former is not.

The growth of nature and the phenomena in it, all often do not follow the fixed law, and in this environment a material transfer model works with inaccurate information such as more or less… How do you think we can model this transfer? To do this, we need knowledge of transmission tools and transmission operators that can work with this inaccurate information. We call these types of operators fuzzy operators. Concerning this modeling, the fractional operators will be more useful and effective.

As far as we know, the different materials and processes in many applied sciences like electrical circuits, biology, bio-mechanics, electrochemistry, electromagnetic processes and, others are widely recognized to be well predicted by using fractional differential operators in accordance with their memory and hereditary properties. For the complex phenomena, the modeling and their results in diverse widespread fields of science and engineering, are also so complicated and for achieving the accurate method the only powerful tool is fractional calculus.

Indeed the fractional calculus is not only a very important and productive topic, it also represents a new point of view that how to construct and apply a certain type of non-local operators to real-world problems. Since the uncertainty in our real environment and data has an important role, this causes us to discuss the uncertainty in our mentioned topics.

In engineering and biological science, many subjects are modeled by fractals and fractional operators like fractional differential operators. The heat, wave, chaos, and other phenomena can be modeled by fractional differential equations and especially if the data and information are uncertain and fuzzy the fuzzy fractional operators are better candidates to model the problems.

Fractional differential computations are one of the branches of mathematical analysis that study and research the properties and applications of integrals and derivatives of arbitrary orders. The fractional derivatives and integrals are not only useful for describing many phenomena and properties of materials, but according to the results, it is clear that new modeling based on fractional derivatives is much more appropriate and accurate than modeling using derivatives with integer order.

This advantage stems from the fact that fractional derivatives are an excellent tool for describing the memory and hereditary properties of different materials and how they are processed, and this is the most important advantage of fractional derivatives compared to derivatives with integer order. In other words, integer order derivatives depend only on the local behavior of the function, while modeling based on fractional derivatives allows all the information of the function to be condensed in a weighted form.

Today, the advantages of fractional derivatives in modeling the electrical and mechanical properties of materials, explaining the properties of rock deformation, nonlinear earthquake oscillations, diffusion equations in mathematics, physics and dynamics of turbulence, quantum mechanics and plasma physics, etc. are well seen. In recent years, due to the frequent emergence of deficit differential equations in flow mechanisms, viscoelastic, biology, electrochemistry, and technical and physical issues, it has attracted the attention of many researchers in engineering and medical sciences to research in this regard. Successful theories can be found in mechanics (Vascular theory Static and viscoelastic) were referred to in biological chemistry (modeling of polymers and proteins), electrical engineering (transmission of ultrasonic waves), and medicine (modeling of human tissue under mechanical loads).

The origins of this branch of mathematics came as Leibniz sought to introduce a new symbol for derivation by playing with symbols in 1695. In his journal, he used the symbol $\frac{d^n x}{dt^n}$ to define the derivative, which led to the controversial question of Hopital by Leibniz: What happens if $n = \frac{1}{2}$? In fact, this question was the beginning of a great change in mathematics, which led to the birth of a new branch of mathematics called fractional arithmetic.

In this regard, scientists and researchers with a new perspective on mathematics began to investigate this issue. In 1730, Euler formed some questions about the fractional derivative and integral, which was a spark for finding the fractional derivative. He published the first work on derivatives of any order. In 1812, Laplace introduced an integral fractional derivative operator, and in 1819 he published the first work on derivatives of any order. Then in the same year, Lacroix, introduced the Gamma function as an extension of factorial, in the following form,

$$\frac{d^n x(t)}{dt^n} = \frac{\Gamma(m+1)}{\Gamma(m-n+1)} t^{m-n}, \quad m \geq n$$

The concepts of fractional differentiation and fractional integration were examined further over the course of the eighteenth and nineteenth centuries. The topic attracted the attention of mathematical giants such as Riemann, Liouville, Abel, Laurent, and Hardy and Littlewood.

The paradoxes described by Leibniz were resolved by others later, but this is not to say that the field of fractional calculus is now wholly free of open problems. One recurring issue over the centuries is the existence of numerous contradictory definitions. By the mid-nineteenth century, several different definitions of the deficit account had been proposed: Liouville defined his definition based on the distinction of exponential functions and another based on an integral formula for inverse power functions, while Lacroix defined differently by differentiation. Had created. The power functions of Liouville and Lacroix's definitions are not equivalent, leading some critics to conclude that one must be right and the other wrong. De Morgan, however, wrote that:

> Therefore, both systems may be part of a more general system.

His words, like those of Leibniz 145 years ago, were prophetic. Liouville's and Lacroix's formulas are, in fact, specific to what is now known as Riemann–Liouville's definition of fractional calculus. This includes a c-integration integration constant, which, when set to zero, adjusts the Lacroix formula and gives Liouville's performance when set as infinity.

1.2 Introduction to Fuzzy Fractional Differential Equations

Several scientists in their earliest works introduced fuzzy fractional calculus as an uncertain fractional calculus to consider fractional-order systems with uncertain initial values or uncertain relationships between parameters. The uniqueness and existence of the solution of fractional differential equations with uncertain initial value are utilized as well. The other ones employed the Riemann–Liouville generalized H-differentiability in order to solve the fuzzy fractional differential equations and presented some new results under this notion. Then they applied the technique of fuzzy Laplace transforms to solve some types of fuzzy fractional differential equations based on the Riemann–Liouville fuzzy derivative.

As an uncertain set-valued problem and considering the delta-Hukuhara derivative in the fuzzy case for uncertain functions, the stability criteria for hybrid fuzzy systems on time scales in the Lyapunov sense were introduced by others. The other scientists switched to the applied topics of fractional calculus and solved a class of time-dependent fuzzy fractional optimal control problems.

In general, the majority of the fuzzy fractional differential equations as same as fuzzy differential equations do not have exact solutions. This is why approximate and numerical procedures are important to be developed. On the other hand, the complicity of many parameters in mathematical modeling of natural phenomena appears as an uncertain fractional model and they play an important role in various disciplines. Hence, it motivates the researchers to investigate effective numerical methods with error analysis to approximate the fuzzy fractional differential equations.

As a result, researchers started to develop numerical techniques for fuzzy fractional differential equations. The first method was introduced as a fuzzy approximate solution using the Euler method to solve fuzzy fractional differential equations. The others adopted the operational Jacobi operational matrix based on the fuzzy Caputo fractional derivative using shifted Jacobi polynomials. The clear advantage of the usage of this method is that the matrix operators have the main role to find the approximate fuzzy solution of fuzzy fractional differential equations instead of considering the methods required the complicated fractional derivatives and their calculations.

Then the spectral numerical method for solving fuzzy fractional kinetic equations was introduced. The simplicity, efficiency, and high accuracy of their method are the main advantages. After a while other scientists exploited a cluster of orthogonal functions, named shifted Legendre functions, to solve fuzzy fractional differential equations under Caputo type. The benefit of the shifted Legendre operational matrices method, over other existing orthogonal polynomials, is its simplicity of execution as well as some other advantages. The achieved solutions present satisfactory results, obtained with only a small number of Legendre polynomials.

Recently, investigated an analytical method (Eigenvalue–Eigenvector) for solving a system of fuzzy differential equations under fuzzy Caputo's derivative. To this end, they exploited generalized H-differentiability and derived the solutions based on this concept. Two definitions of differentiability of type-2 fuzzy number valued functions in sense of Riemann–Liouville and Caputo fractional order are introduced and clearly more accuracy will have more complicity cost.

Then for solving fuzzy fractional differential equations, the notion of revisited Caputo's H-differentiability based on the generalized Hukuhara difference and proposed a novel analytical method entitled fuzzy Laplace transforms are developed. Employing Laplace transforms, they proposed a novel efficient technique for the solution of this type of equation that can efficiently make the original problem easier to achieve the numerical solution. The suggested algorithm for the fuzzy fractional differential equations uses the level-wise fuzzy fractional derivative in Caputo sense.

Moreover, some researchers have investigated the variation iteration method (VIM) to solve the fuzzy linear fractional differential equation under Caputo generalized Hukuhara differentiability. The VIM method is the semi-analytical method and it was introduced in fuzzy sense and modelled as fuzzy VIM in recent years.

Then Semi-Analytical methods for solving fuzzy impulsive fractional differential equations based on fuzzy differential transform method in the fractional case in Caputo fractional derivative sense have been defined and discussed as well. Then two-dimensional Legendre wavelet was studied to approximate the solution of the fuzzy fractional integro-differential equation under Caputo generalized Hukuhara differentiability.

In this research, the existence and uniqueness theorems for a fuzzy fractional integro-differential equation by considering the type of differentiability of solutions were proved. Then the generalized Taylor's expansion was presented for fuzzy-valued functions. To this end, a fuzzy fractional mean value theorem for integral, and some properties of Caputo generalized Hukuhara derivative were discussed. Also, the authors derived the fractional Euler's method for solving fuzzy fractional differential equations in the sense of Caputo's differentiability. Moreover, the fuzzy q-derivative and fuzzy q-fractional derivative in Caputo sense by using generalized Hukuhara difference by means of q-Mittag-Leffler function are provided recently.

Moreover, the characterization theorem between the solutions of the fuzzy Caputo q-fractional initial value problem and system of ordinary Caputo q-fractional differential equations was presented.

The right and left fuzzy fractional Riemann–Liouville integrals and the right and left fuzzy fractional Caputo derivatives are studied in some researches and then the right and left fuzzy fractional Taylor formula has established for a fuzzy fractional Ostrowski type inequality with applications.

1.3 Structure of the Book

In Chap. 2, we are going to cover the fuzzy sets in several forms and discuss their properties and any other related preliminaries including operations. In Chap. 3, fuzzy fractional differential operators are introduced. Several operators including fuzzy fractional integrals are also considered. Moreover, the relations of fractional integrals and also derivatives are considered. In Chap. 4, we study the fuzzy fractional differential equations with the existence and uniqueness of the solutions. In this chapter, each fractional differential equation is explained with the corresponding fractional derivative in detail.

One of the main discussions of this chapter is fuzzy impulsive fractional differential equations that have much application in engineering science. Besides the fuzzy Laplace transforms are expressed to solve the equations theoretically. In Chap. 5 application of these equations is brought. The first application is about the fuzzy optimal control problems and the second one discusses the drug release for Tumor of cancer.

Chapter 2
Fuzzy Sets

2.1 Introduction

In this chapter, we are going to cover the fuzzy sets in several forms and discuss the properties and any other related preliminaries.

2.2 Fuzzy Sets and Variables

Mathematical modeling seeks to describe the phenomenon formally, but it always faces some inconveniences, namely uncertainty, and ambiguity. To be vague, the fuzzy set theory, which is formulated by Prof. Lotfi A. Zadeh in 1965, aims to give the subject matter a mathematical treatment. In addition, it is considered as an important tool for better understanding some of the real situations.

Fuzzy variables are tools for modeling data in an uncertain context that cannot be accurately predicted. For example, coin throwing, dice throwing, poker games, stock pricing, marketing and market demand, longevity and more.

What is the answer to the question, who is handsome or beautiful? Apparently, it depends on the personality of the individual and differs from different personalities. It is clear that we are composed of a fuzzy set of beautiful people. Membership in this set is called membership grade. The degree of membership can be considered as a real number of [0, 1] intervals.

In fact, a fuzzy set is formed by ordered pairs such that the second component is the degree of membership of the first component. Graphically, these sorted pairs are on the function and this function is called the membership function. Sometimes the fuzzy set plays the fuzzy variable role. It depends on the situation. For example, to consider a fuzzy valued function, it must have a fuzzy variable and in fact, is a set of probability measurements and has a membership function. Below, several fuzzy sets and variables are mathematically mapped and modeled.

T. Allahviranloo, *Fuzzy Fractional Differential Operators and Equations*, Studies in Fuzziness and Soft Computing 397, https://doi.org/10.1007/978-3-030-51272-9_2

- Gaussian form membership function on $[a, b]$ (Fig. 2.1).

$$f(x) = \begin{cases} 0, & x \leq a \\ e^{-0.3x^2}, & a \leq x \leq b \\ 0, & x \geq b \end{cases}$$

- Triangular form membership function on $[a, b]$

$$g(x) = \begin{cases} 0, & x \leq a \\ \frac{x-a_1}{c-a_1}, & a \leq x < c \\ 1, & x = c \\ \frac{a_2-x}{a_2-c}, & c < x \leq b \\ 0, & x \geq b \end{cases}$$

If suppose that $a = -1, c = 1, b = 4$ the figure is (Fig. 2.2).

- Trapezoidal form membership function on $[a_1, a_2]$

$$h(x) = \begin{cases} 0, & x \leq a_1 \\ \frac{x-a_1}{\bar{a}-a_1}, & a_1 \leq x < a' \\ 1, & a' \leq x \leq a'' \\ \frac{a_2-x}{a_2-\bar{a}}, & a'' < x \leq a_2 \\ 0, & x \geq a_2 \end{cases}$$

The following figure shows the Trapezoidal fuzzy number for any a_1, a_2, a', a''. It is clear, the functions should define a line between $[a_1, a']$, $[a', a'']$ and $[a'', a_2]$. The Trapezoidal fuzzy number is a general form of a Triangular fuzzy number, because it is sufficient that $a' = a''$. For this reason, in approximation methods, the best approximation of a Triangular fuzzy number is a Trapezoidal fuzzy number (Fig. 2.3).

Consider the following function, in this figure $x(t) \in [0, 1] \subseteq R^+$, the variable x is a fuzzy variable and can be chosen from uncertainty space (Fig. 2.4).

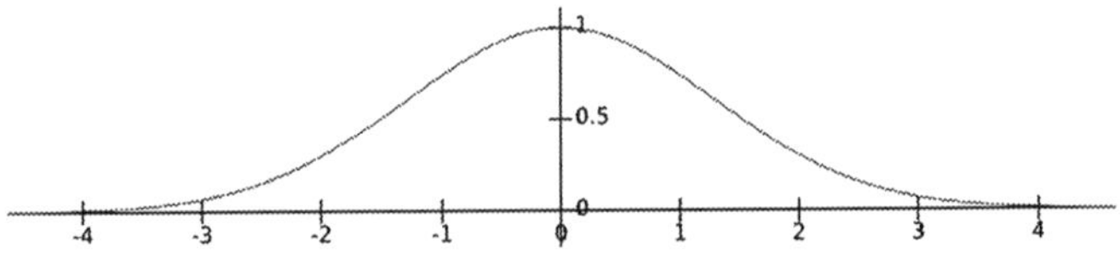

Fig. 2.1 Gaussian membership function on $[a, b] = [-4, 4]$

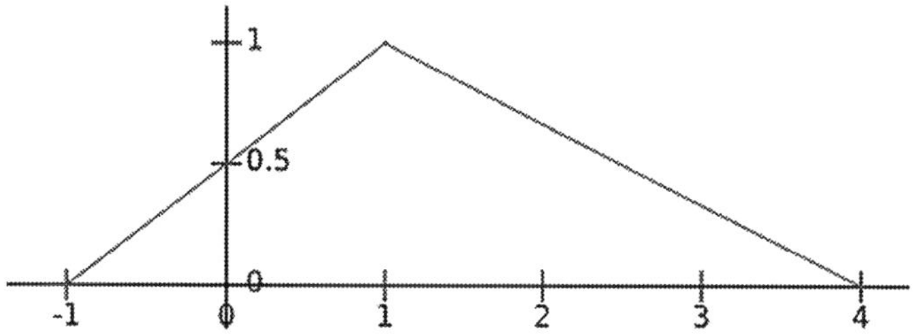

Fig. 2.2 Triangular membership function on $[-1, 4]$

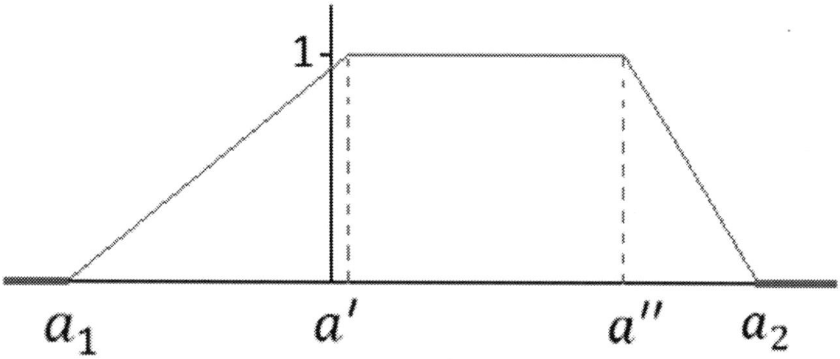

Fig. 2.3 Trapezoidal membership function

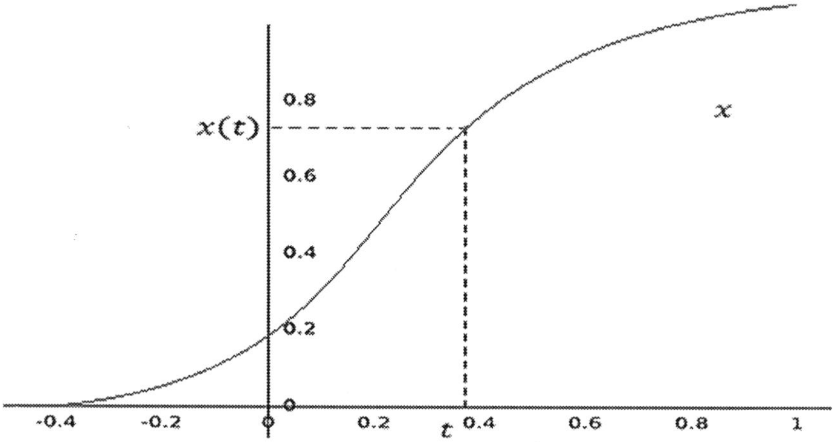

Fig. 2.4 Fuzzy variable

Some examples

$$x_1(t) = t, \quad x_2(t) = \sum_{i=1}^{n} a_i t^i, \quad n \in R^+, \quad x_3(t) = \exp t, \ldots$$

The Zigzag form of an uncertain variable

This example is the shape of zigzag uncertain variables. Suppose we are talking about old people, not everyone who is less than a_1 years old, but all people aged between a_1 to a_2 years old can be called old. The people with the age of a_2 to a_3 years old are almost old and the rest more than a_3 years old is completely old. Suppose that $x(t)$ is the function of oldness of aged people, the following membership function shows us the membership of the people in the function.

$$x(t) = \begin{cases} 0, & t \leq a_1 \\ \frac{\alpha t - a_1}{a_2 - a_1}, & a_1 < t \leq a_2 \\ \frac{\beta t - a_2}{a_3 - a_2}, & a_2 < t \leq a_3 \\ \frac{\alpha t - a_1}{a_2 - a_1} + \frac{\beta t - a_2}{a_3 - a_2} + (t - a_3), & a_3 < t \end{cases}$$

In the following figure some type of a zigzag membership function is shown. (Fig. 2.5)

Experimental uncertain variables

The following membership function is for an experimental uncertain variable with the following figure.

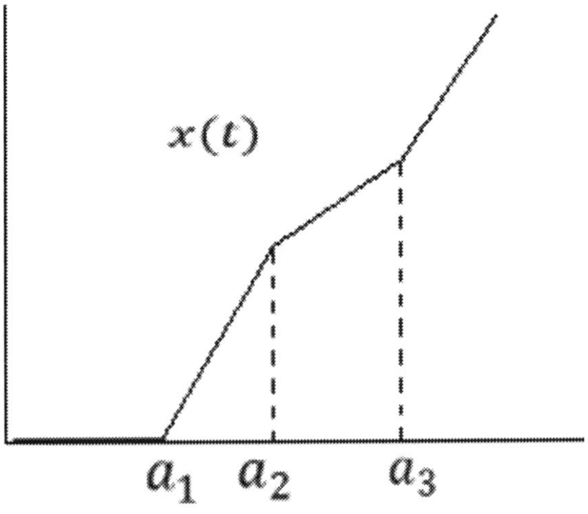

Fig. 2.5 The set of tall kids as an uncertain variable

$$x(t) = \begin{cases} 0, & t \leq t_1 \\ f(\theta), & t = t_i + \theta(t_{i+1} - t_i) \\ 1, & t_n < t \end{cases}$$

where $1 \leq i \leq n-1$, $0 < \theta \leq 1$. The function $f(\theta)$ is any function between the points t_i and t_{i+1} as a jump function. In the following figure the relation between two arbitrary points is not restricted to the linear function. It can be any function even a curve (Fig. 2.6).

Or another case of experimental uncertain variable (Fig. 2.7).

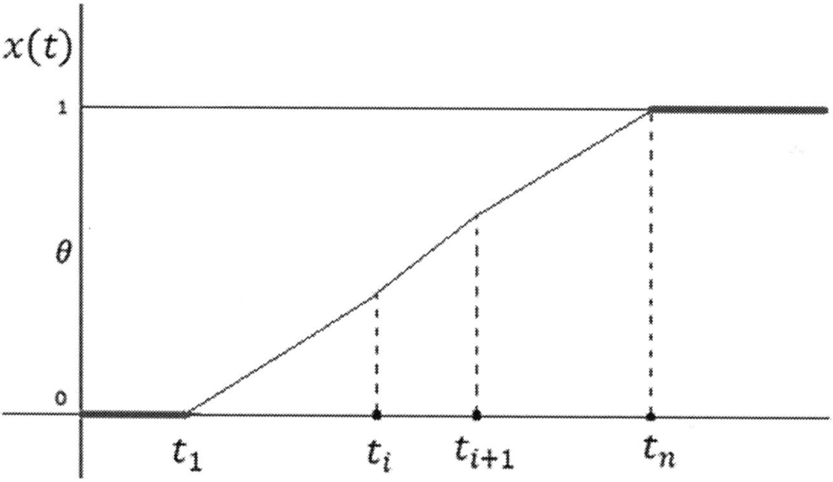

Fig. 2.6 An experimental (piecewise) uncertain variable

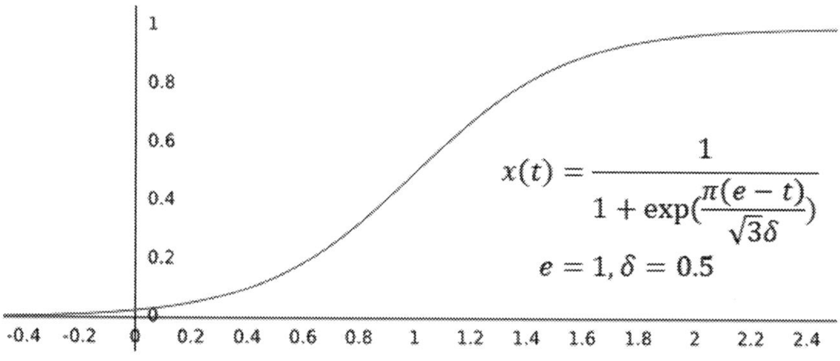

Fig. 2.7 An experimental (non-linear) uncertain variable

$$x(t) = \frac{1}{1 + \exp\left(\frac{\pi(e-t)}{\sqrt{3}\delta}\right)}, \quad e, \delta \in R^{+} x \in R$$

Note. Now we have enough information about the uncertainty and uncertain sets and variables. One of the uncertainties cases is fuzzy sets. Like as other uncertain sets, the fuzzy set is a set of ordered pairs, (member, its membership). It is a membership function as well and their structures and forms are the same as the structures and forms of uncertain sets. See the figures.

2.2.1 *Membership Function*

For a fuzzy set (uncertain set), M, the membership of its elements, $x \in M$, has a membership degree, $M(x) \in [0, 1]$. If the degree is 0, the member does not belong to the set and if the degree is 1, the member belongs to the set completely.

This is the reason that, any real set is a special case of an uncertain set and has the characteristic degree and function. In this case, the range of the membership degree is $\{0, 1\}$. It is obvious than the set M is a fuzzy set and

$$M = \{(x, M(x))|M(x) \in [0, 1], x \in M\}$$

This set is called membership function.

In this concept, if x belongs to the uncertain set with the membership degree $M(x)$, at the same time it does not belong to the set with $1 - M(x)$ membership degree. In the following section, we discuss fuzzy numbers and their properties.

2.3 Fuzzy Numbers and Their Properties

First of all, we should know that why we need to fuzzy number and it is important. Because, we need to computations and ranking the fuzzy sets. This the reason that fuzzy sets must have some additional properties. Now we are going to explain their graphical and mathematical forms and computations. First of all, we should define a fuzzy set as a fuzzy number.

2.3.1 *Definition of a Fuzzy Number*

A fuzzy membership function $M: R \to [0, 1]$, is called a fuzzy number if it has the following conditions:

1. M, is normal. It means there is at least a real member x_0 such that $M(x_0) = 1$.
2. M, is fuzzy convex. It means,

   For two arbitrary real points x_1, x_2 and $\lambda \in [0, 1]$ we have

$$M(\lambda x_1 + (1 - \lambda)x_2) \geq \min\{M(x_1), M(x_2)\}$$

3. M, is upper semi-continuous. It means, if we increase its value at a certain point x_0 to $f(x_0) + \epsilon$ (for some positive constant ϵ), then the result is upper-semicontinuous; if we decrease its value to $f(x_0) - \epsilon$ then the result is lower-semi-continuous.
4. The closure of the set $Supp(M) = \{x \in R | M(x) > 0\}$, as a support set, is a compact set.

As an example; the triangular, trapezoidal fuzzy sets are fuzzy numbers.

The definition of a fuzzy number can be defined as other forms, parametric and level-wise form.

2.3.2 *Level-Wise Form of a Fuzzy Number*

The level-wise membership function is, in fact, an inverse function of membership function that proposes an interval-valued function. In fact, any level in the vertical axis gives us an interval in the horizontal axis. For example, consider one of the following triangular membership functions (Fig. 2.8).

In this figure, all real numbers in the interval [a, b] have degree of membership greater than or equal to the value of "$r - level$" in the fuzzy set M, i.e.

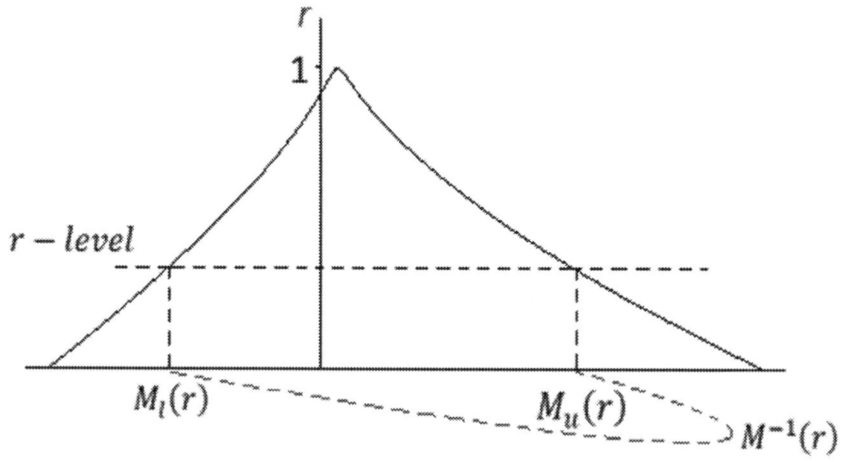

Fig. 2.8 Level-wise form of a fuzzy set

$$\forall x \in [a,b],\ M(x) \geq r, \quad 0 \leq r \leq 1$$

Then the r—cut or r—level set of the membership function can be defined as,

$$M^{-1}(r) = \{x \in R | M(x) \geq r\} = [a,b] = [M_l(r), M_u(r)] := M[r],\ 0 \leq r \leq 1$$

Based on the inequality property in the set, $M[r]$, $M(x) \geq r$, it is clear that the membership function, $M(x)$, can be obtained by,

$$M(x) = \sup\{0 \leq r \leq 1 | x \in M^{-1}(r)\},\ x \in R$$

It means that there is one to one map between two functions, membership function $M^{-1}(r)$, and level wise membership function $M(x)$ (Fig. 2.9).

The figure shows that for each interval there is a degree or level and vice versa. In fact, it can be claimed that,

$$Domain\ of\ M(x) = \bigcup_{0 \leq r \leq 1} [M_l(r), M_u(r)]$$

Now, in general, a fuzzy set M, in level-wise form can be shown as follows,

$$M[r] = [M_l(r), M_u(r)],\ 0 \leq r \leq 1$$

Another form of the definition can be defined as follows.

2.3.3 Definition of a Fuzzy Number in Level-Wise Form

A fuzzy membership function $M : R \to [0,1]$, is called a fuzzy number if its level-wise form $M[r] = [M_l(r), M_u(r)]$, is a compact interval for any $0 \leq r \leq 1$.

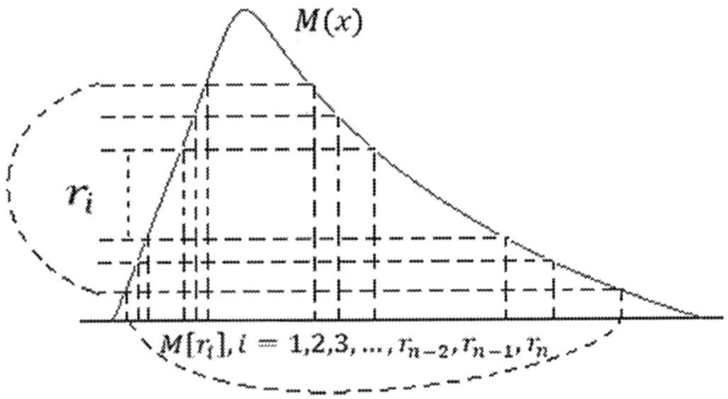

Fig. 2.9 One-to-one corresponding

The following definition shows, when the stacking of levels or cuts in a fuzzy set can establish a fuzzy number.

2.3.4 Definition of a Fuzzy Number in Level-Wise Form

The sufficient and necessary conditions for $M^{-1}(r) = M[r]$ to be a level-wise membership function of a fuzzy number are:

(i) (Nesting property) For any two $r - levels, r_1, r_2$
If $r_1 \leq r_2$ *then* $M[r_1] \supseteq M[r_2]$

(ii) For any monotone increasing sequence of levels, $0 < r_1 < r_2 < \cdots < r_n < 1$,
If $\{r_n\}_n \nearrow r, then\, M[r_n] \to M[r], for\, any\, 0 \leq r \leq 1$ (Fig. 2.10).

A singleton fuzzy number

A real number 'a' is called a singleton fuzzy number, if

$$a[r] = [a_l(r), a_u(r)] = [a, a]$$

It means in the membership function the membership degree at 'a' is 1 and at other values is zero. $a_l(r) = a_u(r) = a$.

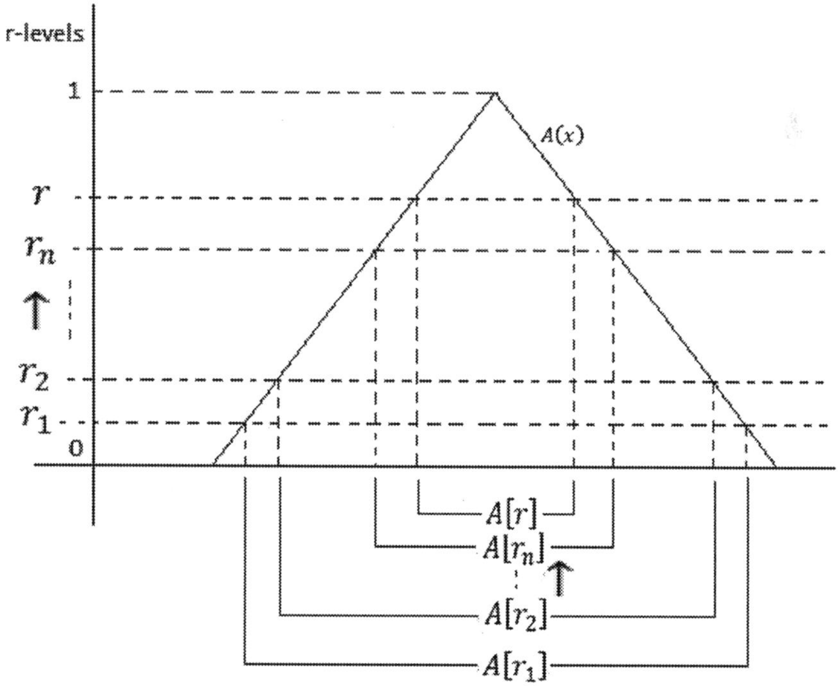

Fig. 2.10 The levels stacking

Based on this definition we have the same for fuzzy zero number or origin and $a_l(r) = a_u(r) = 0$.

2.3.5 Definition of a Fuzzy Number in Parametric Form

Any fuzzy number, M, has the parametric form $M(r) = (M_l(r), M_u(r))$ for any $0 \leq r \leq 1$, if and only if,

(i) $M_l(r) \leq M_u(r)$
(ii) $M_l(r)$ is an increasing and left continuous function on $(0, 1]$ and right continuous at 0 with respect to r.
(iii) $M_u(r)$ is a decreasing and left continuous function on $(0, 1]$ and right continuous at 0 with respect to r.

Note that, in items (ii) and (iii), both functions can be bounded.

In both forms of fuzzy numbers—level-wise form and parametric form—both functions, lower, $M_l(r)$ and upper, $M_u(r)$, are the same. But the differences can be listed as follow:

1. In level-wise form, the values of both functions for any arbitrary but fixed r, are real numbers. But in parametric form, they act the role of function with respect to r.
2. In level-wise form, the level is an interval for any arbitrary but fixed r. But in parametric form, it is a couple of functions with respect to r.

Now we are going to introduce other forms of a fuzzy number in linear, non-linear cases. The following figure shows that it doesn't matter which fuzzy number we do we consider for analyzing.

Indeed, a membership function which corresponds to a fuzzy number is a piece-wise function. For examples,

$$M_1(x) = \begin{cases} x, & 0 \leq x \leq 1 \\ 2 - x, & 1 \leq x \leq 2 \\ 0, & otherwise \end{cases}$$

$$M_2(x) = \begin{cases} 1 - |x|, & -1 \leq x \leq 1 \\ 0, & otherwise \end{cases}$$

$$M_3(x) = \begin{cases} |x|, & -1 \leq x \leq 1 \\ 0, & otherwise \end{cases}$$

$$M_4(x) = \begin{cases} \sqrt{x}, & -1 \leq x \leq 1 \\ 1, & otherwise \end{cases}$$

In general form, the function can be shown as follow,

$$M(x) = \begin{cases} L\left(\frac{x-a}{b-a}\right), & a \le x \le b \\ 1, & b \le x \le c \\ R\left(\frac{d-x}{d-c}\right), & c \le x \le d \\ 0, & otherwise \end{cases}$$

where $L, R: [0, 1] \rightarrow [0, 1]$ are two non-decreasing shape functions and

$$R(0) = L(0) = 0, R(1) = L(1) = 1.$$

To obtain the level-wise or parametric forms of the fuzzy number in the general form, the relations are,

$$M_l(r) = a + (b - a)L^{-1}(r),$$
$$M_u(r) = d - (d - c)R^{-1}(r), \qquad r \in [0, 1]$$

Clearly, if the functions L and R are linear then we will have Trapezoidal and Triangular membership functions as fuzzy numbers. If they are non-linear then they will appear as curves look like Trapezoidal or Triangular.

The general form of a fuzzy number ın linear or non-linear cases can be shown as the following figure (Fig. 2.11).

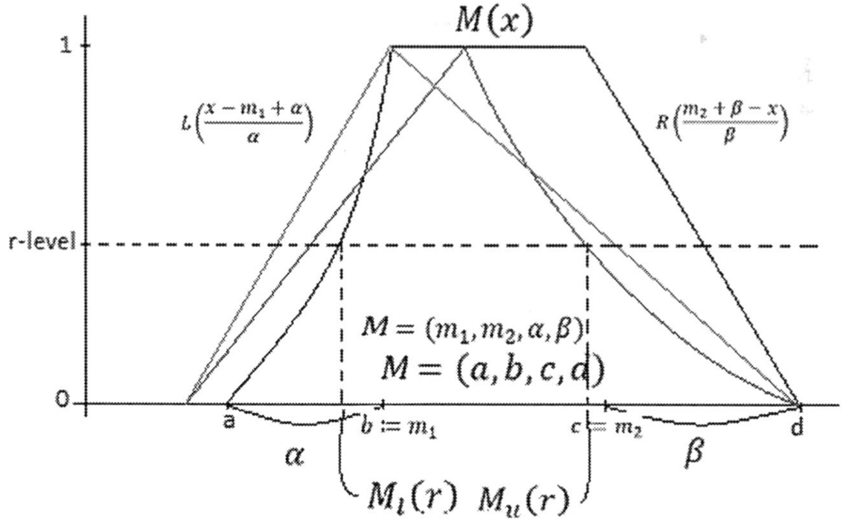

Fig. 2.11 General from of a fuzzy number

2.3.6 Non-linear Fuzzy Number

For instance, in the following cases, the fuzzy numbers do not have any linear lower and upper functions.

$$L(x) = \frac{1}{1+x^2}, \quad R(x) = \frac{1}{1+2|x|}, \quad \alpha = 3, \quad \beta = 2, \quad m = 4$$

Then the membership function is,

$$M(x) = \begin{cases} L\left(\frac{4-x}{3}\right) = \frac{1}{1+\left(\frac{4-x}{3}\right)^2}, & x \leq 4 \\ R\left(\frac{x-4}{2}\right) = \frac{1}{1+2\left|\frac{x-4}{2}\right|}, & 4 \leq x \\ 0, & otherwise \end{cases}$$

The figure of this membership function is (Fig. 2.12).

Generally, the r-level set of a fuzzy set M is,

$$M[r] = [M_l(r), M_u(r)] = [a + (b-a)L^{-1}(r), d - (d-c)R^{-1}(r)]$$

Note that, members of an r-level set as an interval is included in the membership function with membership degree or uncertain measure, r.

2.3.7 Trapezoidal Fuzzy Number

The general level-wise or parametric forms are,

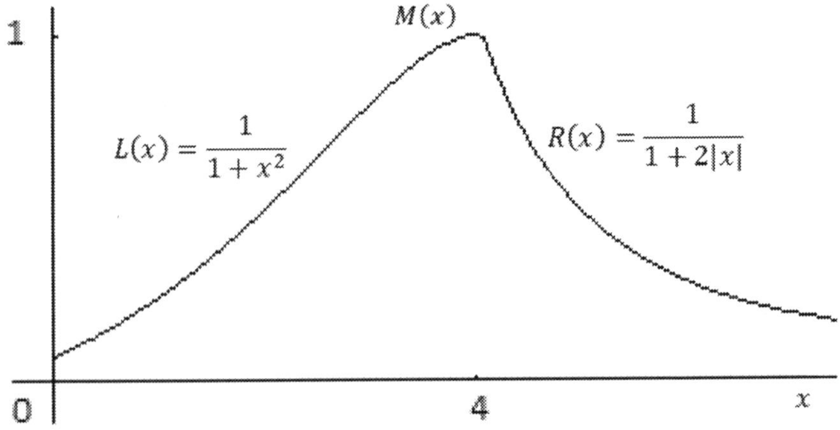

Fig. 2.12 Non-linear fuzzy number

$$M_l(r) = b - (b-a)L^{-1}(r),$$
$$M_u(r) = c + (d-c)R^{-1}(r), \qquad r \in [0,1]$$

where in linear case the left and right functions can be replaced by linear functions like $R(r) = L(r) = r$ or others.

Now suppose that $b - a = \alpha, d - c = \beta, b = m_1, c = m_2$ then the membership function will be as,

$$M(x) = \begin{cases} L\left(\frac{m_1 - x}{\alpha}\right), & m_1 - \alpha \leq x \leq m_1 \\ R\left(\frac{x - m_2}{\beta}\right), & m_2 \leq x \leq m_2 + \beta \\ 1, & m_1 \leq x \leq m_2 \\ 0, & otherwise \end{cases}$$

where α, β are left and right spreads, and m_1, m_2 are cores of a trapezoidal fuzzy number.

$$M_l(r) = m_1 - \alpha L^{-1}(r),$$
$$M_u(r) = m_2 + \beta R^{-1}(r), \qquad r \in [0,1]$$

The formal formats to show this number can be written as,

$$M = (a,b,c,d) := M = (m_1, m_2, \alpha, \beta) := M[r] = [M_l(r), M_u(r)] := M(r)$$
$$= (M_l(r), M_u(r))$$

Please note that these formats are only for presenting the fuzzy number and each one has own properties and calculations. The calculations on them will be explained in the next sections.

The r-level set of a trapezoidal fuzzy set $M = (a,b,c,d)$ is,

$$M[r] = [M_l(r), M_u(r)] = [b + (b-a)(r-1), c + (d-c)(1-r)]$$

2.3.8 *Triangular Fuzzy Number*

The only difference of triangular fuzzy number from trapezoidal is the number of the cores in membership functions. In the trapezoidal, if the cores are the same, i.e. $m_1 = m_2 = m$ then it will be triangular fuzzy number. The general level-wise or parametric forms are,

$$M_l(r) = b - (b-a)L^{-1}(r),$$
$$M_u(r) = b + (c-b)R^{-1}(r), \qquad r \in [0,1]$$

where in linear case the left and right functions can be replaced by linear functions like $R(r) = L(r) = r$ or others.

Now suppose that $b - a = \alpha, c - b = \beta, b = m$ then the membership function will be as,

$$M(x) = \begin{cases} L\left(\frac{x-m+\alpha}{\alpha}\right), & m - \alpha \leq x \leq m \\ R\left(\frac{m+\beta-x}{\beta}\right), & m \leq x \leq m + \beta \\ 0, & otherwise \end{cases}$$

where α, β are left and right spreads, and m is the core of a triangular fuzzy number.

$$M_l(r) = m - \alpha L^{-1}(r),$$
$$M_u(r) = m + \beta R^{-1}(r), \qquad r \in [0, 1]$$

The formal formats to show this number can be written as,

$$M = (a, b, c) := M = (m, \alpha, \beta) := M[r] = [M_l(r), M_u(r)] := M(r) \\ = (M_l(r), M_u(r))$$

The r-level set of a triangular fuzzy set $M = (a, b, c)$ is,

$$M[r] = [M_l(r), M_u(r)] = [b + (b - a)(r - 1), b + (c - b)(1 - r)] \ .$$

For both format of the r-level sets of triangular and trapezoidal fuzzy numbers the lower and upper functions can be obtained practically as follow,

Consider the following figure of fuzzy set M. First, the line segment between two points A and B is defined as,

$$\frac{y - y_A}{x - x_A} = \frac{y_B - y_A}{x_B - x_A} \Rightarrow y = \frac{1}{\alpha}(x - m + \alpha)$$

Then after finding the equation of line, we have this system,

$$y = \frac{1}{\alpha}(x - m + \alpha) \quad \& \quad y = r$$

So,

$$r = \frac{1}{\alpha}(x - m + \alpha)$$

The inverse is reflective function on $x = r$ and it is as,

$$M_l(r) = x = m + \alpha(r - 1)$$

The same procedure will be true for the points B and C. Then it will be obtained as,

$$\frac{y - y_B}{x - x_B} = \frac{y_C - y_B}{x_C - x_B} \Rightarrow y = \frac{1}{\beta}(m - x) + 1$$

Again, we will have

$$r = \frac{1}{\beta}(m - x) + 1$$

And the inverse function will be $M_u(r)$ (Fig. 2.13).

$$M_u(r) = m + \beta(1 - r)$$

2.3.9 Operations on Level-Wise Form of Fuzzy Numbers

Here, the main calculations on fuzzy sets in parametric or level-wise form are going to be defined and discussed. Since the concept of the difference is different and needs more attention, so this operation will be discussed more. Because here the subtraction has the meaning of difference between two sets or two functions.

Before the discussion on the operations, we need some explanations. Please note that the fuzziness will be growing under these operators. It means, for two arbitrary fuzzy numbers M and N, and a continuous measurable function like $f \in \{\oplus, \odot, ø\}$, the fuzziness of M and fuzziness of N is less than or equal to the fuzziness of $f(M, N)$. One of the concepts of the fuzziness is diameter of a fuzzy number. The diameter of an interval in level-wise form of a fuzzy number can be called as fuzziness of M in any level of r.

$$Fuzz(M_r) = diam([M_l(r), M_u(r)]) = M_u(r) - M_l(r), \quad 0 \le r \le 1.$$

Now it is proven that

$$Fuzz(M_r) \le Fuzz\big(f(M, N)_r\big) \;\&\; Fuzz(N_r) \le Fuzz\big(f(M, N)_r\big)$$

also,

$$K = f(M, N) \Rightarrow K[r] = f(M[r], N[r])$$

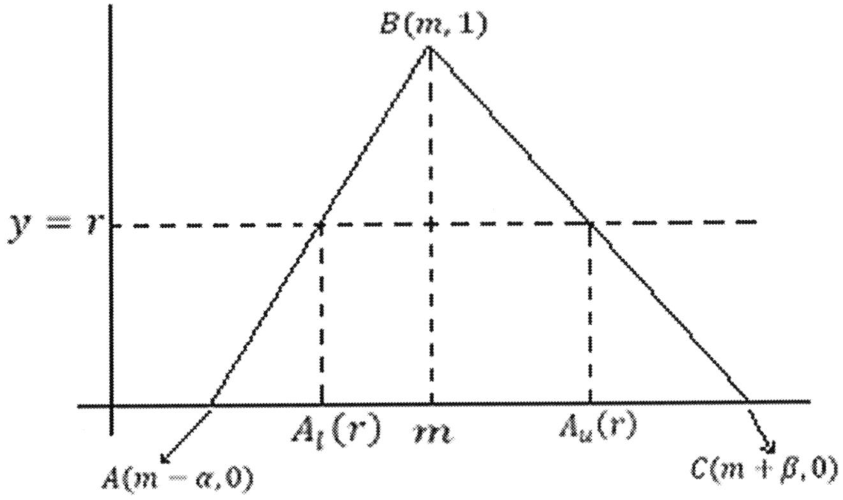

Fig. 2.13 Fuzzy number in triangular form

This means that if the function f is continuous and measurable then

$$\begin{aligned} &\text{if } M \oplus N = K \text{ then } M[r] + N[r] = K[r], \\ &\text{if } M \odot N = K \text{ then } M[r] \odot N[r] = K[r], \\ &\text{if } M \o N = K \text{ then } M[r] \o N[r] = K[r]. \end{aligned}$$

Now we can discuss on the operators separately.

For any two arbitrary fuzzy numbers M and N and any arbitrary but fixed $0 \le r \le 1$, if $M \oplus N = K$, we have,

$$K[r] = [K_l(r), K_u(r)] = M[r] + N[r] = [M_l(r), M_u(r)] + [N_l(r), N_u(r)]$$

Then

$$K_l(r) = M_l(r) + N_l(r), \qquad K_u(r) = M_u(r) + N_u(r)$$

An example

Suppose that

$$M = [r - 1, 1 - r], \qquad N = [r, 2 - r^2],$$

To compute the summation $M \oplus N = K$ we need to,

$$M_l(r) = r - 1, \quad M_u(r) = 1 - r, \quad N_l(r) = r, \quad N_u(r) = 2 - r^2$$

So,

$$K_l(r) = 2r - 1, \qquad K_u(r) = 3 - r - r^2.$$

The following figure shows the summation of two fuzzy numbers in the example (Fig. 2.14).

As it is seen we have

$$M_u(r) - M_l(r) = Fuzz(M_r) \le Fuzz(M \oplus N)_r = K_u(r) - K_l(r)$$
$$\&$$
$$N_u(r) - N_l(r) = Fuzz(N_r) \le Fuzz(M \oplus N)_r = K_u(r) - K_l(r)$$

For all levels.

2.3.9.1 Multiplication

For any two arbitrary fuzzy numbers M and N and any arbitrary but fixed $0 \le r \le 1$, if $M \odot N = \mathrm{K}$, we have,

$$K[r] = [K_l(r), K_u(r)] = M[r] \cdot N[r] = [M_l(r), M_u(r)] \cdot [N_l(r), N_u(r)]$$

Then

$$K_l(r) = \min\{M_l(r) \cdot N_l(r), M_l(r) \cdot N_u(r), M_u(r) \cdot N_l(r), M_u(r) \cdot N_u(r)\},$$

$$K_u(r) = \max\{M_l(r) \cdot N_l(r), M_l(r) \cdot N_u(r), M_u(r) \cdot N_l(r), M_u(r) \cdot N_u(r)\}$$

In the previous example

$$K_l(r) = \min\left\{(r-1) \cdot r, (r-1) \cdot \left(2 - r^2\right), (1-r) \cdot r, (1-r) \cdot \left(2 - r^2\right)\right\}$$

$$K_u(r) = \max\left\{(r-1) \cdot r, (r-1) \cdot \left(2 - r^2\right), (1-r) \cdot r, (1-r) \cdot \left(2 - r^2\right)\right\}$$

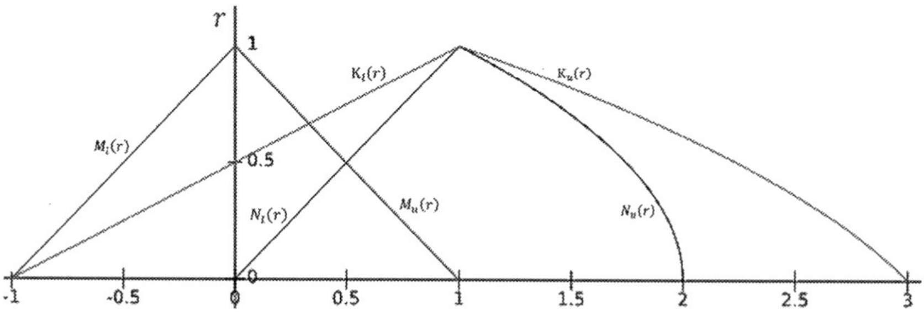

Fig. 2.14 Summation

We will see that

$$K_l(r) = (r-1)(2-r^2), \quad K_u(r) = (1-r)(2-r^2)$$

The following figure shows the multiplication of two fuzzy numbers in the example (Fig. 2.15).

2.3.9.2 Scalar Multiplication

The difference of two fuzzy numbers is indeed the difference of two membership functions. Two different differences in the sense of standard and non-standard cases will be discussed in level-wise form.

First of all, we should consider the multiplying a membership function by a scalar in level-wise form. Suppose that $\lambda \in R$ is a scalar. Then

In triple form of fuzzy number:

$$\lambda \odot M = \lambda \cdot (a,b,c) = \begin{cases} (\lambda a, \lambda b, \lambda c), & \lambda \geq 0 \\ (\lambda c, \lambda b, \lambda a), & \lambda < 0 \end{cases}$$

In level-wise form of fuzzy number:

$$\lambda M[r] = \begin{cases} [\lambda M_l(r), \lambda M_u(r)], & \lambda \geq 0 \\ [\lambda M_u(r), \lambda M_l(r)], & \lambda < 0 \end{cases}$$

For any $0 \leq r \leq 1$.

The concept of scalar multiplication is the same as the multiplication of the scalar to each member of the interval. It means,

$$M[r] = [M_l(r), M_u(r)] = \{z_t | z_t = M_l(r) + t(M_u(r) - M_l(r)), 0 \leq t \leq 1\}$$

For any $0 \leq r \leq 1$.

So, in $r - level$,

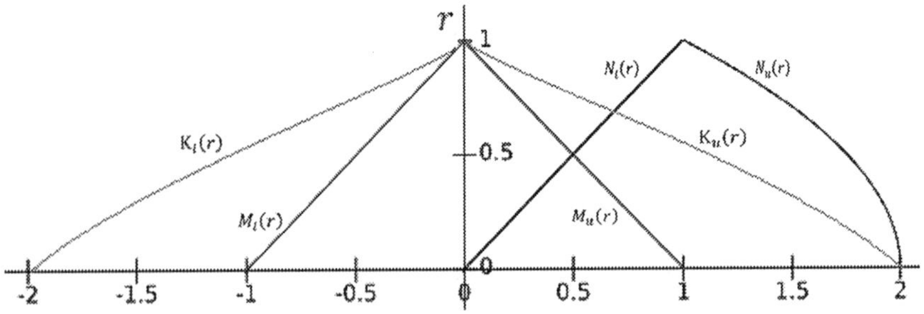

Fig. 2.15 Multiplication

$$\lambda M[r] = \{\lambda z_t | 0 \le t \le 1\} = \begin{cases} [\lambda M_l(r), \lambda M_u(r)], & \lambda \ge 0 \\ [\lambda M_u(r), \lambda M_l(r)], & \lambda < 0 \end{cases}$$

For instance, if consider one of the previous fuzzy numbers like,

$$M[r] = [M_l(r), M_u(r)] = \left[r, 2 - r^2\right]$$

And $\lambda = -1$, then

$$(-1)M[r] = [-M_u(r), -M_l(r)] = \left[r^2 - 2, -r\right]$$

For more illustration see the following figure. (Figure 2.16)

For using this scalar multiplication in the definition of difference of two fuzzy numbers, let $M[r] = [M_l(r), M_u(r)]$ and $B[r] = [N_l(r), N_u(r)]$ are two fuzzy numbers in level-wise form. In this case the difference is defined as follow,

$$\begin{aligned} M[r] - N[r] &= M[r] + (-1)N[r] \\ &= [M_l(r), M_u(r)] + (-1)[N_l(r), N_u(r)] \\ &= [M_l(r), M_u(r)] + [-N_u(r), -N_l(r)] \\ &= [M_l(r) - N_u(r), N_u(r) - M_l(r)] \end{aligned}$$

Or in the format of convex combination, for any arbitrary but fixed $0 \le r \le 1$,

$$\begin{aligned} M[r] &= \{z'_t | z'_t = M_l(r) + t(M_u(r) - M_l(r)), 0 \le t \le 1\} \\ (-1)N[r] &= \{z''_t | z''_t = -N_u(r) + t(N_u(r) - N_l(r)), 0 \le t \le 1\} \\ M[r] + (-1)N[r] &= \{z'_t - z''_t | 0 \le t \le 1\} = [M_l(r) - N_u(r), N_u(r) - M_l(r)] \end{aligned}$$

In this definition,

$$M \oplus (-1) \odot M \neq 0$$

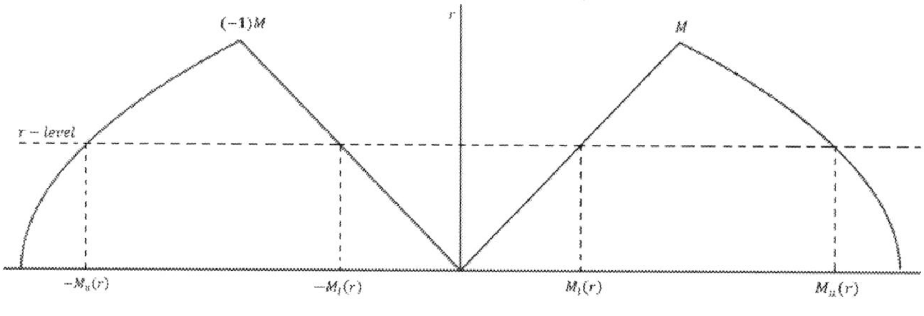

Fig. 2.16 Multiplication of (-1)

Because, based on the definition the result is a symmetric interval centred at zero and it is always non-zero interval.

$$M[r] - M[r] = [M_l(r) - M_u(r), M_u(r) - M_l(r)] \neq 0 = [0, 0]$$

Please note that the symmetric interval centered at zero is called as a zero interval. It came from the concept of equivalency class. If we consider an equivalency class of zero as a set of all symmetric intervals centered at zero, then all of the members of the class are called as a zero interval. Moreover,

$$Fuzz(M_r) \leq Fuzz\left((\lambda \odot M)_r\right), \quad 0 \leq r \leq 1.$$

2.3.9.3 Hukuhara Difference

Suppose that M and N are two fuzzy numbers in level-wise form. The Hukuhara difference of $M \ominus_H N$ is defined as,

$$\exists K; M \ominus_H N = K \Leftrightarrow M = N \oplus K$$

It is clear, the existence of the difference is conditional and depends on the existence of fuzzy number K.

Note. For the existence of H-difference, all M, N and K must be fuzzy numbers.

It means that if the fuzzy set B can be transformed by C then it will be fallen into A.

Now consider $M = N \oplus N$, and level-wise form of both side of the equality, we have,

$$M[r] = N[r] + K[r]$$
$$[M_l(r), M_u(r)] = [N_l(r), N_u(r)] + [K_l(r), K_u(r)] = [N_l(r) + K_l(r), N_u(r) + K_u(r)]$$
$$M_l(r) = N_l(r) + K_l(r), M_u(r) = N_u(r) + K_u(r)$$

Finally,

$$K_l(r) = M_l(r) - N_l(r), K_u(r) = M_u(r) - N_u(r)$$

The level-wise form of the Hukuhara difference or H-difference is defined as subtractions of two endpoints of two intervals respectively.

$$\begin{aligned}\left[(M \ominus_H N)_l(r), (M \ominus_H N)_u(r)\right] &= [K_l(r), K_u(r)] \\ &= [M_l(r) - N_l(r), M_u(r) - N_u(r)]\end{aligned}$$

Please note that the difference

$$M \ominus_H N \neq M \oplus (-1) \odot N$$

Because in level-wise form the differences between intervals in both sides are not the same.

$$\begin{aligned}(M \ominus_H N)[r] &= [M_l(r) - N_l(r), M_u(r) - N_u(r)] \\ &\neq [M_l(r) - N_u(r), N_u(r) - M_l(r)] = (M \oplus (-1) \odot N)[r]\end{aligned}$$

An example

Consider the following two fuzzy numbers in parametric forms,

$$M[r] = [M_l(r), M_u(r)] = [2r, 4 - 2r], N[r] = [N_l(r), N_u(r)] = [r - 1, 1 - r],$$

Now to obtain $M \ominus_H N = K$,

$$K_l(r) = r + 1, \quad K_u(r) = 3 - r$$

So, the Hukuhara difference in parametric form is,

$$K[r] = [r + 1, 3 - r]$$

In the following figure the H-difference is shown (Fig. 2.17).

As it is seen in the figure and based on the definition of H-difference,

$$M_u(r) - M_l(r) = Fuzz(M_r) \geq Fuzz(M \ominus_H N)_r = K_u(r) - K_l(r)$$
$$\&$$
$$N_u(r) - N_l(r) = Fuzz(N_r) \leq M_u(r) - M_l(r) = Fuzz(M_r)$$

Because the difference K is a shift for extending of N to be into M.

Now we are going to find some sufficient conditions for existence of H-difference. To this purpose let us consider,

$$M_r = (M_{1,r}, M_{2,r}, M_{3,r}), N_r = (N_{1,r}, N_{2,r}, N_{3,r}), K_r = (K_{1,r}, K_{2,r}, K_{3,r})$$

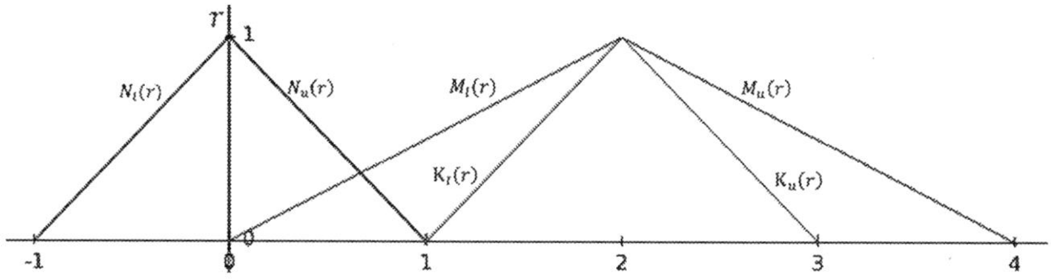

Fig. 2.17 $M \ominus_H N$

where

$$M_{1,r} \leq M_{2,r} \leq M_{3,r}, \;\; N_{1,r} \leq N_{2,r} \leq N_{3,r}, \;\; K_{1,r} \leq K_{2,r} \leq K_{3,r}$$

are representation of fuzzy numbers in triple forms for each level r.

It means in each level the intervals are satisfied the conditions of a real interval.

Lemma The sufficient condition for the existence of the H-difference $M \ominus_H N = K$ is

$$Fuzz(N_r) = N_{3,r} - N_{1,r} \leq \min\{M_{2,r} - M_{1,r}, M_{3,r} - M_{2,r}\}$$

To show $M = N \oplus K$ with condition $K_{1,r} \leq K_{2,r} \leq K_{3,r}$, our assertion for the existence is only proving $K_{1,r} \leq K_{2,r} \leq K_{3,r}$ or $M_{1,r} - N_{1,r} \leq M_{2,r} - N_{2,r} \leq M_{3,r} - N_{3,r}$. Because,

$$M = N \oplus K \Leftrightarrow M_{1,r} = N_{1,r} + K_{1,r}, M_{2,r} = N_{2,r} + K_{2,r}, M_{3,r} = N_{3,r} + K_{3,r}$$

To show $M_{1,r} - N_{1,r} \leq M_{3,r} - N_{3,r}$ it is enough to show $N_{3,r} - N_{1,r} \leq M_{3,r} - M_{1,r}$. In the first case suppose that, $\min\{M_{2,r} - M_{1,r}, M_{3,r} - M_{2,r}\} = M_{2,r} - M_{1,r} > 0$. Now we have $N_{3,r} - N_{1,r} \leq M_{2,r} - M_{1,r} \leq M_{3,r} - M_{1,r}$ and the proof is completed.

In the second case suppose that, $\min\{M_{2,r} - M_{1,r}, M_{3,r} - M_{2,r}\} = M_{3,r} - M_{2,r} > 0$. Now we have $N_{3,r} - N_{1,r} \leq M_{3,r} - M_{2,r} \leq M_{3,r} - M_{1,r}$ and the proof is also completed.

To show $M_{1,r} - N_{1,r} \leq M_{2,r} - N_{2,r}$ it is enough to show $N_{2,r} - N_{1,r} \leq M_{2,r} - M_{1,r}$. In the first case suppose that, $\min\{M_{2,r} - M_{1,r}, M_{3,r} - M_{2,r}\} = M_{2,r} - M_{1,r} > 0$. Now we have $N_{2,r} - N_{1,r} \leq N_{3,r} - N_{1,r} \leq M_{2,r} - M_{1,r}$ and the proof is completed. In the second case suppose that, $\min\{M_{2,r} - M_{1,r}, M_{3,r} - M_{2,r}\} = M_{3,r} - M_{2,r} > 0$. Now we have $N_{2,r} - N_{1,r} \leq N_{3,r} - N_{1,r} \leq M_{3,r} - M_{2,r} \leq M_{2,r} - M_{1,r}$ and the proof is also completed.

To show $M_{2,r} - N_{2,r} \leq M_{3,r} - N_{3,r}$ it is enough to show $N_{3,r} - N_{2,r} \leq M_{3,r} - M_{2,r}$. In the first case suppose that, $\min\{M_{2,r} - M_{1,r}, M_{3,r} - M_{2,r}\} = M_{2,r} - M_{1,r} > 0$. Now we have $N_{3,r} - N_{2,r} \leq N_{3,r} - N_{1,r} \leq M_{2,r} - M_{1,r} \leq M_{3,r} - M_{2,r}$ and the proof is completed. In the second case suppose that, $\min\{M_{2,r} - M_{1,r}, M_{3,r} - M_{2,r}\} = M_{3,r} - M_{2,r} > 0$. Now we have $N_{3,r} - N_{2,r} \leq N_{3,r} - N_{1,r} \leq M_{3,r} - M_{2,r}$ and the proof is also completed.

An example As we mentioned the existence of the difference is conditional. Now in this example we will see that it does not always exist. Suppose that in $M \ominus_H N$

$$Fuzz(N_r) = N_{3,r} - N_{1,r} > \min\{M_{2,r} - M_{1,r}, M_{3,r} - M_{2,r}\}$$

For instance, $M[r] = [1 - r, r - 1]\, and\, N[r] = [2 - r, r - 2]$, now

$$(M \ominus_H N)[r] = [M_l(r) - N_l(r), M_u(r) - N_u(r)] = [-1, -3]$$

As you see it is not an interval and for any r. So, the difference does not exist.

2.3.9.4 Generalized Hukuhara Difference

In this case, we can define the difference in another way. Suppose that we want to try $N \ominus_H M = (-1)K$ and may the difference K does exist. Considering the level-wise forms of two sides,

$$\begin{aligned}(N \ominus_H M)[r] &= [N_l(r) - M_l(r), N_u(r) - M_u(r)] = ((-1)K)[r] \\ &= [-K_u(r), -K_l(r)]\end{aligned}$$

We have,

$$N_l(r) - M_l(r) = -K_u(r), \quad N_u(r) - M_u(r) = -K_l(r)$$

Or

$$M_l(r) - N_l(r) = K_u(r), \quad M_u(r) - N_u(r) = K_l(r)$$

In general,

$$\begin{aligned}\left[(N \ominus_H M)_l(r), (N \ominus_H M)_u(r)\right] &= [K_l(r), K_u(r)] \\ &= [M_u(r) - N_u(r), M_l(r) - N_l(r)] \\ &= \ominus(-1)\left[(M \ominus_H N)_l(r), (M \ominus_H N)_u(r)\right]\end{aligned}$$

Now to define an almost right definition for the difference, we have two cases to consider.

$$M \ominus_{gH} N = K \Leftrightarrow \begin{cases} (i) & M = N \oplus K \\ & or \\ (ii) & N = M \oplus (-1)K \end{cases}$$

The generalized Hukuhara difference is defined in two cases. If the case (i) exist so there is no need to consider the case (ii). Otherwise we will need the second case. The relation between two cases can be explained as follow,

- In case (i),

$$M \ominus_{gH} N = K$$

- In case (ii),

$$N \ominus_{gH} M = (-1)K$$

The relationship is,

$$(M \ominus_{gH} N)_i[r] := 0 \ominus_H (-1)((M \ominus_{gH} N)_{ii}[r])$$

In case both exist then, $K = (-1)K$ and it is concluded that both types of the difference are the same and equal.

The level-wise form of generalized difference

As we found the level-wise form in case (i) is as,

$$\begin{aligned}\left[(M \ominus_{gH} N)_l(r), (M \ominus_{gH} N)_u(r)\right] &= [K_l(r), K_u(r)] \\ &= [M_l(r) - N_l(r), M_u(r) - N_u(r)]\end{aligned}$$

And in case (ii) it is as follow,

$$\begin{aligned}\left[(N \ominus_{gH} M)_l(r), (N \ominus_{gH} M)_u(r)\right] &= [K_l(r), K_u(r)] \\ &= [M_u(r) - N_u(r), M_l(r) - N_l(r)]\end{aligned}$$

So to define the endpoints of the difference,

$$K_l(r) = \min\{M_l(r) - N_l(r), M_u(r) - N_u(r)\}$$

$$K_u(r) = \max\{M_l(r) - N_l(r), M_u(r) - N_u(r)\}$$

To show two cases at the same time, we use gH-difference notation and define it in the following form,

$$(M \ominus_{gH} N)[r] = [\min\{M_l(r) - N_l(r), M_u(r) - N_u(r)\}, \max\{M_l(r) - N_l(r), M_u(r) - N_u(r)\}]$$

Some properties of gH-difference

Please note that all of the following properties can be proved in level-wise form easily.

1. If the gH-difference exist, it is unique.
2. $M \ominus_{gH} M = 0$
3. If $M \ominus_{gH} N$ exists in case (i) then $N \ominus_{gH} M$ exists in case (ii) and vise-versa.
4. In both cases $(M \oplus N) \ominus_{gH} N = M$. (It is easy to show in level-wise form.)
5. If $M \ominus_{gH} N$ and $N \ominus_{gH} M$ exist, then $0 \ominus_{gH} \left(M \ominus_{gH} N\right) = N \ominus_{gH} M$
6. If $M \ominus_{gH} N = N \ominus_{gH} M = K$ if and only if $K = -K$ and $M = N$.

The difference even in the gH-difference case may not exist. It can be say that the gH-difference of two fuzzy numbers are not always a fuzzy number.

An example This example shows that the generalized Hukuhara difference does not exist for each arbitrary level of difference.

Suppose that one of the numbers is triangular and another one is in trapezoidal forms. $M = (0, 2, 4)$ or in parametric form $M[r] = [2r, 4 - 2r]$ and $N = (0, 1, 2, 3)$ or in parametric form $N[r] = [r, 3 - r]$.

In case (i),

$$\left(M \ominus_{gH} N\right)[r] = [r, 1 - r]$$

If $r = 1$, the difference is as $[1, 0]$ that is not an interval.

In case (ii),

$$\left(M \ominus_{gH} N\right)[r] = [1 - r, r]$$

If $r = 0$, the difference is as $[1, 0]$ that is not an interval. So as we see the gH-difference does not exist for all $r \in [0, 1]$.

Note. In all methods of this book, we will suppose that the gH-difference always exists.

2.3.9.5 Partial Ordering

For two fuzzy numbers $M, N \in \mathbb{F}_R$, we call $\preccurlyeq$ as a partial order notation and

$$M \preccurlyeq N \text{ if and only if } M_l(r) \leq N_l(r) \text{ and } M_u(r) \leq N_u(r)$$

also we have the same definition for the strict inequality,

$$M \prec N \text{ if and only if } M_l(r) < N_l(r) \text{ and } M_u(r) < N_u(r)$$

For any $r \in [0, 1]$.

Some properties of partial ordering

- If $M \preccurlyeq N$ then $-N \preccurlyeq -M$
- If $M \preccurlyeq N$ and $N \preccurlyeq M$ then $M = N$.

To prove the properties, we use the level-wise form and they are very clear. For instance, we prove the first property,

$$M \preccurlyeq N \text{ if and only if } M_l(r) \leq N_l(r) \text{ and } M_u(r) \leq N_u(r)$$
$$-N \preccurlyeq -M \text{ if and only if } -N_l(r) \leq -M_l(r) \text{ and } -N_u(r) \leq -M_u(r)$$

So the proof is completed. The second on is obtained in similar way.

Absolute value of a fuzzy number

The absolute value of a fuzzy number M is defined as follow,

$$|M| = \begin{cases} M, & M \succcurlyeq 0 \\ -M, & M \prec 0 \end{cases}$$

where 0 fuzzy number is called a singleton fuzzy zero number.

Some properties of partial ordering in gH-difference

- If $M \preccurlyeq N$ then $M \ominus_{gH} N \preccurlyeq 0$
- If $M \succcurlyeq N$ then $M \ominus_{gH} N \succcurlyeq 0$

They are very easy to prove in level-wise form.

2.3.9.6 Approximately Generalized Hukuhara Difference

In case of the gH-diiference does not exist or $(M \ominus_{gH} N)[r]$ don't define a fuzzy number for any $r \in [0, 1]$, we can use the nested property of the fuzzy numbers and define a proper fuzzy number as a difference. We call this approximately gH-difference and denoted by $\ominus_g$ and it is defined in level-wise form as follows,

$$(M \ominus_g N)[r] = cl\left(\bigcup_{\beta \geq r} (M \ominus_{gH} N)[\beta]\right), \quad r \in [0, 1]$$

If the gH-difference $(M \ominus_{gH} N)[\beta]$ exist or define a proper fuzzy number for any $\beta \in [0, 1]$ then $(M \ominus_g N)[r]$ is exactly same as gH-difference $(M \ominus_{gH} N)[r]$ and it is exactly same as $(M \ominus_H N)[r]$ Huhuhara difference.

where

$$(M \ominus_g N)[r] = \left[\underset{\beta \geq r}{inf} \min\{M_l(\beta) - N_l(\beta), M_u(\beta) - N_u(\beta)\}, \underset{\beta \geq r}{sup} \max\{M_l(\beta) - N_l(\beta), M_u(\beta) - N_u(\beta)\}\right]$$

Proposition g.1 For any two fuzzy numbers $M, N \in R_F$, the two of $M \ominus_g N$ and $N \ominus_g M$ exist for any $r \in [0, 1]$ and $M \ominus_g N = -(N \ominus_g M)$ where

$$(M \ominus_g N)[r] = [D_l(r), D_u(r)]$$

$$D_l(r) = \inf\{\{M_l(\beta) - N_l(\beta) | \beta \geq r\} \cup \{M_u(\beta) - N_u(\beta) | \beta \geq r\}\}$$
$$D_u(r) = \sup\{\{M_l(\beta) - N_l(\beta) | \beta \geq r\} \cup \{M_u(\beta) - N_u(\beta) | \beta \geq r\}\}$$

Please note that in case of finite numbers of level or discretized levels, the mentioned above interval is as same as gH-difference.

$$(M \ominus_g N)[r] = [D_l(r), D_u(r)]$$

$$D_l(r) = \min\{M_l(r) - N_l(r), M_u(r) - N_u(r)\}$$
$$D_u(r) = \max\{M_l(r) - N_l(r), M_u(r) - N_u(r)\}]$$

Proposition g.2 For any two fuzzy numbers $M, N \in R_F$, $M \ominus_g N$ always exist or is a fuzzy number. Because it satisfies the conditions of fuzzy numbers.

1. $\underset{\beta \geq r}{inf} \min\{M_l(\beta) - N_l(\beta), M_u(\beta) - N_u(\beta)\} \leq \underset{\beta \geq r}{sup} \max\{M_l(\beta) - N_l(\beta), M_u(\beta) - N_u(\beta)\}$
2. $\underset{\beta \geq r}{inf} \min\{M_l(\beta) - N_l(\beta), M_u(\beta) - N_u(\beta)\}$ is non-decreasing, left continuous and bounded function for any $r \in [0, 1]$.
3. $\underset{\beta \geq r}{sup} \max\{M_l(\beta) - N_l(\beta), M_u(\beta) - N_u(\beta)\}$ is non-increasing, left continuous and bounded function for any $r \in [0, 1]$.

Some properties of g-difference

For any two fuzzy numbers $M, N \in R_F$,

1. $M \ominus_g N = M \ominus_{gH} N$ subject to $M \ominus_{gH} N$ exists.
2. $M \ominus_g M = 0$
3. $(M \oplus N) \ominus_g N = M$
4. $0 \ominus_g (M \ominus_g N) := \ominus_g (M \ominus_g N) = N \ominus_g M$

5. $M \ominus_g N = N \ominus_g M = K$ if and only if $K = -K$, the immediate conclusion is $K = 0$ and $M = N$. In conclusion,

$$M \ominus_g N = N \ominus_g M \Leftrightarrow M = N$$

All the properties can be proved very easily based on the definition of the g-difference in level-wise form.

Let us consider some examples when the gH-difference does not exist, while the g-difference exists.

Example Consider two trapezoidal fuzzy numbers as follow,

$$M = (0, 2, 2, 4), \quad N = (0, 1, 2, 3)$$

where

$$M_l(\beta) = 2\beta, \quad M_u(\beta) = 4 - 2\beta, \quad N_l(\beta) = \beta, \quad N_u(\beta) = 3 - \beta$$

And

$$\inf_{\beta \geq r} \min\{M_l(\beta) - N_l(\beta), M_u(\beta) - N_u(\beta)\} = \inf_{\beta \geq r} \min\{\beta, 1 - \beta\} = 0$$
$$\sup_{\beta \geq r} \max\{M_l(\beta) - N_l(\beta), M_u(\beta) - N_u(\beta)\} = \sup_{\beta \geq r} \max\{\beta, 1 - \beta\} = 1$$

So the g-difference is,

$$\left(M \ominus_g N\right)[r] = [0, 1]$$

And the fuzzy number for g-difference $(0, 0, 1, 1)$ is as a fuzzy number.

In the next example we will see that the gH-difference does not exist.

Example Consider two trapezoidal fuzzy numbers as follow,

$$M = (2, 3, 5, 6), \ N = (0, 4, 4, 8)$$

where

$$M_l(\beta) = 2 + \beta, \quad M_u(\beta) = 6 - \beta, \quad N_l(\beta) = 4\beta, \quad N_u(\beta) = 8 - 4\beta$$

And

$$\inf_{\beta \geq r} \min\{M_l(\beta) - N_l(\beta), M_u(\beta) - N_u(\beta)\} = \inf_{\beta \geq r} \min\{2 - 3\beta, -2 + 3\beta\} =$$
$$\sup_{\beta \geq r} \max\{M_l(\beta) - N_l(\beta), M_u(\beta) - N_u(\beta)\} = \sup_{\beta \geq r} \max\{2 - 3\beta, -2 + 3\beta\} =$$

Based on the definition of gH-difference we **cannot claim** that

$$\min\{2-3r,-2+3r\}\le\max\{2-3r,-2+3r\}$$

For any $r\in[0,1]$. Because,

$$M_u(r)-N_u(r)=-(M_l(r)-N_l(r))$$

And
If $M_l(r)-N_l(r)\ge 0$ then $M_u(r)-N_u(r)\le M_l(r)-N_l(r)$ then

$$\begin{aligned}\min\{M_l(r)-N_l(r),M_u(r)-N_u(r)\}&=M_u(r)-N_u(r)\\ \max\{M_l(r)-N_l(r),M_u(r)-N_u(r)\}&=M_l(r)-N_l(r)\end{aligned}$$

If $M_l(r)-N_l(r)<0$ then $M_u(r)-N_u(r)>M_l(r)-N_l(r)$ then

$$\begin{aligned}\min\{M_l(r)-N_l(r),M_u(r)-N_u(r)\}&=M_l(r)-N_l(r)\\ \max\{M_l(r)-N_l(r),M_u(r)-N_u(r)\}&=M_u(r)-N_u(r)\end{aligned}$$

Now here for $0\le r\le\frac{2}{3}$,

$$\begin{aligned}\min\{M_l(r)-N_l(r),M_u(r)-N_u(r)\}&=-2+3r\\ \max\{M_l(r)-N_l(r),M_u(r)-N_u(r)\}&=2-3r\end{aligned}$$

and otherwise for $\frac{2}{3}<r\le 1$,

$$\begin{aligned}\min\{M_l(r)-N_l(r),M_u(r)-N_u(r)\}&=2-3r\\ \max\{M_l(r)-N_l(r),M_u(r)-N_u(r)\}&=-2+3r\end{aligned}$$

Then the gH-difference doesn't exist.
But the g-difference is as
For $0\le r\le 0.32$,

$$(M\ominus_g N)[r]=\left[\frac{100}{32}r-2,\frac{-100}{32}r+2\right]$$

And for $0.32<r\le 1$,

$$(M\ominus_g N)[r]=[-1,1]$$

2.3.9.7 Generalized Division

Now like generalized difference, for any two arbitrary fuzzy numbers *M* and *N*, the generalized division can be defined as follows:

$$M \o_g N = K \Leftrightarrow \begin{cases} (i) & M = N \odot K \\ & or \\ (ii) & N = M \odot K^{-1} \end{cases}$$

It is clear if both cases are true then $M = N \odot K = M \odot K^{-1} \odot K$ and in this case the only option is $K^{-1} \odot K = K \odot K^{-1} = \{1\}$ where 1 is a singleton fuzzy number or a real scalar is and *K* is also a non-zero real scalar.

The generalized division in the level-wise or interval form can be defined as follow,

$$M[r]/N[r] = K[r] \Leftrightarrow \begin{cases} (i) & M[r] = N[r] \cdot K[r] \\ & or \\ (ii) & N[r] = M[r] \cdot K^{-1}[r] \end{cases}$$

where $K^{-1}[r] := \left[\frac{1}{K_u(r)}, \frac{1}{K_l(r)}\right]$ and doesn't contain zero for all $r \in [0, 1]$.

In case (i),

$$[M_l(r), M_u(r)] = [N_l(r), N_u(r)] \cdot [K_l(r), K_u(r)]$$

And indeed,

$$[K_l(r), K_u(r)] = [M_l(r), M_u(r)]/[N_l(r), N_u(r)]$$

Then

$$K_l(r) = \min\left\{\frac{M_l(r)}{N_l(r)}, \frac{M_l(r)}{N_u(r)}, \frac{M_u(r)}{N_l(r)}, \frac{M_u(r)}{N_u(r)}\right\},$$

$$K_u(r) = \max\left\{\frac{M_l(r)}{N_l(r)}, \frac{M_l(r)}{N_u(r)}, \frac{M_u(r)}{N_l(r)}, \frac{M_u(r)}{N_u(r)}\right\}$$

Subject to $0 \notin [N_l(r), N_u(r)]$ it means $N_u(r) < 0,\ or\ 0 < N_l(r)$ for all $r \in [0, 1]$.

In case (ii),

$$[N_l(r), N_u(r)] = [M_l(r), M_u(r)] \cdot \left[\frac{1}{K_u(r)}, \frac{1}{K_l(r)}\right]$$

And indeed,

$$\left[\frac{1}{K_u(r)}, \frac{1}{K_l(r)}\right] = [N_l(r), N_u(r)]/[M_l(r), M_u(r)]$$

So,

$$\frac{1}{K_u(r)} = \min\left\{\frac{N_l(r)}{M_l(r)}, \frac{N_l(r)}{M_u(r)}, \frac{N_u(r)}{M_l(r)}, \frac{N_u(r)}{M_u(r)}\right\},$$

$$\frac{1}{K_l(r)} = \max\left\{\frac{N_l(r)}{M_l(r)}, \frac{N_l(r)}{M_u(r)}, \frac{N_u(r)}{M_l(r)}, \frac{N_u(r)}{M_u(r)}\right\}$$

In fact similar to two cases of generalized difference, two endpoints changed the roles.

$$K_u(r) = \min\left\{\frac{M_l(r)}{N_l(r)}, \frac{M_l(r)}{N_u(r)}, \frac{M_u(r)}{N_l(r)}, \frac{M_u(r)}{N_u(r)}\right\},$$

$$K_l(r) = \max\left\{\frac{M_l(r)}{N_l(r)}, \frac{M_l(r)}{N_u(r)}, \frac{M_u(r)}{N_l(r)}, \frac{M_u(r)}{N_u(r)}\right\}$$

Remark The following results in both cases of division are easy to investigate.

If $M_l(r) > 0, N_l(r) > 0$ then $K_l(r) = \frac{M_l(r)}{N_u(r)}, K_u(r) = \frac{M_u(r)}{N_l(r)}$

If $M_u(r) < 0, N_u(r) < 0$ then $K_l(r) = \frac{M_u(r)}{N_l(r)}, K_l(r) = \frac{M_l(r)}{N_u(r)}$

In general, if

$$K_l(r) = \frac{M_i(r)}{N_i(r)}, K_u(r) = \frac{M_j(r)}{N_j(r)}, i, j \in \{l, u\}$$

Then the enough condition to reach $K_l(r) \leq K_u(r)$ is

$$M_i(r) \cdot N_j(r) \leq M_j(r) \cdot N_i(r), i, j \in \{l, u\} \quad \text{for all } r \in [0, 1]$$

And $K_l(r)$ and $K_u(r)$ must be left continuous increasing and decreasing functions respectively.

Some properties of division

Please note that all the following properties can be proved in level-wise form easily. Suppose that M and N are fuzzy number and 1 is $\{1\}$.

1. If $0 \notin M[r], \forall r \in [0, 1]$ then $M \o_g M = 1$
2. If $0 \notin N[r], \forall r \in [0, 1]$ then $M \odot N \o_g N = M$
3. If $0 \notin M[r], \forall r \in [0, 1]$ then $1 \o_g M = M^{-1}$ and $1 \o_g M^{-1} = M$

4. If $M ø N$ exists, then either $N \odot (M ø_g N) = M$ or $M \odot (M ø_g N)^{-1} = N$ and both equalities hold if and only if $M ø_g N$ is real number.

Note. In all methods of this book, we will suppose that the division always exists.

Examples Here we explain the generalized division based on its level-wise definition.

Suppose that $M[r] = [1 + 2r, 7 - 4r]$ and $N[r] = [-3 + r, -1 - r]$ now according to case (i),

$$M[r]/N[r] = K[r] \Leftrightarrow M[r] = N[r] \cdot K[r]$$

And is $K[r] = \left[\frac{7-4r}{-3+r}, \frac{1+2r}{-1-r}\right]$.

Suppose that $M[r] = [-7 + 2r - 4 - r]$ and $N[r] = [12 + 5r, -4 - 3r]$ now according to case (ii),

$$M[r]/N[r] = K[r] \Leftrightarrow N[r] = M[r] \cdot K^{-1}[r]$$

And is $K[r] = \left[\frac{-7+2r}{-12+5r}, \frac{-5+r}{11-5r}\right]$.

In some cases the g-division doesn't exist. For instance consider

$$M[r] = [1 + 0.5r, 5 - 3.5r] \text{ and } N[r] = [-4 + 2r, -1 - r]$$

In this case like as the g-difference, we have to define an approximate g-division as follows,

2.3.9.8 Approximately Generalized Division

In case of the g-division does not exist or $(M ø_{Ag} N)[r]$ don't define a fuzzy number for any $r \in [0, 1]$, we can use the nested property of the fuzzy numbers and define a proper fuzzy number as a division. We call this approximately g-division and denoted by $ø_{Ag}$ and it is defined in level-wise form as follows,

$$(M ø_{Ag} N)[r] = cl\left(\bigcup_{\beta \geq r} (M ø_g N)[\beta]\right), \quad r \in [0, 1]$$

If the g-division $(M ø_g N)[\beta]$ exist or define a proper fuzzy number for any $\beta \in [0, 1]$ then $(M ø_{Ag} N)[r]$ is exactly same as gH-difference $(M ø_g N)[r]$ subject to $0 \notin N[r]$, for any $r \in [0, 1]$.

Suppose that $M\text{ø}_{Ag}N = K$ on a discrete partition $0 \le r_0 \le r_1 \le \cdots \le r_n \le 1$ of $[0, 1]$. A discretize version of $M\text{ø}_{Ag}N = K$ is obtained using

$$M[r_i]/N[r_i] = W[r_i] = [W_{l,i}, W_{u,i}], 0 \le i \le 1$$

And $K_{l,n} = W_{l,n}, K_{u,n} = W_{u,n}$ also for $i = n-1, \ldots, 0$

$$K_{l,i} = \min\{K_{l,i+1}, W_{l,i}\}$$

$$K_{u,i} = \max\{K_{u,i+1}, W_{u,i}\}$$

Example Suppose that $M[r] = [1+0.5r, 5-3.5r]$ and $N[r] = [-4+2r, -1-r]$. The g-division $M\text{ø}_g N = \left[\frac{5-3.5r}{-4+2r}, \frac{1+0.5r}{-1-r}\right]$ do not define a proper fuzzy number for any $r \in [0, 1]$. But using the Ag-division the result is as $M\text{ø}_{Ag}N = \left[\frac{5-3.5r}{-4+2r}, -0.75\right]$.

2.3.10 Piece Wise Membership Function

Sometimes the lower and upper functions are piece wise. For instance (Fig. 2.18).

$$\begin{aligned}
M_1(r) &= x_1 + (x_2 - x_1)L_1^{-1}(r), && r \in [x_1, x_2],\\
M_2(r) &= x_2 + (x_3 - x_2)L_2^{-1}(r), && r \in [x_2, x_3],\\
M_3(r) &= x_4 - (x_4 - x_3)R_1^{-1}(r), && r \in [x_3, x_4],\\
M_4(r) &= x_5 - (x_5 - x_4)R_2^{-1}(r), && r \in [x_4, x_5].
\end{aligned}$$

The point is, in any format including piece-wise form, the conditions of a fuzzy number should be satisfied by the function.

2.3.11 Some Properties of Addition and Scalar Product on Fuzzy Numbers

In this section we need to consider and prove the following properties. Here it is supposed that the zero fuzzy number is a singleton fuzzy number. In general, a singleton fuzzy number is defined as,

Definition—Singleton fuzzy number A fuzzy number like a is called a singleton fuzzy number if the membership degree of a is one and the membership degrees for the other members are zero. See the following figure (Fig. 2.19).

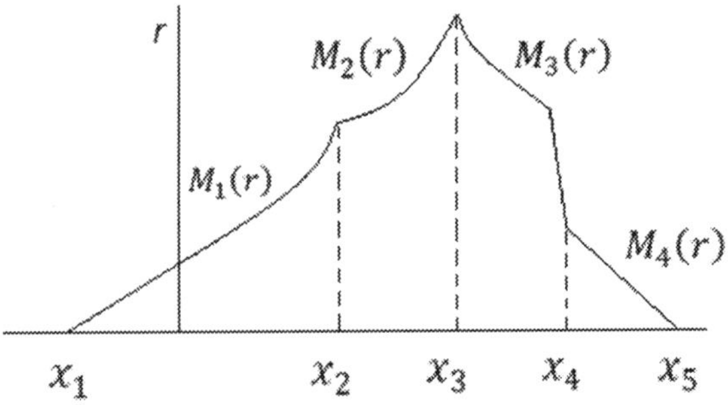

Fig. 2.18 Multi functional membership function

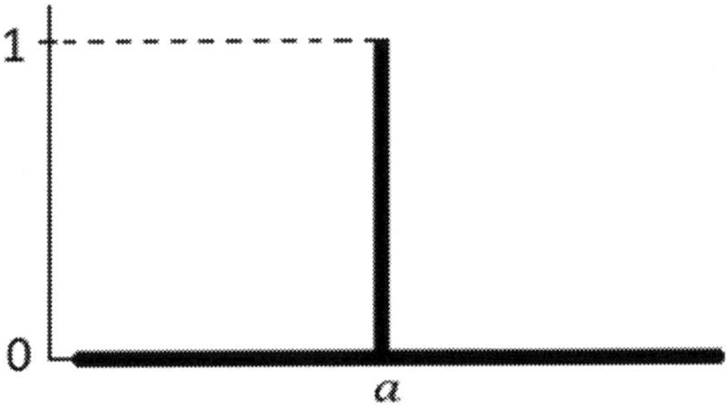

Fig. 2.19 Singleton fuzzy number

Some of the scalar product and addition properties on fuzzy numbers are as follows. For all properties $0, M \in F_R$ are singleton zero number and any fuzzy number respectively which F_R is the set of all fuzzy numbers those are defined on real numbers.

1. $0 \oplus M = M \oplus 0$
2. There is no inverse with respect to $\oplus$. It means $M \oplus (-1)M \neq 0$
3. For two arbitrary real numbers $a, b \in R$ in which $a \cdot b \geq 0$,

$$(a+b)M = a \odot M \oplus b \odot M$$

4. For any $\lambda \in R$ and $M, N \in R_F$,

$$\lambda \odot (M \oplus N) = \lambda \odot M \oplus \lambda \odot N$$

5. For ant $\lambda, \mu \in R$ and $M \in R_F$,

$$\lambda \odot (\mu \odot M) = (\lambda \cdot \mu) \odot M$$

All mentioned above properties can be proved very easily by using the computations in level-wise form.

2.4 Some Operators on Fuzzy Numbers

In this section, some operators like Distance, Limit and, Riemann integral are introduced for the fuzzy numbers and fuzzy number valued functions.

2.4.1 Distance

Basically, the distance is a type of operators from the space all fuzzy numbers to the set of real numbers. If we suppose that the $\mathbb{F}_R$ is the set of all fuzzy numbers then,

$$D: \mathbb{F}_R \times \mathbb{F}_R \rightarrow R$$

Clearly, this operator should have some conditions or properties like as following.

For any arbitrary fuzzy numbers $M,\ N, K\ and\ P$,

- $D(M,N) > 0$
- $D(M,N) = 0 \Leftrightarrow M = N$
- $D(M \oplus K, N \oplus K) = D(M,K)$
- $D(M \oplus N, K \oplus P) \leq D(M,K) + D(N,P)$
- $D(\lambda M, \lambda N) = |\lambda| D(M,N), \quad \lambda \in R$

Hausdorff distance

Again, for two fuzzy numbers $M\ and\ N$, $D_H(M,N) \in R^{\geq 0}$ as Hausdorf distance is defined as follow,

$$D_H(M,N) = sup_{0 \leq r \leq 1} \max\{|M_l(r) - N_l(r)|, |M_u(r) - N_u(r)|\}$$

All the properties can be verified easily. Here some of the properties are proved as follow,

- $D_H(M \ominus_H N, K \ominus_H P) \leq D_H(M,K) + D_H(N,P)$, subject that $M \ominus_H N$ and $K \ominus_H P$ exist.

To prove it, we have the following relations,

$$|M_l(r) - N_l(r)| \leq \max\{|M_l(r) - N_l(r)|, |M_u(r) - N_u(r)|\}$$
$$|M_u(r) - N_u(r)| \leq \max\{|M_l(r) - N_l(r)|, |M_u(r) - N_u(r)|\}$$

And

$$|K_l(r) - P_l(r)| \leq \max\{|K_l(r) - P_l(r)|, |K_u(r) - P_u(r)|\}$$
$$|K_u(r) - P_u(r)| \leq \max\{|K_l(r) - P_l(r)|, |K_u(r) - P_u(r)|\}$$

$$\begin{aligned}|M_l(r) - N_l(r) - K_l(r) + P_l(r)| &\leq |M_l(r) - N_l(r)| + |K_l(r) - P_l(r)| \\ &\leq \max\{|M_l(r) - N_l(r)|, |M_u(r) - N_u(r)|\} + \max\{|K_l(r) - P_l(r)|, |K_u(r) - P_u(r)|\}\end{aligned}$$

$$\begin{aligned}|M_u(r) - N_u(r) - K_u(r) + P_u(r)| &\leq |M_u(r) - N_u(r)| + |K_u(r) - P_u(r)| \\ &\leq \max\{|M_l(r) - N_l(r)|, |M_u(r) - N_u(r)|\} + \max\{|K_l(r) - P_l(r)|, |K_u(r) - P_u(r)|\}\end{aligned}$$

So,

$$\begin{aligned}&\max\{|M_l(r) - N_l(r) - K_l(r) + P_l(r)|, |M_u(r) - N_u(r) - K_u(r) + P_u(r)|\} \\ &\leq \max\{|M_l(r) - N_l(r)|, |M_u(r) - N_u(r)|\} + \max\{|K_l(r) - P_l(r)|, |K_u(r) - P_u(r)|\}\end{aligned}$$

Then

$$sup_{0 \leq r \leq 1} \max\{|M_l(r) - N_l(r) - K_l(r) + P_l(r)|, |M_u(r) - N_u(r) - K_u(r) + P_u(r)|\}$$

$$\begin{aligned}&\leq sup_{0 \leq r \leq 1} \max\{|M_l(r) - N_l(r)|, |M_u(r) - N_u(r)|\} \\ &\quad + sup_{0 \leq r \leq 1} \max\{|K_l(r) - P_l(r)|, |K_u(r) - P_u(r)|\}\end{aligned}$$

And the property is now proved and

$$D_H(M \ominus_H N, K \ominus_H P) \leq D_H(M, K) + D_H(N, P)$$

- $D_H\big(M \ominus_{gH} N, 0\big) = D_H(M, N)$

Based on the definition of Hausdorff distance the proof is easy. Because the left side can be written as,

$$\begin{aligned}D_H\big(M \ominus_{gH} N, 0\big) &= sup_{0 \leq r \leq 1} \max\{|M_l(r) - N_l(r) - 0|, |M_u(r) - N_u(r) - 0|\} \\ &= sup_{0 \leq r \leq 1} \max\{|M_l(r) - N_l(r)|, |M_u(r) - N_u(r)|\} = D_H(M, N)\end{aligned}$$

- $D_H\big(\lambda \odot M \ominus_{gH} \mu \odot M, 0\big) = |\lambda - \mu| D_H(M, 0), M \in \mathbb{F}_R, \lambda, \mu \geq 0.$

To prove the property, in accordance with the property of the distance the left side can be appeared as,

$$D_H\big(\lambda \odot M \ominus_{gH} \mu \odot M, 0\big) = D_H(\lambda \odot M, \mu \odot M)$$

Now, considering the definition of the distance,

$$\begin{aligned} D_H(\lambda \odot M, \mu \odot M) &= sup_{0 \le r \le 1} \max\{|\lambda M_l(r) - \mu M_l(r)|, |\lambda M_u(r) - \mu M_u(r)|\} \\ &= sup_{0 \le r \le 1} \max\{|\lambda - \mu|M_l(r), |\lambda - \mu|M_u(r)\} \\ &= |\lambda - \mu| sup_{0 \le r \le 1} \max\{M_l(r), M_u(r)\} = |\lambda - \mu| D_H(M, 0) \end{aligned}$$

The proof is now completed.

2.4.2 *Limit of Fuzzy Number Valued Functions*

In this section, we are going to display some preliminarily definitions and theorems about fuzzy set-number valued functions.

Definition—Fuzzy set valued function Any function like $x(t)$ is called a fuzzy set valued function, if it is a fuzzy set for any $t \in R$.

Definition—Fuzzy Number Valued Function Any function like $x(t)$ is called a fuzzy number valued function, if it is a fuzzy number for any $t \in R$.

Definition—The Limit of Fuzzy number valued function Suppose that $x(t)$ is a fuzzy number valued function and is defined from any real interval like $[a, b]$ to $\mathbb{F}_R$. If for any positive ϵ there is a positive δ such that

$$\forall \epsilon > 0 \exists \delta > 0 \forall t (|t - t_0| < \delta \Rightarrow D_H(x(t), L) < \epsilon)$$

where D is the Hausdorff of $x(t)$ and $L \in \mathbb{F}_R$. It is equivalent to the following limit

$$\lim_{t \to t_0} x(t) = L$$

Note. The limit of a fuzzy number valued function $x(t)$ exist whenever the value L is a fuzzy number not fuzzy set.

Also, if we have

$$\lim_{t \to t_0} x(t) = x(t_0)$$

Then the function $x(t)$ is called continuous at the point t_0.

Please note that the same function is continuous on the its real domain $[a, b]$ if it is continuous at all points of the interval.

Now we are going to consider some properties on the limit of a fuzzy number valued function.

Theorem—Limit of Summation of Functions Suppose that $x(t), y(t) : [a, b] \rightarrow \mathbb{F}_R$ are two fuzzy number valued functions. If

$$\lim_{t \to t_0} x(t) = L_1, \lim_{t \to t_0} y(t) = L_2$$

Such that two values $L_1, L_2 \in \mathbb{F}_R$. Then

$$\lim_{t \to t_0} [x(t) \oplus y(t)] = L_1 \oplus L_2$$

Proof For the function $x(t)$,

$$\forall \frac{\epsilon}{2} > 0 \exists \delta_1 > 0 \forall t \left(|t - t_0| < \delta_1 \Rightarrow D_H(x(t), L_1) < \frac{\epsilon}{2} \right)$$

Because $\lim_{t \to t_0} x(t) = L_1$. And also, for another one $y(t)$, we have

$$\forall \frac{\epsilon}{2} > 0 \exists \delta_2 > 0 \forall t \left(|t - t_0| < \delta_2 \Rightarrow D_H(y(t), L_2) < \frac{\epsilon}{2} \right)$$

Now $|t - t_0| < \delta_1$ and $|t - t_0| < \delta_2$ clearly $|t - t_0| < \min\{\delta_1, \delta_2\} = \delta$. Now for any positive $\frac{\epsilon}{2}$ there is a positive δ and based on the property of the Hausdorff distance,

$$D_H(x(t) \oplus y(t), L_1 \oplus L_2) \leq D_H(x(t), L_1) + D_H(y(t), L_2) < \frac{\epsilon}{2} + \frac{\epsilon}{2} = \epsilon$$

So finally, we prove that

$$\forall \epsilon > 0 \exists \delta > 0 \forall t (|t - t_0| < \delta \Rightarrow D_H(x(t) \oplus y(t), L_1 \oplus L_2) < \epsilon)$$

And the proof is completed.

Theorem—Limit of Difference of Functions Considering all assumptions of the previous theorem then we can prove that

$$\lim_{t \to t_0} \left[x(t) \ominus_{gH} y(t) \right] = L_1 \ominus_{gH} L_2$$

Subject to $L_1 \ominus_{gH} L_2$ exists.

Proof Suppose that $x(t) \ominus_{gH} y(t) = z(t)$ so based on the definition of gH-difference,

$$x(t) \ominus_{gH} y(t) = z(t) \Leftrightarrow \begin{cases} (i) & x(t) = y(t) \oplus z(t) \\ & or \\ (ii) & y(t) = x(t) \oplus (-1)z(t) \end{cases}$$

- In case (i),

Since $\lim_{t\to t_0} x(t) = L_1$ then $\lim_{t\to t_0}(y(t) \oplus z(t)) = L_1$ and based on the previous theorem we have,

$$\lim_{t\to t_0}(y(t) \oplus z(t)) = \lim_{t\to t_0} y(t) \oplus \lim_{t\to t_0} z(t) = L_1$$

Now it is concluded that,

$$\lim_{t\to t_0} z(t) = L_1 \ominus \lim_{t\to t_0} y(t) = L_1 \ominus L_2$$

- In case (ii),

Since $\lim_{t\to t_0} y(t) = L_2$ then $\lim_{t\to t_0}(x(t) \oplus (-1)z(t)) = L_2$ and again based on the previous theorem we have,

$$\lim_{t\to t_0}(x(t) \oplus (-1)z(t)) = \lim_{t\to t_0} x(t) \oplus (-1) \lim_{t\to t_0} z(t) = L_2$$

Now it is concluded that,

$$L_1 \oplus (-1) \lim_{t\to t_0} z(t) = L_2$$

Moreover, based on the second case of the gH-difference

$$\lim_{t\to t_0} z(t) = L_1 \ominus L_2$$

and the proof is completed.

Theorem—Limit of Multiplication Suppose that $x(t) : [a, b] \to \mathbb{F}_R$ is a fuzzy number valued functions and $y(t)$ is a non-negative real function. If

$$\lim_{t\to t_0} x(t) = L_1, \lim_{t\to t_0} y(t) = L_2$$

Such that two values $L_1 \in \mathbb{F}_R, L_2 \in R^+$. Then

$$\lim_{t\to t_0}[x(t) \odot y(t)] = L_1 \odot L_2$$

Proof To prove it, first we investigate the Hausdorff distance of a fuzzy number valued function $x(t)$ and zero.

$$D_H(x(t),0) = D_H\big(x(t)\ominus_{gH} L_1 \oplus L_1, 0\big) \le D_H\big(x(t)\ominus_{gH} L_1, 0\big) + D_H(L_1,0)$$
$$= D_H(x(t),L_1) + D_H(L_1,0) < \epsilon + D_H(L_1,0)$$

The ϵ is any arbitrary positive number and it can be considered as $\epsilon = 1$, so

$$D_H(x(t),0) < 1 + D_H(L_1,0)$$

For the function $x(t)$ we have the same limit definition as,

$$\forall \epsilon_1 = \frac{\epsilon}{2L_2} > 0 \exists \delta_1 > 0 \forall t(|t-t_0| < \delta_1 \Rightarrow D_H(x(t),L_1) < \epsilon_1)$$

The function $y(t)$ is real and non-negative function and we have,

$$\forall \epsilon_2 = \frac{\epsilon}{2(1+L_1)} > 0 \exists \delta_2 > 0 \forall t(|t-t_0| < \delta_2 \Rightarrow |y(t) - L_2| < \epsilon_2)$$

By all assumptions we have,

$$D_H(x(t)\odot y(t), L_1 \odot L_2) = D_H\big(x(t)\odot y(t) \ominus_{gH} L_1 \odot L_2, 0\big)$$
$$= D_H\big(x(t)\odot y(t) \ominus_{gH} x(t)\odot L_2 \oplus x(t)\odot L_2 \ominus_{gH} L_1 \odot L_2, 0\big)$$

As we know $y(t)$ and L_2 are non-negative and let consider $(y(t) - L_2) \geq 0$. So the following result is immediately concluded.

$$x(t)\odot y(t) \ominus_{gH} x(t) \odot L_2 = x(t) \odot (y(t) - L_2)$$

Considering the level cuts of both sides it is proved easily.
Moreover

$$x(t)\odot L_2 \ominus_{gH} L_1 \odot L_2 = \big(x(t)\ominus_{gH} L_1\big)\odot L_2$$

Now the proof is continued as

$$\begin{aligned} D_H\big(x(t)\odot y(t) \ominus_{gH} x(t)\odot L_2 \oplus x(t)\odot L_2 \ominus_{gH} L_1 \odot L_2, 0\big) \\ = D_H\big(x(t)\odot (y(t)-L_2) \oplus \big(x(t)\ominus_{gH} L_1\big)\odot L_2, 0\big) \\ \le D_H(x(t)\odot (y(t)-L_2), 0) + D_H\big(\big(x(t)\ominus_{gH} L_1\big)\odot L_2, 0\big) \\ = |y(t) - L_2| D_H(x(t),0) + |L_2| D_H\big(x(t)\ominus_{gH} L_1, 0\big) \\ = |y(t) - L_2| D_H(x(t),0) + |L_2| D_H(x(t), L_1) \\ < \tfrac{\epsilon}{2(1+D_H(L_1,0))}(1 + D_H(L_1,0)) + |L_2|\tfrac{\epsilon}{2L_2} = \tfrac{\epsilon}{2} + \tfrac{\epsilon}{2} = \epsilon \end{aligned}$$

The proof now is completed.

Other properties of limit

Suppose that

$$\lim_{n\to\infty} x_n = x, \lim_{n\to\infty} y_n = y$$

Such that $x_n, y_n, x, y \in \mathbb{F}_R$, then

- $\lim_{n\to\infty} D_H(x_n, y_n) = D_H(\lim_{n\to\infty} x_n, \lim_{n\to\infty} y_n) = D_H(x, y)$
- $\lim_{n\to\infty}(c_n \odot x_n) = c \odot x$ where $\lim_{n\to\infty} c_n = c$ and c, c_n are two non-negative real numbers.

The proofs are very clear.

2.4.3 *Fuzzy Riemann Integral Operator*

Suppose that the function $f(t)$ is a fuzzy number valued function, $f(t) \in \mathbb{F}_R$. The Fuzzy Riemann integral is defined as the following summation and denoted by

$$J = FR\int_a^b f(t)dt := \oplus \sum_{i=0}^{n} \Delta t_i \odot f(t_i)$$

where $FR\int_a^b f(t)dt$ is denoted as fuzzy Riemann integral and $\oplus \sum_{i=0}^{n} \Delta t_i \odot f(t_i)$ points out fuzzy summations of fuzzy numbers $\Delta t_i \odot f(t_i)$ for any $i = 0, 1, \ldots, n$. Moreover $\Delta t_i = t_{i+1} - t_i$ and the set of points $\{a = t_0 < t_1 < \cdots < t_n = b\}$ is a partition on interval $[a, b]$.

To define the fuzzy Riemann integral, since both sides are fuzzy numbers and for the equality of two fuzzy numbers, we use the Hausdorff distance so,

$$\forall \epsilon > 0, \exists \delta > 0, D_H\left(\oplus \sum_{i=0}^{n} \Delta t_i \odot f(t_i), J\right) < \epsilon$$

Note. The important point is, J should be a fuzzy number not a fuzzy set.
In the level-wise form of integral, it can be displayed as,

$$J[r] = \left(FR\int_a^b f(t)dt\right)[r] := \left(\oplus \sum_{i=0}^{n} \Delta t_i \odot f(t_i)\right)[r]$$

For any $0 \le r \le 1$. Indeed, the definition of the Hausdorff distance needs the level-wise form and it can be brought as

$$[J_l(r), J_u(r)] = \left[\sum_{i=0}^{n} \Delta t_i \cdot f_l(t_i, r), \sum_{i=0}^{n} \Delta t_i \cdot f_u(t_i, r)\right]$$
$$= \left[R\int_a^b f_l(t, r)dt, R\int_a^b f_u(t, r)dt\right]$$

On the hand,

$$\left[\sum_{i=0}^{n} \Delta t_i \cdot f_l(t_i, r), \sum_{i=0}^{n} \Delta t_i \cdot f_u(t_i, r)\right] = \sum_{i=0}^{n} \Delta t_i [f_l(t_i, r), f_u(t_i, r)]$$

Then

$$\left[R\int_a^b f_l(t, r)dt, R\int_a^b f_u(t, r)dt\right] = R\int_a^b [f_l(t, r), f_u(t, r)]dt$$

In summary,

$$J_l(r) = R\int_a^b f_l(t, r)dt = \sum_{i=0}^{n} \Delta t_i \cdot f_l(t_i, r),$$
$$J_u(r) = R\int_a^b f_u(t, r)dt = \sum_{i=0}^{n} \Delta t_i \cdot f_u(t_i, r)$$

Some properties of fuzzy Riemann integral

- Suppose that the functions $f(t), g(t) \in \mathbb{F}_R$ and are Riemann integrable functions. Then

$$FR\int_a^b (f(t) \oplus g(t))dt = \oplus \sum_{i=0}^{n} \Delta t_i \odot (f(t_i) \oplus g(t_i))$$

This property is so easy to prove.

- For the previous function and any $c \in [a, b]$,

$$FR\int_a^b f(t)dt = FR\int_a^c f(t)dt \oplus FR\int_c^b f(t)dt$$

To prove the property, since $c \in [a, b]$ the partition for the integral will be as

$$\{a = t_0 < t_1 < \cdots < t_k = c < \cdots < t_n = b\}, \exists k \in \{0, 1, \ldots, n\}$$

Now based on the definition,

$$FR\int_a^c f(t)dt = \oplus\sum_{i=0}^{k} \Delta t_i \odot f(t_i)$$

$$FR\int_c^b f(t)dt = \oplus\sum_{i=k+1}^{n} \Delta t_i \odot f(t_i)$$

It is clear that

$$\oplus\sum_{i=0}^{k} \Delta t_i \odot f(t_i) \oplus \sum_{i=k+1}^{n} \Delta t_i \odot f(t_i) = \oplus\sum_{i=0}^{n} \Delta t_i \odot f(t_i)$$

The property is now proved.

2.4.4 Some Additional Properties of gH-Difference

Suppose that $A, B, C, D \in \mathbb{F}_R$ then

1. $(A \oplus B) \ominus_{gH} (C \oplus D) = (A \ominus_{gH} C) \oplus (B \ominus_{gH} D)$
2. $(A \ominus_{gH} B) \oplus (C \ominus_{gH} D) = (A \ominus_{gH} D) \oplus (C \ominus_{gH} B)$
3. $(A \ominus_{gH} B) \ominus_{gH} (C \ominus_{gH} D) = (A \ominus_{gH} C) \ominus_{gH} (B \ominus_{gH} D)$
4. $A \oplus (B \ominus_{gH} C) = \begin{cases} A \oplus B \ominus_{gH} C \\ A \ominus_{gH} (-1)B \oplus (-1)C \end{cases}$
5. $A \ominus_{gH} (B \oplus C) = \begin{cases} A \ominus_{gH} B \ominus_{gH} C \\ A \ominus_{gH} B \oplus (-1)C \end{cases}$
6. $A \ominus_{gH} (B \ominus_{gH} C) = \begin{cases} A \ominus_{gH} B \oplus C \\ A \oplus (-1)B \ominus_{gH} (-1)C \end{cases}$

To show the properties, we can use the definition of gH-difference and all are easily proved. For instance, we prove the properties 1 and 5. To this end, we use the level-wise form of both sides in two types of gH-difference,

$$
\begin{aligned}
&((A \oplus B) \ominus_{gH} (C \oplus D))[r] \\
&= \begin{cases} [(A_l(r) + B_l(r)) - (C_l(r) + D_l(r)), (A_u(r) + B_u(r)) - (C_u(r) + D_u(r))], & i - gH \\ [(A_u(r) + B_u(r)) - (C_u(r) + D_u(r)), (A_l(r) + B_l(r)) - (C_l(r) + D_l(r))], & ii - gH \end{cases} \\
&= \begin{cases} [(A_l(r) - C_l(r)) + (B_l(r) - D_l(r)), (A_u(r) - C_u(r)) + (B_u(r) - D_u(r))], & i - gH \\ [(A_u(r) - C_u(r)) + (B_u(r) - D_u(r)), (A_l(r) - C_l(r)) + (B_l(r) - D_l(r))], & ii - gH \end{cases} \\
&= \begin{cases} ((A \ominus_{gH} C) \oplus (B \ominus_{gH} D))[r] \\ ((A \ominus_{gH} C) \oplus (B \ominus_{gH} D))[r] \end{cases} \\
&= ((A \ominus_{gH} C) \oplus (B \ominus_{gH} D))[r]
\end{aligned}
$$

Now the property 5,

$$
\begin{aligned}
&(A \ominus_{gH} (B \oplus C))[r] \\
&\quad = \begin{cases} [A_l(r) - (B_l(r) + C_l(r)), A_u(r) - (B_u(r) + C_u(r))], & i - gH \\ [A_u(r) - (B_u(r) + C_u(r)), A_l(r) - (B_l(r) + C_l(r))], & ii - gH \end{cases} \\
&\quad = \begin{cases} [(A_l(r) - B_l(r)) - C_l(r), (A_u(r) - B_u(r)) - C_u(r)], & i - gH \\ [(A_u(r) - B_u(r)) - C_u(r), (A_l(r) - B_l(r)) - C_l(r)], & ii - gH \end{cases} \\
&\quad = \begin{cases} (A \ominus_{gH} B \ominus_{gH} C)[r] \\ (A \ominus_{gH} B \oplus (-1)C)[r] \; or \; (A \ominus_{gH} B \ominus_{gH} C)[r] \end{cases} = \begin{cases} A \ominus_{gH} B \ominus_{gH} C \\ A \ominus_{gH} B \oplus (-1)C \end{cases}
\end{aligned}
$$

To continuo the discussion we need some more definitions and properties about the derivatives of fuzzy valued functions.

Remark Suppose that $f(x) = a \odot x \oplus b$ such that $a, b \in \mathbb{F}_R$ then $f'_{gH}(x) = a$.

To show it, we use the definition of the derivative,

$$
f'_{gH}(x) = \lim_{h \to 0} \frac{f(x+h) \ominus_{gH} f(x)}{h} = \lim_{h \to 0} \frac{(a \odot (x+h) \oplus b) \ominus_{gH} (a \odot x \oplus b)}{h}
$$

Based on the item 1 of the properties of gH,

$$
\begin{aligned}
f'_{gH}(x) &= \lim_{h \to 0} \frac{(a \odot (x+h) \ominus_{gH} a \odot x) \oplus (b \ominus_{gH} b)}{h} \\
&= \lim_{h \to 0} \frac{(a \odot (x+h) \ominus_{gH} a \odot x)}{h}
\end{aligned}
$$

Let us discuss $a \odot (x+h) \ominus_{gH} a \odot x$,
If $x > 0$ then we have the first case of gH-difference,

$$
a \odot (x+h) \ominus_H a \odot x
$$

And its level-wise form is

$$[a_l(r)(x+h) - a_l(r)x, a_u(r)(x+h) - a_u(r)x] = h[a_l(r), a_u(r)]$$

If $x+h<0$ then we have to consider the second case of gH-difference,

$$(-1)(a \odot x \ominus_H a \odot (x+h)) = \ominus_H(-1)a \odot (x+h) \oplus (-1)a \odot x$$

And its level-wise form is again as,

$$[a_l(r)(x+h) - a_l(r)x, a_u(r)(x+h) - a_u(r)x]$$

So in two cases,

$$a \odot (x+h) \ominus_{gH} a \odot x = a \odot h$$

Then

$$f'_{gH}(x) = \lim_{h\to 0} \frac{(a \odot (x+h) \ominus_{gH} a \odot x)}{h} \lim_{h\to 0} \frac{a \odot h}{h} = a$$

This can be proved even $x \in \mathbb{F}_R$.

Remark—Summation in gH-differentiability For two gH-differentiable fuzzy number valued functions like $f(x), g(x)$ it is clearly proved that if

$$h(x) = f(x) \oplus g(x)$$

Then $h'(x) = f'(x) \oplus g'(x)$ and,

$$h'_{i-gH}(x) = f'_{i-gH}(x) \oplus g'_{i-gH}(x)$$

$$h'_{ii-gH}(x) = f'_{ii-gH}(t) \oplus g'_{ii-gH}(x)$$

where

$$f'_{i-gH}(x) = \lim_{h\to 0} \frac{f(x+h) \ominus_H f(x)}{h},$$
$$f'_{ii-gH}(x) = \lim_{h\to 0} \frac{\ominus_H (-1)f(x+h) \oplus (-1)f(x)}{h}$$

The proof can be done by Hausdorff distance. Based on the properties of D_H and the property 1 of gH-difference, $(A \oplus B) \ominus_{gH} (C \oplus D) = (A \ominus_{gH} C) \oplus (B \ominus_{gH} D)$ we have,

$$
\begin{aligned}
& D_H\left(\frac{(f(x+h)\oplus g(x+h))\ominus_{gH}(f(x)\oplus g(x))}{h}, f'(x)\oplus g'(x)\right)\\
& = D_H\left(\frac{(f(x+h)\ominus_{gH} f(x))\oplus(g(x+h)\ominus_{gH} g(x))}{h}, f'(x)\oplus g'(x)\right)\\
& = D_H\left(\frac{f(x+h)\ominus_{gH} f(x)}{h}\oplus\frac{g(x+h)\ominus_{gH} g(x)}{h}, f'(x)\oplus g'(x)\right)\\
& \le D_H\left(\frac{f(x+h)\ominus_{gH} f(x)}{h}, f'(x)\right)+D_H\left(\frac{g(x+h)\ominus_{gH} g(x)}{h}, g'(x)\right)
\end{aligned}
$$

Which leads to the proof when $h \to 0$.

Remark—Difference in gH-differentiability For two gH-differentiable fuzzy number valued functions like $f(x), g(x)$ it is clearly proved that if

$$h(x) = f(x) \ominus_{gH} g(x)$$

Then $h'_{gH}(x) = f'_{gH}(x) \ominus_{gH} g'_{gH}(x)$ and,

$$
\begin{aligned}
h'_{i-gH}(x) &= f'_{i-gH}(x) \ominus_H g'_{i-gH}(x)\\
h'_{ii-gH}(x) &= f'_{ii-gH}(t) \ominus_H g'_{ii-gH}(x)
\end{aligned}
$$

These two equations are proved separately as follows in accordance with the definition of gH-difference,

$$h(x) = f(x) \ominus_{gH} g(x) \Leftrightarrow \begin{cases} i-gH & f(x) = h(x) \oplus g(x)\ or \\ ii-gH & g(x) = f(x) \oplus (-1)h(x) \end{cases}$$

First consider the case (i) and based on the Remark—summation in gH-differentiability,

$$f'_{gH}(x) = h'_{gH}(x) \oplus g'_{gH}(x) \Leftrightarrow h'_{gH}(x) = f'_{gH}(x) \ominus_{gH} g'_{gH}(x)$$

Now in case (ii), we have $g(x) = f(x) \oplus (-1)h(x)$ and again,

$$g'_{gH}(x) = f'_{gH}(x) \oplus (-1)h'_{gH}(x) \Leftrightarrow h'_{gH}(x) = f'_{gH}(x) \ominus_{gH} g'_{gH}(x)$$

$$h'_{i-gH}(x) = f'_{i-gH}(x) \oplus (-1)g'_{i-gH}(x), h'_{ii-gH}(x) = f'_{ii-gH}(x) \oplus (-1)g'_{ii-gH}(x)$$

Because for any real number λ we have,

$$\lambda(A \ominus_{i-gH} B) = (\lambda A \ominus_{i-gH} \lambda B)$$

$$\lambda\left(A \ominus_{ii-gH} B\right) = \left(\lambda A \ominus_{ii-gH} \lambda B\right)$$

Remark—Production in gH-differentiability
For a gH-differentiable fuzzy functions $f(x)$ and real function $g(x)$,

$$(f(x) \odot g(x))'_{gH} = f'_{gH}(x) \odot g(x) \oplus f(x) \odot g'_{gH}(x)$$

The left side is as

$$(f(x) \odot g(x))'_{gH} = \lim_{h \to 0} \frac{f(x+h) \odot g(x+h) \ominus_{gH} f(x) \odot g(x))}{h}$$

Now considering the distance of two sides of the equation, it is enough to show that the distance intends zero.

$$D_H\left(\frac{f(x+h) \odot g(x+h) \ominus_{gH} f(x) \odot g(x))}{h}, f'_{gH}(x) \odot g(x) \oplus f(x) \odot g'_{gH}(x)\right)$$

$$= D_H\left(\frac{f(x+h) \odot g(x+h) \oplus f(x) \odot g(x+h) \ominus_{gH} f(x) \odot g(x+h) \ominus_{gH} f(x) \odot g(x))}{h}, f'_{gH}(x) \odot g(x) \oplus f(x) \odot g'_{gH}(x)\right)$$

$$= D_H\left(\frac{\left(f(x+h) \ominus_{gH} f(x)\right) \odot g(x+h) \oplus f(x) \odot \left(g(x+h) \ominus_{gH} \odot g(x)\right)}{h}, f'_{gH}(x) \odot g(x) \oplus f(x) \odot g'_{gH}(x)\right)$$

$$\leq D_H\left(\frac{\left(f(x+h) \ominus_{gH} f(x)\right) \odot g(x+h)}{h}, f'_{gH}(x) \odot g(x)\right) + D_H\left(\frac{f(x) \odot \left(g(x+h) \ominus_{gH} \odot g(x)\right)}{h}, f(x) \odot g'_{gH}(x)\right)$$

Now limit of two sides when $h \to 0$ are,

$$\lim_{h \to 0} D_H\left(\frac{f(x+h) \odot g(x+h) \ominus_{gH} f(x) \odot g(x))}{h}, f'_{gH}(x) \odot g(x) \oplus f(x) \odot g'_{gH}(x)\right)$$

$$= \lim_{h \to 0} D_H\left(\frac{\left(f(x+h) \ominus_{gH} f(x)\right) \odot g(x+h)}{h}, f'_{gH}(x) \odot g(x)\right) + \lim_{h \to 0} D_H\left(\frac{f(x) \odot \left(g(x+h) \ominus_{gH} \odot g(x)\right)}{h}, f(x) \odot g'_{gH}(x)\right)$$

Based on the properties of the limit and distance (as we expressed before) the proof is completed.

2.4.5 Proposition—Minimum and Maximum

For any gH-differentiable fuzzy number function $f(x)$ at the inner point like $x_0 \in (a,b)$, if the function has local minimum or maximum then $f'_{gH}(x_0) = 0$.

To prove the proposition, we use the definition of the minimum and maximum value.

The function at the point x_0 has the local maximum value if there is a positive δ and for all x from the neighborhood $|x - x_0| < \delta$ then $f(x) \preccurlyeq f(x_0)$. Then $f(x) \ominus_{gH} f(x_0) \preccurlyeq 0$. On the other hand for all $x \in (a, x_0)$, $x - x_0 < 0$. Now it can be concluded that

$$f'_{gH}(x) = \lim_{h \to 0} \frac{f(x) \ominus_{gH} f(x_0)}{x - x_0} \succcurlyeq 0$$

Also similarly, the function at the point c has the local minimum value if there is a positive δ and for all x from the neighborhood $|x - x_0| < \delta$ then $f(x_0) \preccurlyeq f(x)$. Finally we have, .

$$f'_{gH}(x) = \lim_{h \to 0} \frac{f(x) \ominus_{gH} f(x_0)}{x - x_0} \preccurlyeq 0$$

Then

$$f'_{gH}(x) \succcurlyeq 0 \; and \; f'_{gH}(x) \preccurlyeq 0 \; then \; f'_{gH}(x) = 0$$

2.4.6 Proposition—Cauchy's Fuzzy Mean Value Theorem

Assume that $f(x)$ is a continuous and gH-differentiable fuzzy number valued function on closed and open intervals respectively and $g(x)$ is a real valued continuous and differentiable function on the same intervals. Then there is $x_0 \in (a,b)$ such that

$$[f(b) \ominus_{gH} f(a)] \odot g'(x_0) = [g(b) - g(a)] \odot f'_{gH}(x_0)$$

where the gH-difference exists.

Proof Let us consider a new function as follow,

$$\phi(t) = [f(b) \ominus_{gH} f(a)] \odot g(x) \ominus_{gH} [g(b) - g(a)] \odot f(x)$$

This function is continuous because,

Let us consider $f(b) \ominus_{gH} f(a) = k$ and $g(b) - g(a) = l$ so we have

$$\phi(t) = k \odot g(x) \ominus_{gH} l \odot f(x)$$

Now for any $\epsilon > 0 \exists \delta > 0$ if for any x in $|x - \bar{x}| < \delta$ at an arbitrary point x we prove,

$$D_H(\phi(x), \phi(\bar{x})) = D_H\big(k \odot g(x) \ominus_{gH} l \odot f(x), k \odot g(\bar{x}) \ominus_{gH} l \odot f(\bar{x})\big)$$

Based on the properties of the Hausdorff distance,

$$\leq D_H(k \odot g(x), k \odot g(\bar{x})) + D_H(l \odot f(x), l \odot f(\bar{x}))$$

Based on the definition of Hausdorff distance and definition of absolute value of a fuzzy number we can write,

$$\leq |k||g(x) - g(\bar{x})| + |l| D_H(f(x), f(\bar{x}))$$

And finally, this summation is less than ϵ.

On the other hand, $\phi(a) = \phi(b)$ so it is gH-differentiable at the inner point like x_0 and equals to zero. Then

$$\phi'_{gH}(x) = \big[f(b) \ominus_{gH} f(a)\big] \odot g'(x) \ominus_{gH} [g(b) - g(a)] \odot f'_{gH}(x)$$

And $\phi'_{gH}(x_0) = 0$ then

$$\big[f(b) \ominus_{gH} f(a)\big] \odot g'(x_0) \ominus_{gH} [g(b) - g(a)] \odot f'_{gH}(x_0) = 0$$

The proof is completed.

As a corollary if we take the real function $g(x) = x$ then we will have the fuzzy mean value theorem.

2.4.7 *Corollary—Fuzzy Mean Value Theorem*

Considering all the previous assumptions and $f(x) = x$ we have

$$\big[f(b) \ominus_{gH} f(a)\big] \odot f'(x_0) = [f(b) - f(a)] \odot f'_{gH}(x_0) \Rightarrow$$

$$\big[f(b) \ominus_{gH} f(a)\big] = (b - a) \odot f'_{gH}(x_0)$$

Then

$$f'_{gH}(x_0) = \frac{f(b) \ominus_{gH} f(a)}{b-a}$$

2.4.8 Integral Relation

Let $f_{gH}^{(n)}(x)$ is a continuous fuzzy number valued function for any $s \in (a, b)$ the following integral equations are valid.

Item 1. Consider $f_{gH}^{(n)}(x), n = 1, 2, \ldots, m$ is $(i - gH)$—differentiable and the type of differentiability do not change on (a, b).

$$\begin{aligned}
\left(FR\int_a^s f_{i-gH}^{(n)}(x)dx\right)[r] &= FR\int_a^s \left(f_{i-gH}^{(n)}(x)\right)[r]dx \\
&= \left[R\int_a^s f_l^{(n)}(x,r)dx, R\int_a^s f_u^{(n)}(x,r)dx\right] \\
&= \left[f_l^{(n-1)}(s,r) - f_l^{(n-1)}(a,r), f_u^{(n-1)}(s,r) - f_u^{(n-1)}(a,r)\right] \\
&= \left[f_l^{(n-1)}(s,r), f_u^{(n-1)}(s,r)\right] - \left[f_l^{(n-1)}(a,r), f_u^{(n-1)}(a,r)\right] \\
&= \left(f_{i-gH}^{(n-1)}(s)\right)[r] - \left(f_{i-gH}^{(n-1)}(a)\right)[r]
\end{aligned}$$

Then it is concluded,

$$FR\int_a^s f_{i-gH}^{(n)}(x)dt = f_{i-gH}^{(n-1)}(s) \ominus_H f_{i-gH}^{(n-1)}(a)$$

$$f_{i-gH}^{(n-1)}(s) = f_{i-gH}^{(n-1)}(a) \oplus FR\int_a^s f_{i-gH}^{(n)}(x)dx$$

If $n = 1, s = b$ then

$$f(b) = f(a) \oplus FR\int_a^b f'_{i-gH}(x)dx$$

Item 2. Consider $f_{gH}^{(n)}(x), n = 1, 2, \ldots, m$ is $(ii - gH)$—differentiable and the type of differentiability do not change on (a, b).

$$\left(FR\int_a^s f_{ii-gH}^{(n)}(x)dx\right)[r] = FR\int_a^s f_{ii-gH}^{(n)}(x, r)dx$$
$$= \left[R\int_a^s f_u^{(n)}(x, r)dt, R\int_a^s f_l^{(n)}(x, r)dx\right]$$
$$= \left[f_u^{(n-1)}(s, r) - f_u^{(n-1)}(a, r), f_l^{(n-1)}(s, r) - f_l^{(n-1)}(a, r)\right]$$
$$= \left[f_u^{(n-1)}(s, r), f_l^{(n-1)}(s, r)\right] - \left[f_u^{(n-1)}(a, r), f_l^{(n-1)}(a, r)\right]$$
$$= f_{ii-gH}^{(n-1)}(s, r) - f_{ii-gH}^{(n-1)}(a, r)$$
$$= (-1)\left[f_l^{(n-1)}(a, r), f_u^{(n-1)}(a, r)\right] \ominus_H (-1)\left[f_l^{(n-1)}(s, r), f_u^{(n-1)}(s, r)\right]$$

Then it is concluded,

$$FR\int_a^s f_{ii-gH}^{(n)}(x)dx = f_{ii-gH}^{(n-1)}(s) \ominus_H f_{ii-gH}^{(n-1)}(a)$$
$$f_{ii-gH}^{(n-1)}(s) = f_{ii-gH}^{(n-1)}(a) \oplus FR\int_a^s f_{ii-gH}^{(n)}(x)dx$$

Or

$$FR\int_a^s f_{ii-gH}^{(n)}(x)dx = \ominus_H (-1)f_{ii-gH}^{(n-1)}(a) \ominus_H (-1)f_{ii-gH}^{(n-1)}(s)$$

If $n = 1, s = b$ then

$$f(b) = f(a) \oplus FR\int_a^b f'_{ii-gH}(x)dx$$

Item 3. Consider $f_{gH}^{(n)}(x)$ is $(i - gH)$—differentiable and $f_{gH}^{(n-1)}(x)$ is $(ii - gH)$—differentiable then

$$\begin{aligned}
\ominus_H(-1)\left(FR\int_a^s f_{i-gH}^{(n)}(x)dx\right)[r] &= \ominus_H(-1)FR\int_a^s f_{i-gH}^{(n)}(x,r)dx \\
&= \left[R\int_a^s f_u^{(n)}(x,r)dx, R\int_a^s f_l^{(n)}(x,r)dx\right] \\
&= \left[f_u^{(n-1)}(s,r) - f_u^{(n-1)}(a,r), f_l^{(n-1)}(s,r) - f_l^{(n-1)}(a,r)\right] \\
&= \left[f_u^{(n-1)}(s,r), f_l^{(n-1)}(s,r)\right] - \left[f_u^{(n-1)}(a,r), f_l^{(n-1)}(a,r)\right] \\
&= f_{ii-gH}^{(n-1)}(s,r) - f_{ii-gH}^{(n-1)}(a,r)
\end{aligned}$$

Then it is concluded,

$$\ominus_H(-1)FR\int_a^s f_{i-gH}^{(n)}(x)dx = f_{ii-gH}^{(n-1)}(s) \ominus_H f_{ii-gH}^{(n-1)}(a)$$

$$f_{ii-gH}^{(n-1)}(s) = f_{ii-gH}^{(n-1)}(a) \ominus_H(-1)FR\int_a^s f_{i-gH}^{(n)}(x)dx$$

If $n = 1, s = b$ then

$$f(b) = f(a) \ominus_H(-1)FR\int_a^b f'_{i-gH}(x)dx$$

Item 4. Consider $f_{gH}^{(n)}(x)$ is $(ii-gH)-$ differentiable and $f_{gH}^{(n-1)}(x)$ is $(i-gH)-$ differentiable then

$$\begin{aligned}
\ominus_H(-1)\left(FR\int_a^s f_{ii-gH}^{(n)}(x)dx\right)[r] &= \ominus_H(-1)FR\int_a^s f_{ii-gH}^{(n)}(x,r)dx \\
&= \left[R\int_a^s f_l^{(n)}(x,r)dx, R\int_a^s f_u^{(n)}(x,r)dx\right] \\
&= \left[f_l^{(n-1)}(s,r)-f_l^{(n-1)}(a,r), f_u^{(n-1)}(s,r)-f_u^{(n-1)}(a,r)\right] \\
&= \left[f_l^{(n-1)}(s,r), f_u^{(n-1)}(s,r)\right]-\left[f_l^{(n-1)}(a,r), f_u^{(n-1)}(a,r)\right] \\
&= f_{i-gH}^{(n-1)}(s,r)-f_{i-gH}^{(n-1)}(a,r)
\end{aligned}$$

Then it is concluded,

$$\ominus_H(-1)FR\int_a^s f_{ii-gH}^{(n)}(x)dx = f_{i-gH}^{(n-1)}(s)\ominus_H f_{i-gH}^{(n-1)}(a)$$

$$f_{i-gH}^{(n-1)}(s) = f_{i-gH}^{(n-1)}(a)\ominus_H(-1)FR\int_a^s f_{ii-gH}^{(n)}(x)dx$$

If $n=1, s=b$ then

$$f(b) = f(a)\ominus_H(-1)FR\int_a^b f'_{ii-gH}(x)dx$$

2.4.9 *Fuzzy Taylor Expansion of Order One*

Let us assume $f(x)$ is a continuous and gH-differentiable fuzzy number valued function.

Item 1. $f(x)$,$f'(x)$ both are $i-gH$ differentiable without changing in type of differentiability. Then based on the integral relation we have the following relation.

$$f(s) = f(a) \oplus FR\int_a^s f'_{i-gH}(s_1)ds_1, f'_{i-gH}(s_1) = f'_{i-gH}(a) \oplus FR\int_a^{s_1} f''_{i-gH}(s_1)ds_1$$

Taking fuzzy Riemann integral of two sides,

$$FR\int_a^s f'_{i-gH}(s_1)ds_1 = FR\int_a^s f'_{i-gH}(a)ds_1 \oplus FR\int_a^s FR\int_a^{s_1} f''_{i-gH}(s_2)ds_2ds_1$$

$$= f'_{i-gH}(a) \odot (s-a) \oplus FR\int_a^s \left(\int_a^{s_1} f''_{i-gH}(s_2)ds_2\right)ds_1$$

On the other hand the right side is as,

$$FR\int_a^s f'_{i-gH}(s_1)ds_1 = f(s) \ominus_H f(a)$$

Then

$$f(s) = f(a) \oplus f'_{i-gH}(a) \odot (s-a) \oplus FR\int_a^s \left(\int_a^{s_1} f'_{i-gH}(s_2)ds_2\right)ds_1$$

Item 2. Both $f(x), f'(x)$ are $ii-gH$ differentiable without changing in type of differentiability. Then based on the integral relation we have the following relation.

$$f(s) = f(a) \ominus_H (-1)FR\int_a^s f'_{ii-gH}(s_1)ds_1,$$

$$f'_{ii-gH}(s_1) = f'_{ii-gH}(a) \oplus FR\int_a^{s_1} f'_{ii-gH}(s_1)ds_1$$

Taking integral of two sides,

$$FR\int_a^s f'_{ii-gH}(s_1)ds_1 = FR\int_a^s f'_{ii-gH}(a)ds_1 \oplus FR\int_a^s FR\int_a^{s_1} f''_{ii-gH}(s_2)ds_2ds_1$$
$$= f'_{ii-gH}(a)\odot(s-a)\oplus FR\int_a^s\left(\int_a^{s_1} f''_{ii-gH}(s_2)ds_2\right)ds_1$$

On the other hand the right side is as,

$$\ominus_H(-1)FR\int_a^s f'_{ii-gH}(s_1)ds_1 = f(s)\ominus f(a)$$

Then

$$f(s) = f(a)\ominus_H(-1)f'_{ii-gH}(a)\odot(s-a)\ominus(-1)FR\int_a^s\left(\int_a^{s_1} f''_{ii-gH}(s_2)ds_2\right)ds_1$$

Item 3. Suppose that $f(x)$ is $i-gH$ differentiable and $f'(x)$ is $ii-gH$ differentiable,

$$f(s) = f(a)\ominus_H(-1)FR\int_a^s f'_{ii-gH}(s_1)ds_1$$

According to hypothesis $f'_{gH}(t)$ is $i-gH$ differentiable and we have

$$f'_{ii-gH}(s) = f'_{ii-gH}(a)\ominus_H(-1)FR\int_a^s f''_{i-gH}(t)dt$$

Taking integral of two sides,

$$FR\int_a^s f'_{ii-gH}(s_1)ds_1 = FR\int_a^s f'_{ii-gH}(a)ds_1\ominus_H(-1)FR\int_a^s FR\int_a^{s_1} f''_{ii-gH}(s_2)ds_2ds_1$$
$$= f'_{ii-gH}(a)\odot(s-a)\ominus_H(-1)FR\int_a^s\left(\int_a^{s_1} f''_{ii-gH}(s_2)ds_2\right)ds_1$$

So

$$f(s) = f(a) \ominus_H (-1) f'_{ii-gH}(a) \odot (s-a) \oplus FR \int_a^s \left(\int_a^{s_1} f''_{ii-gH}(s_2) ds_2 \right) ds_1$$

Item 4. Suppose that $f(x)$ is $ii - gH$ differentiable and $f'(x)$ is $i - gH$ differentiable. So

$$f(\xi) = f(a) \oplus FR \int_a^{\xi} f'_{i-gH}(s_1) ds_1$$

And it is $ii - gH$ differentiable in interval $[\xi, b]$

$$f(s) = f(\xi) \ominus_H (-1) FR \int_{\xi}^{s} f'_{ii-gH}(t_1) dt_1$$

By replacement

$$f(s) = f(a) \oplus FR \int_a^{\xi} f'_{i-gH}(s_1) ds_1 \ominus_H (-1) FR \int_{\xi}^{s} f'_{ii-gH}(t_1) dt_1$$

Now we are going to find the first integral in the right hand side. Let us consider ζ_1 as a switching point for the second gH-derivative. And suppose that f'_{i-gH} is $ii - gH$ differentiable on $[a, \zeta_1]$, then the type of differentiability changes and

$$f'_{i-gH}(\zeta_1) = f'_{i-gH}(a) \ominus_H (-1) FR \int_a^{\zeta_1} f''_{ii-gH}(s_2) ds_2$$

Now for $s_1 \in [\zeta_1, \xi]$, f'_{i-gH} is $i - gH$ differentiable on $[\zeta_1, \xi]$, then the type of differentiability changes and

$$f'_{i-gH}(s_1) = f'_{i-gH}(\zeta_1) \oplus FR \int_{\zeta_1}^{s_1} f''_{i-gH}(s_3) ds_3$$

By substituting,

$$f'_{i-gH}(s_1) = f'_{i-gH}(a) \ominus_H (-1)FR\int_a^{\zeta_1} f''_{ii-gH}(s_2)ds_2 \oplus FR\int_{\zeta_1}^{s_1} f''_{i-gH}(s_3)ds_3$$

On the other hand,

$$f''_{ii-gH}(s_2) = f''_{ii-gH}(a) \oplus FR\int_{\zeta_1}^{s_2} f'''_{ii-gH}(s_4)ds_4$$

Using the FR integral on $[a, \zeta_1]$

$$FR\int_a^{\zeta_1} f''_{ii-gH}(s_2)ds_2 = f''_{ii-gH}(a) \odot (\zeta_1 - a) \oplus FR\int_a^{\zeta_1}\left(\int_a^{s_2} f'''_{ii-gH}(s_4)ds_4\right)ds_2$$

And

$$f''_{i-gH}(s_3) = f''_{i-gH}(\zeta_1) \oplus FR\int_{\zeta_1}^{s_3} x^{(3)}_{i-gH}(s_5)ds_5$$

Using the FR integral operator on $[\zeta_1, s_1]$

$$FR\int_{\zeta_1}^{s_1} f''_{i-gH}(s_3)ds_3 = f''_{i-gH}(\zeta_1) \odot (s_1 - \zeta_1) \oplus FR\int_{\zeta_1}^{s_1}\left(\int_{\zeta_1}^{s_3} f'''_{i-gH}(s_5)ds_5\right)ds_3$$

To find $f'_{i-gH}(s_1)$, we insert two of last equations in $f'_{i-gH}(s_1)$.

$$\begin{aligned}
f'_{i-gH}(s_1) &= f'_{i-gH}(a) \ominus_H (-1)\left(f''_{ii-gH}(a) \odot (\zeta_1 - a) \oplus FR\int_a^{\zeta_1}\left(\int_a^{s_2} f'''_{ii-gH}(s_4)ds_4\right)ds_2\right) \\
&\quad \oplus f''_{i-gH}(\zeta_1) \odot (s_1 - \zeta_1) \oplus FR\int_{\zeta_1}^{s_1}\left(\int_{\zeta_1}^{s_3} f'''_{i-gH}(s_5)ds_5\right)ds_3 \\
&= f'_{i-gH}(a) \ominus_H f''_{ii-gH}(a) \odot (a - \zeta_1) \oplus f''_{i-gH}(\zeta_1) \odot (s_1 - \zeta_1) \\
&\quad \ominus_H (-1)FR\int_a^{\zeta_1}\left(\int_a^{s_2} f'''_{ii-gH}(s_4)ds_4\right)ds_2 \oplus FR\int_{\zeta_1}^{s_1}\left(\int_{\zeta_1}^{s_3} f'''_{i-gH}(s_5)ds_5\right)ds_3
\end{aligned}$$

Using the FR integral operator on $[a, \xi]$,

$$FR\int_a^{\xi} f''_{i-gH}(s_1)ds_1 = f'_{i-gH}(a) \odot (\xi - a) \ominus_H f''_{ii-gH}(a) \odot (a - \zeta_1) \odot (\xi - a)$$

$$\oplus f''_{i-gH}(\zeta_1) \odot \left(\frac{(\xi - \zeta_1)^2}{2} - \frac{(a - \zeta_1)^2}{2}\right) \ominus_H (-1)FR\int_a^{\xi}\left(\int_a^{\zeta_1}\left(\int_a^{s_2} f'''_{ii-gH}(s_4)ds_4\right)ds_2\right)ds_1$$

$$\oplus FR\int_a^{\xi}\left(\int_{\zeta_1}^{s_1}\left(\int_{\zeta_1}^{s_3} f'''_{i-gH}(s_5)ds_5\right)ds_3\right)ds_1$$

We had this equation:

$$f(s) = f(a) \oplus FR\int_a^{\xi} f'_{i-gH}(s_1)ds_1 \ominus_H (-1)FR\int_{\xi}^{s} f'_{ii-gH}(t_1)dt_1$$

2.4.10 Integration by Part

Consider the function $f(x)$ as gH-differentiable fuzzy number valued function on $[a, b]$ and $g(x)$ as real valued function on the same interval. Then

$$\int_a^b f'_{gH}(x) \odot g(x)dx = ((f(b) \odot g(b)) \ominus_H (f(a) \odot g(a))) \ominus_{gH} \int_a^b f(x) \odot g'(x)dx$$

To prove the relation, we use the derivative of combination of these two functions.

$$(f \odot g)'_{gH}(x) = f'_{gH}(x) \odot g(x) \oplus f(x) \odot g'(x)$$

Integrating both side respect to t over the interval $[a, b]$. We will have

$$\int_a^b (f \odot g)'_{gH}(x)dx = \int_a^b \left(f'_{gH}(x) \odot g(x)\right)dx \oplus \int_a^b (f(x) \odot g'(x))dx$$

The left side can be obtained immediately as

$$((f(b) \odot g(b)) \ominus_H (f(a) \odot g(a))) = \int_a^b \left(f'_{gH}(x) \odot g(x)\right)dx \oplus \int_a^b (f(x) \odot g'(x))dx$$

Now based on the definition of the H-difference the proof is completed.

2.4.11 Definition—gH-Partial Differentiability

The fuzzy number valued function of two variables $f(x,y) \in \mathbb{F}_R$ is called gH-partial differentiable $(gH - p)$ at the point $(x_0, y_0) \in \mathbb{D}$ with respect to x and y and denoted by $\partial_{xgH}f(x_0, y_0)$ and $\partial_{ygH}f(x_0, y_0)$ if

$$\partial_{xgH}f(x_0, y_0) = \lim_{h \to 0} \frac{f(x_0 + h, y_0) \ominus_{gH} f(x_0, y_0)}{h}$$

$$\partial_{ygH}f(x_0, y_0) = \lim_{k \to 0} \frac{f(x_0, y_0 + k) \ominus_{gH} f(x_0, y_0)}{k}$$

Provided to both derivatives $\partial_{xgH}f(x_0, y_0)$ and $\partial_{ygH}f(x_0, y_0)$ are fuzzy number valued functions not fuzzy sets.

Definition—Level-Wise form of gH-Partial differentiability Suppose that the fuzzy number valued function $f(x,y) \in \mathbb{F}_R$ is $gH - p$ differentiable at the point $(x_0, y_0) \in \mathbb{D}$ with respect to t and $f_l(x,y,r)$, $f_u(x,y,r)$ are real valued functions and partial differentiable with respect to t. We say,

- $f(x,y)$ is $(i - gH - p)$ differentiable w.r.t. x at (x_0, y_0) if

$$\partial_{x,i-gH}f(x_0, y_0, r) = [\partial_x f_l(x_0, y_0, r), \partial_x f_u(x_0, y_0, r)]$$

- $f(x,y)$ is $(ii - gH - p)$ differentiable w.r.t. x at (x_0, y_0) if

$$\partial_{x,ii-gH}f(x_0, y_0, r) = [\partial_x f_u(x_0, y_0, r), \partial_x f_l(x_0, y_0, r)]$$

Please note that in each cases the conditions of the definition in level-wise form should be satisfied.

Definition—Switching point in gH-Partial differentiability For any fixed ξ_0 we say the point $(\xi_0, y) \in \mathbb{D}$ is a switching point for the gH-differentiability of $f(x, y)$ w.r.t x, if in any neighborhood V of (ξ_0, y) there exist points $(x_1, y) < (\xi_0, y) < (x_2, y)$ for any fixed y such that,

Type I at the point (x_1, y) is $(i - gH - p)$ differentiable and not $(ii - gH - p)$ differentiable and at the point (x_2, y) is $(ii - gH - p)$ differentiable and not $(i - gH - p)$ differentiable.
Type II at the point (x_1, y) is $(ii - gH - p)$ differentiable and not $(i - gH - p)$ differentiable and at the point (x_2, y) is $(i - gH - p)$ differentiable and not $(ii - gH - p)$ differentiable.

Definition—Higher order of gH-Partial differentiability Suppose that the fuzzy number valued function $\partial_{xgH} f(x, y) \in \mathbb{F}_R$ is $gH - p$ differentiable at the point $(x_0, y_0) \in \mathbb{D}$ with respect to x and there is no switching point. Moreover suppose that $\partial_{xx} f_l(x, y, r), \partial_{xx} f_u(x, y, r)$ are real valued functions and partial differentiable with respect to x. We say,

- $\partial_x f(x, y)$ is $(i - gH - p)$ differentiable w.r.t. x at (x_0, y_0) if

$$\partial_{xx,i-gH} f(x_0, y_0, r) = [\partial_{xx} f_l(x_0, y_0, r), \partial_{xx} f_u(x_0, y_0, r)]$$

- $\partial_x f(x, y)$ is $(ii - gH - p)$ differentiable w.r.t. x at (x_0, y_0) if

$$\partial_{xx,ii-gH} f(x_0, y_0, r) = [\partial_{xx} f_u(x, y, r), \partial_{xx} f_l(x, y, r)]$$

Note. Please note that in each case the conditions of the definition in level-wise form should be satisfied and the type of gH-partial differentiability for both functions $f(x, y)$ and $\partial_x f(x, y)$ are the same.

Note. The same definition is defined for the partial derivatives of first and second order for the second variable y.

Bivariate fuzzy chain rule in gH- Partial differentiability

Let $x_i(t)$ is defined on $\mathbb{I}_i := [a, b] \subseteq R, i = 1, 2$ and are strictly increasing and differentiable functions. Consider U is an open set of R^2 such that $\prod_{i=1}^{2} \mathbb{I}_i \subseteq R$. Let us assume that the function $f : U \rightarrow \mathbb{F}_R$ is a continuous fuzzy function. Suppose that $\partial_{x_i gH} f : U \rightarrow \mathbb{F}_R, i = 1, 2$ the $gH - p$ derivative of f exist and are fuzzy continuous functions. Call $x_i := x_i(t)$ and $z := z(t) := f(x_1, x_2)$. Then $\partial_{tgH} z$ exist and

$$\partial_{tgH}z = \partial_{x_1gH}f(x_1, x_2) \odot \partial_t x_1(t) \oplus \partial_{x_2gH}f(x_1, x_2) \odot \partial_t x_2(t)$$

where $\partial_t x_i(t), i = 1, 2$ are derivatives of $x_i(t)$ with respect to t.

To show the assertion let $t \in (a, b)$ and $(x_1, x_2) \in U$ are fixed and $\Delta x_i > 0, i = 1, 2$ are enough small. Now set

$$\begin{aligned} \alpha_1 &= f(x_1 + \Delta x_1, x_2 + \Delta x_2) \ominus_{gH} f(x_1, x_2 + \Delta x_2) \in \mathbb{F}_R \\ \alpha_2 &= f(x_1, x_2 + \Delta x_2) \ominus_{gH} f(x_1, x_2) \in \mathbb{F}_R \end{aligned}$$

So we have

$$f(x_1 + \Delta x_1, x_2 + \Delta x_2) \ominus_{gH} f(x_1, x_2) = \oplus \sum_{i=1}^{2} \alpha_i \in \mathbb{F}_R$$

Since the partial gH-derivative $\partial_{x_igH}f$ exist the above gH-differences in α_i exist for $i = 1, 2$ when $\Delta x_i \to 0$. Here $\Delta x_i = x_i(t + \Delta t) - x_i(t) := x_i + \Delta x_i, i = 1, 2$.

Now

$$\begin{aligned}
&\lim_{\Delta t \to 0} D_H\left(\frac{f(x_1 + \Delta x_1, x_2 + \Delta x_2) \ominus_{gH} f(x_1, x_2)}{\Delta t}, \oplus \sum_{i=1}^{2} \partial_{x_igH}f(x_1, x_2) \odot \partial_t x_i(t)\right) \\
&= \lim_{\Delta t \to 0} D_H\left(\frac{\oplus \sum_{i=1}^{2} \alpha_i}{\Delta t}, \oplus \sum_{i=1}^{2} \partial_{x_igH}f(x_1, x_2) \odot \partial_t x_i(t)\right) \\
&\leq \lim_{\Delta t \to 0} D_H\left(\frac{f(x_1 + \Delta x_1, x_2 + \Delta x_2) \ominus_{gH} f(x_1, x_2 + \Delta x_2)}{\Delta x_1} \odot \frac{\Delta x_1}{\Delta t}, \quad \partial_{x_1gH}f(x_1, x_2) \odot \partial_t x_1(t)\right) \\
&+ \lim_{\Delta t \to 0} D_H\left(\frac{f(x_1, x_2 + \Delta x_2) \ominus_{gH} f(x_1, x_2)}{\Delta x_2} \odot \frac{\Delta x_2}{\Delta t}, \partial_{x_2gH}f(x_1, x_2) \odot \partial_t x_2(t)\right) \\
&\leq \lim_{\Delta t \to 0} D_H\left(\frac{\int_{x_1}^{x_1 + \Delta x_1} \partial_{x_1gH}f(t, x_2 + \Delta x_2)dt}{\Delta x_1} \odot \frac{\Delta x_1}{\Delta t}, \partial_{x_1gH}f(x_1, x_2) \odot \partial_t x_1(t)\right) \\
&+ \lim_{\Delta t \to 0} D_H\left(\frac{\int_{x_2}^{x_2 + \Delta x_2} \partial_{x_2gH}f(x_1, x_2)dt}{\Delta x_2} \odot \frac{\Delta x_2}{\Delta t}, \partial_{x_2gH}f(x_1, x_2) \odot \partial_t x_2(t)\right)
\end{aligned}$$

If the limit operator goes to inside of the distance then $\lim_{\Delta t \to 0} \frac{\Delta x_i}{\Delta t} = \partial_t x_i(t) := x_i'(t)$. Moreover in each term the Δx_i is an constant with respect to the integral variable. So

$$\leq \partial_t x_1(t) \lim_{\Delta t\to 0} \frac{1}{\Delta x_1} D_H\left(\int_{x_1}^{x_1+\Delta x_1} \partial_{x_1 gH} f(t, x_2+\Delta x_2)dt, \partial_{x_1 gH} f(x_1, x_2)\right)$$

$$+ \partial_t x_2(t) \lim_{\Delta t\to 0} \frac{1}{\Delta x_2} D_H\left(\int_{x_2}^{x_2+\Delta x_2} \partial_{x_2 gH} f(x_1, x_2)dt, \partial_{x_2 gH} f(x_1, x_2)\right) + 0$$

$$\leq \frac{\partial_t x_1(t)}{\Delta x_1} \lim_{\Delta t\to 0}\left(\int_{x_1}^{x_1+\Delta x_1} D_H\left(\partial_{x_1 gH} f(t, x_2+\Delta x_2), \partial_{x_1 gH} f(x_1, x_2)\right)dt\right)$$

$$+ \frac{\partial_t x_2(t)}{\Delta x_2} \lim_{\Delta t\to 0}\left(\int_{x_2}^{x_2+\Delta x_2} D_H\left(\partial_{x_2 gH} f(x_1, x_2), \partial_{x_2 gH} f(x_1, x_2)\right)dt\right)$$

$$\leq \frac{\partial_t x_1(t)}{\Delta x_1} \lim_{\Delta t\to 0}\left(\sup_{\tau\in[x_1, x_1+\Delta x_1]} D_H\left(\partial_{x_1 gH} f(\tau, x_2+\Delta x_2), \partial_{x_1 gH} f(x_1, x_2)\right)\right)\Delta x_1$$

$$+ \frac{\partial_t x_2(t)}{\Delta x_2} \lim_{\Delta t\to 0}\left(\sup_{\tau\in[x_1, x_1+\Delta x_1]} D_H\left(\partial_{x_1 gH} f(x_1, x_2), \partial_{x_1 gH} f(x_1, x_2)\right)\right)\Delta x_2 \to 0$$

As $\Delta t \to 0$ then all $\Delta x_i \to 0$ ant thus $\tau_i \to x_i$ for all $i = 1, 2$. Then by continuity of $\partial_{x_i gH}$ two of the terms intend to the zero. The proof is completed.

2.4.12 *Fuzzy Fubini—Theorem*

Let $I_1, I_2 \subset R$ and $f : I_1 \times I_2 \to \mathbb{F}_R$ be a fuzzy integrable function. Then

$$\int_a^b \int_a^x f(x, y)dydx = \int_a^b \int_y^b f(x, y)dxdy$$

Proof. To show the assertion, we use the level-wise form and it can be easily shown. The interval form of left side is as,

$$\left[\int_a^b \int_a^x f_l(x, y, r)dydx, \int_a^b \int_a^x f_u(x, y, r)dydx\right], \quad 0 \leq r \leq 1$$

The lower and upper functions are real functions with the parameter r, and we have

$$\int_a^b \int_a^x f_l(x,y,r)dydx = \int_a^b \int_y^b f_l(x,y,r)dxdy$$

$$\int_a^b \int_a^x f_u(x,y,r)dydx = \int_a^b \int_y^b f_u(x,y,r)dxdy$$

The theorem is in fact proved.

2.4.13 First Order Fuzzy Taylor Expansion for Two Variables Function

For a fuzzy continuous function $f(x,y)$, let assume that the function has continuous partial gH-derivatives on the interval containing $a = (a_1, a_2)$ and there is no switching point. Then the first order fuzzy Taylor expansion of the function around a is defined as follows.

Item 1.

$$f(a_1+h_1, a_2+h_2) = f(a_1,a_2) \oplus \left(\partial_{xgH}(a_1,a_2) \odot h_1 \oplus \partial_{ygH}(a_1,a_2) \odot h_2\right) \oplus R_2(h_1,h_2)$$

Item 2.

$$f(a_1+h_1, a_2+h_2) = f(a_1,a_2) \ominus_H (-1)\left(\partial_{xgH}(a_1,a_2) \odot h_1 \oplus \partial_{ygH}(a_1,a_2) \odot h_2\right) \ominus_H (-1)R_2(h_1,h_2)$$

Item 3.

$$f(a_1+h_1, a_2+h_2) = f(a_1,a_2) \ominus_H (-1)\left(\partial_{xgH}(a_1,a_2) \odot h_1 \oplus \partial_{ygH}(a_1,a_2) \odot h_2\right) \oplus R_2(h_1,h_2)$$

Item 4.

$$f(a_1+h_1, a_2+h_2) = f(a_1,a_2) \oplus \left(\partial_{xgH}(a_1,a_2) \odot h_1 \oplus \partial_{ygH}(a_1,a_2) \odot h_2\right) \ominus_H (-1)R_2(h_1,h_2)$$

Item 5.

$$f(a_1+h_1, a_2+h_2) = f(a_1,a_2) \ominus_H (-1)\left(\partial_{xgH}(a_1,a_2) \odot h_1 \oplus \partial_{ygH}(a_1,a_2) \odot h_2\right) \oplus (-1)R_2(h_1,h_2)$$

where

$$R_2(h_1,h_2)=\int_0^1\left(\int_0^{t_1} f''_{gH}(a_1+t_2h_1,a_2+t_2h_2)dt_2\right)dt_1$$

Proof. For the single variable fuzzy functions we have,

Item 1. $f(x)=f(a)\oplus f'_{gH}(a)\odot(x-a)\oplus R_2(x)$
Item 2. $f(x)=f(a)\ominus_H(-1)f'_{gH}(a)\odot(x-a)\ominus_H(-1)R_2(x)$
Item 3. $f(x)=f(a)\ominus_H(-1)f'_{gH}(a)\odot(x-a)\oplus R_2(x)$
Item 4. $f(x)=f(a)\oplus f'_{gH}(a)\odot(x-a)\ominus_H(-1)R_2(x)$
Item 5. $f(x)=f(a)\ominus_H(-1)f'_{gH}(a)\odot(x-a)\oplus(-1)R_2(x)$

where $R_2(x)=\int_a^x\left(\int_a^{t_1} f''_{gH}(t_2)dt_2\right)dt_1$. Indeed in $R_2(x)$,

$$R_2(x)=\int_a^x\left(\int_a^{t_1} f''_{gH}(t_2)dt_2\right)dt_1=f(x)\ominus_{gH}f(a)\ominus_{gH}(x-a)\odot f'_{gH}(a)$$

For any case of gH-difference we can defined the mentioned above items. Now, if $a=0$ then we have the following items,

Item 1. $f(x)=f(0)\oplus f'_{gH}(0)\odot x\oplus R_2(x)$
Item 2. $f(x)=f(0)\ominus_H(-1)f'_{gH}(0)\odot x\ominus_H(-1)R_2(x)$
Item 3. $f(x)=f(0)\ominus_H(-1)f'_{gH}(0)\odot x\oplus R_2(x)$
Item 4. $f(x)=f(0)\oplus f'_{gH}(0)\odot x\ominus_H(-1)R_2(x)$
Item 5. $f(x)=f(0)\ominus_H(-1)f'_{gH}(0)\odot x\oplus(-1)R_2(x)$

where,

$$R_2(x)=\int_0^x\left(\int_0^{t_1} f''_{gH}(t_2)dt_2\right)dt_1=f(x)\ominus_{gH}f(0)\ominus_{gH}x\odot f'_{gH}(0)$$

Now suppose that $L(t)=f(a_1+th_1,a_2+th_2)$, based on the bivariate fuzzy chain rule in gH- Partial differentiability, we have

$$L'_{gH}(t)=\partial_{xgH}f(x(t),y(t))\odot x'(t)\oplus\partial_{ygH}f(x(t),y(t))\odot y'(t)$$

where $x'(t) = h_1, y'(t) = h_2$ so,

$$L'_{gH}(t) = \partial_{xgH}f(x(t), y(t)) \odot h_1 \oplus \partial_{ygH}f(x(t), y(t)) \odot h_2$$

Suppose that $t = 0$ because it is arbitrary. Then we can obtain the following property,

$$L'_{gH}(0) = \partial_{xgH}f(a_1, a_2) \odot h_1 \oplus \partial_{ygH}f(a_1, a_2) \odot h_2$$

You suppose that we discuss the item 1. Based on this item, the first order Taylor expansion for this single valued fuzzy function is as,

$$L(t) = L(0) \oplus L'_{gH}(0) \odot t \oplus R_2(t)$$

By substituting $L'_{gH}(0)$,

$$L(t) = L(0) \oplus \left(\partial_{xgH}f(a_1, a_2) \odot h_1 \oplus \partial_{ygH}f(a_1, a_2) \odot h_2\right) \odot t \oplus R_2(t)$$

Such that $L(0) = f(a_1, a_2)$, and $R_2(t) := R_2(a_1 + th_1, a_2 + th_2)$. Finally we obtained,

$$f(a_1 + th_1, a_2 + th_2) = f(a_1, a_2) \oplus \left(\partial_{xgH}f(a_1, a_2) \odot h_1 \oplus \partial_{ygH}f(a_1, a_2) \odot h_2\right) \oplus R_2(a_1 + th_1, a_2 + th_2)$$

Now we can set $t = 1$,

$$f(a_1 + h_1, a_2 + h_2) = f(a_1, a_2) \oplus \left(\partial_{xgH}f(a_1, a_2) \odot h_1 \oplus \partial_{ygH}f(a_1, a_2) \odot h_2\right) \oplus R_2(a_1 + h_1, a_2 + h_2)$$

The item 1, is now proved. The rest can be proved in a similar way.

Chapter 3
Fuzzy Fractional Operators

3.1 Introduction

The objectives of this chapter are to consider operators such as differentials and integral operators on fuzzy number valued functions. In this chapter, several types of fractional differentiability on fuzzy number valued functions and their relations are explained.

3.2 Fuzzy Grunwald-Letnikov Derivative—Fuzzy GL Derivative

The main discussion of this chapter is related to the fractional differential operator. As far as we know any differential (Natural or fractional) uses the difference. To this purpose, the fuzzy gH-difference and differentiability should be defined shortly.

In this section, first we discuss on the first order derivative shortly. Because it is necessary for defining the fractional derivative. As before, suppose that the function $x(t)$ is fuzzy number valued function and t is an inner point of its domain $[t_0, T]$. So, we consider $t \in (t_0, T)$. Also, we suppose that for any enough small number $h \longrightarrow 0$, $t + h \in (t_0, T)$.

Now the gH-differential of the $x(t)$ at the point t is denoted by $x'_{gH}(t)$ and defined as,

$$x'_{gH}(t) = \lim_{h \to 0} \frac{x(t+h) \ominus_{gH} x(t)}{h} := \lim_{h \to 0} \frac{\Delta_h x(t)}{h}$$

Subject to the gH-difference $x(t+h) \ominus_{gH} x(t)$ exist.

The operator Δ_h is called as forward operator which is defined by shift operator E_h as follow,

T. Allahviranloo, *Fuzzy Fractional Differential Operators and Equations*, Studies in Fuzziness and Soft Computing 397, https://doi.org/10.1007/978-3-030-51272-9_3

$$\Delta_h = E_h - 1, \quad E_h x(t) = x(t+h),$$
$$\Delta_h x(t) = (E_h - 1)x(t) = x(t+h) \ominus_{gH} x(t)$$

Note Please note that the gH-differential exists at the point t if $x'_{gH}(t)$ is a fuzzy number not a fuzzy set.

The level wise form of the derivative as a fuzzy number-valued function can be explained in the following two cases. Because the gH-difference in its definition, is defined in two cases too.

- Case (i), x is $[i - gH]$-differentiable at t if

$$x'_{i-gH}(t, r) = [x'_l(t, r), x'_u(t, r)]$$

- Case (ii), x is $[ii - gH]$-differentiable at t if

$$x'_{ii-gH}(t, r) = [x'_u(t, r), x'_l(t, r)]$$

Subject to the functions $x'_l(t, r)$ and $x'_u(t, r)$ are two real valued differentiable functions with respect to t and uniformly with respect to $r \in [0, 1]$. Please note that both functions are left continuous on $r \in (0, 1]$ and right continuous at $r = 0$. Moreover, the following conditions should be satisfied.

- The function $x'_l(t, r)$ is non-decreasing and the function $x'_u(t, r)$ is non-increasing as functions of r and, $x'_l(t, r) \leq x'_u(t, r)$. Or
- The function $x'_l(t, r)$ is non-increasing and the function $x'_u(t, r)$ is non-decreasing as functions of r and, $x'_u(t, r) \leq x'_l(t, r)$.
- $\left(x'_{gH}(t)\right)[r] = \left[\min\{x'_l(t, r), x'_u(t, r)\}, \max\{x'_l(t, r), x'_u(t, r)\}\right]$.

Note The relation between two forms of differentiability can be expressed as follow,

$$x'_{i-gH}(t) := \ominus_H(-1)x'_{ii-gH}(t)$$

or

$$x'_{ii-gH}(t) := \ominus_H(-1)x'_{i-gH}(t)$$

Because, in the level-wise form it is verified very easily for any $0 \leq r \leq 1$.

$$x'_{i-gH}(t), r = \ominus_H(-1)x'_{ii-gH}(t, r)$$
$$[x'_l(t, r), x'_u(t, r)] = \ominus_H(-1)[x'_u(t, r), x'_l(t, r)]$$

or

$$\left[x_u'(t,r), x_l'(t,r)\right] = \ominus_H(-1)\left[x_l'(t,r), x_u'(t,r)\right]$$

For an interval in level-wise form, for each level, the length operator or a a function can be defined.

3.2.1 Definition—Length of Fuzzy Function

For a fuzzy function $x(t)$, in the level-wise form $x(t,r) = [x_l(t,r), x_u(t,r)]$, the length is defined as,

$$length(x(t,r)) = x_u(t,r) - x_l(t,r)$$

Now consider $ii - gH$ differentiability,

$$x_{ii-gH}'(t,r) = \left[x_u'(t,r), x_l'(t,r)\right], \quad x_u'(t,r) \leq x_l'(t,r)$$

It means,

$$x_u(t+h,r) - x_u(t,r) \leq x_l(t+h,r) - x_l(t,r)$$

Then

$$\begin{aligned} length(x(t+h,r)) &= x_u(t+h,r) - x_l(t+h,r) \leq x_u(t,r) - x_l(t,r) \\ &= length(x(t,r)) \end{aligned}$$

Now it is claimed that the length operator is a decreasing operator.
Also, in $i - gH$ differentiabilty,

$$x_{i-gH}'(t,r) = \left[x_l'(t,r), x_u'(t,r)\right], \quad x_l'(t,r) \leq x_u'(t,r)$$

It means,

$$x_l(t+h,r) - x_l(t,r) \leq x_u(t+h,r) - x_u(t,r)$$

Then

$$\begin{aligned} length(x(t,r)) &= x_u(t,r) - x_l(t,r) \leq x_u(t+h,r) - x_l(t+h,r) \\ &= length(x(t+h,r)) \end{aligned}$$

In this case the length is increasing.

Now, the higher order of differentiability can be covered in this way, but it may more difficult and complicated when the derivatives are switched.

Second order of gH-differentiability
For the same function at the same point, the second order of gH-differential is defined as,

$$x''_{gH}(t) = \lim_{h\to 0} \frac{x'(t+h)\ominus_{gH} x'(t)}{h} = \lim_{h\to 0} \frac{\Delta_h^2 x(t)}{h}$$

Subject to the gH-difference $x'(t+h)\ominus_{gH} x'(t)$ exists.

The operator Δ_h^2 is called as forward operator with order two which is defined by shift operator of order two, E_h^2 as follow,

$$\Delta_h^2 = (E_h - 1)^2, E_h x(t) = x(t+h),$$
$$\Delta_h^2 x(t) = (E_h - 1)^2 x(t) = x(t+2h)\ominus_{gH} 2 \odot x(t+h) \oplus x(t)$$

Based on the previous discussion the same relation is true between two cases of differentiability's for the second order.

$$x''_{i-gH}(t) := \ominus_H(-1)x''_{ii-gH}(t)$$

or

$$x''_{ii-gH}(t) = \ominus_H(-1)x''_{i-gH}(t)$$

Basically, the differential is an operator and here we have two types of operators, the ${}_{i-gH}\boldsymbol{D}_{GL}$ operator and ${}_{ii-gH}\boldsymbol{D}_{GL}$ operator. So, the following four cases can happen.

$${}_{i-gH}\boldsymbol{D}_{GL}\left({}_{i-gH}\boldsymbol{D}_{GL}\right) := {}_{i,i}\boldsymbol{D}^2_{GL}, \quad {}_{ii-gH}\boldsymbol{D}_{GL}\left({}_{i-gH}\boldsymbol{D}_{GL}\right) := {}_{ii,i}\boldsymbol{D}^2_{GL}$$
$${}_{i-gH}\boldsymbol{D}_{GL}\left({}_{ii-gH}\boldsymbol{D}_{GL}\right) := {}_{i,ii}\boldsymbol{D}^2_{GL}, \quad {}_{ii-gH}\boldsymbol{D}_{GL}\left({}_{ii-gH}\boldsymbol{D}_{GL}\right) := {}_{ii,ii}\boldsymbol{D}^2_{GL}$$

All the following cases can be proved in level-wise form very easily.

1. In case ${}_{i-gH}\boldsymbol{D}_{GL}\left({}_{i-gH}\boldsymbol{D}_{GL}\right) := {}_{i,i}\boldsymbol{D}^2_{GL}$,

$${}_{i,i}\boldsymbol{D}^2_{GL}x(t) = \lim_{h\to 0} \frac{1}{h^2}\left[x(t+2h)\ominus_H 2 \odot x(t+h) \oplus x(t)\right]$$

Subject to the difference $x(t+2h)\ominus_H 2 \odot x(t+h)$ exists. To show it, we use the $i-gH$ differentiability operator on x for two times. To do this, it is enough to use the difference operator for two times.

$$\begin{aligned}&(x(t+2h)\ominus_H x(t+h))\ominus_H(x(t+h)\ominus_H x(t))\\&= x(t+2h)\ominus_H 2\odot x(t+h)\oplus x(t)\end{aligned}$$

Subject to the difference $(x(t+2h)\ominus_H x(t+h))\ominus_H(x(t+h)\ominus_H x(t))$ exists.

Note In this case both differential operators are increasing and there is no chance for switching to another case. Moreover, in the level-wise form,

$$x''_{i-gH}(t,r) = \left[x''_l(t,r), x''_u(t,r)\right]$$

2. In case ${}_{ii-gH}\boldsymbol{D}_{GL}\left({}_{i-gH}\boldsymbol{D}_{GL}\right) := {}_{ii,i}\boldsymbol{D}^2_{GL}$,

$${}_{ii,i}\boldsymbol{D}^2_{GL}x(t) = \lim_{h\to 0}\frac{1}{h^2}\left[2\odot x(t+h)\ominus_H x(t+2h)\ominus x(t)\right]$$

Since we have the $i-gH$ differentiability so we use the $ii-gH$ differentiability on it and the nominator of the derivative is,

$$\begin{aligned}&(x(t+h)\ominus_H x(t))\ominus_H(x(t+2h)\ominus_H x(t+h))\\&= 2\odot x(t+h)\ominus_H x(t+2h)\ominus x(t)\end{aligned}$$

Subject to the difference $(x(t+h)\ominus_H x(t))\ominus_H(x(t+2h)\ominus_H x(t+h))$ exists because both $x(t+h)\ominus_H x(t)$ and $x(t+2h)\ominus_H x(t+h)$ are already exist because of the definition of the second gH-derivative.

Note Here the first derivative operator is increasing and the second one is decreasing. Also, in the level-wise form,

$$\begin{aligned}{}_{i-gH}\boldsymbol{D}_{GL}x(t,r) &= \left[x'_l(t,r), x'_u(t,r)\right]\\{}_{ii-gH}\boldsymbol{D}_{GL}\left({}_{i-gH}\boldsymbol{D}_{GL}\right)x(t,r) &= \left[x''_u(t,r), x''_l(t,r)\right]\end{aligned}$$

3. In case ${}_{i-gH}\boldsymbol{D}_{GL}\left({}_{ii-gH}\boldsymbol{D}_{GL}\right) := {}_{i,ii}\boldsymbol{D}^2_{GL}$, based on the definition of H-difference in the nominator of the derivative, we have,

$${}_{ii,i}\boldsymbol{D}^2_{GL}x(t) = \lim_{h\to 0}\frac{1}{h^2}\left[\ominus_H x(t+2h)\oplus 2\odot x(t+h)\ominus_H x(t)\right]$$

Since we have the $ii-gH$ differentiability so we use the $i-gH$ differentiability on it and the nominator is,

$$(x(t+h)\ominus_H x(t+2h))\ominus_H (x(t)\ominus_H x(t+h))$$
$$= \ominus_H x(t+2h) \oplus 2 \odot x(t+h)\ominus_H x(t)$$

Subject to the difference $(x(t+h)\ominus_H x(t+2h))\ominus_H (x(t)\ominus_H x(t+h))$ exists.

Note Here the first length operator is decreasing and the second one is increasing. Also, in the level-wise form,

$$_{ii-gH}\boldsymbol{D}_{\boldsymbol{GL}}x(t,r) = \left[x_u'(t,r), x_l'(t,r)\right]$$
$$_{i-gH}\boldsymbol{D}_{\boldsymbol{GL}}\left(_{ii-gH}\boldsymbol{D}_{\boldsymbol{GL}}\right)x(t,r) = \left[x_l''(t,r), x_u''(t,r)\right]$$

4. In case $_{ii-gH}\boldsymbol{D}_{\boldsymbol{GL}}\left(_{ii-gH}\boldsymbol{D}_{\boldsymbol{GL}}\right) := {}_{ii,ii}\boldsymbol{D}^2_{\boldsymbol{GL}}$, since,

$$x''_{ii-gH}(t) = \ominus_H(-1)x''_{i-gH}(t)$$

Then

$$_{ii,ii}\boldsymbol{D}^2_{\boldsymbol{GL}}x(t) = \lim_{h\to 0}\frac{1}{h^2}\left[\ominus_H(-1)x(t+2h) \oplus (-2) \odot x(t+h)\ominus_H(-1)x(t)\right]$$

Note In this case both length operators are decreasing and there is no chance for switching to another case. Moreover

$$x''_{ii-gH}(t,r) = \left[x_u'(t,r), x_l'(t,r)\right]$$

Remark Taking to account all the results, in conclusion, it can be discovered that two cases $_{i-gH}\boldsymbol{D}_{\boldsymbol{GL}}\left(_{i-gH}\boldsymbol{D}_{\boldsymbol{GL}}\right) := {}_{i}\boldsymbol{D}^2_{\boldsymbol{GL}}$ and $_{ii-gH}\boldsymbol{D}_{\boldsymbol{GL}}\left(_{ii-gH}\boldsymbol{D}_{\boldsymbol{GL}}\right) := {}_{ii}\boldsymbol{D}^2_{\boldsymbol{GL}}$ are so logical and comfortable to use.

Third order of gH-differentiability

For the same function at the same point, the third order of gH-differential is defined as,

$$x'''_{gH}(t) = \lim_{h\to 0}\frac{x''(t+h)\ominus_{gH}x''(t)}{h} = \lim_{h\to 0}\frac{\Delta_h^3 x(t)}{h}$$

Subject to the gH-difference $x''(t+h)\ominus_{gH}x''(t)$ exists.

The operator Δ_h^3 is called as forward operator with order three which is defined by shift operator of order three, E_h^3 as follow,

$$\begin{aligned}\Delta_h^3 &= (E_h - 1)^3, E_h x(t) = x(t+h),\\ \Delta_h^3 x(t) &= (E_h - 1)^3 x(t)\\ &= x(t+3h)\ominus_{gH} 3\odot x(t+2h)\oplus 3\odot x(t+h)\ominus_H x(t)\end{aligned}$$

As it is mentioned, here are two cases of differentiability for the third order.

$$x'''_{i-gH}(t) = \ominus_H(-1)x'''_{ii-gH}(t)$$

or

$$x'''_{ii-gH}(t) = \ominus_H(-1)x'''_{i-gH}(t)$$

So, in case of ${}_i\boldsymbol{D}_{GL}^3$ we have,

$${}_i\boldsymbol{D}_{GL}^1\left({}_i\boldsymbol{D}_{GL}^2\right)x(t) = {}_i\boldsymbol{D}_{GL}^3 x(t),$$

$${}_i\boldsymbol{D}_{GL}^3 x(t) = \lim_{h\to 0}\frac{1}{h^3}[x(t+3h)\ominus_H 3\odot x(t+2h)\oplus 3\odot x(t+h)\ominus_H x(t)]$$

or

$${}_i\boldsymbol{D}_{GL}^3 x(t) = \lim_{h\to 0}\frac{1}{h^3}[x(t+3h)\ominus_H(-1)(-3)\odot x(t+2h)\oplus 3\odot x(t+h)\ominus_H(-1)(-1)x(t)]$$

It is supposed that, the first, second and third order of derivatives are in $i-gH$ form and those length functions are increasing. In level-wise form,

$${}_i\boldsymbol{D}_{GL}^3 x(t,r) = \left[x_l'''(t,r), x_u'''(t,r)\right]$$

Also, in case of ${}_{ii}\boldsymbol{D}_{GL}^3$ we have,

$$\begin{aligned}&{}_{ii}\boldsymbol{D}_{GL}^1\left({}_{ii}\boldsymbol{D}_{GL}^2\right)x(t) = {}_{ii}\boldsymbol{D}_{GL}^3 x(t),\\ &{}_{ii}\boldsymbol{D}_{GL}^3 x(t)\\ &= \lim_{h\to 0}\frac{1}{h^3}[\ominus_H(-1)x(t+3h)\oplus(-3)\odot x(t+2h)\ominus_H(-3)\odot x(t+h)\oplus(-1)x(t)]\end{aligned}$$

In level-wise form,

$${}_{ii}\boldsymbol{D}_{GL}^3 x(t,r) = \left[x_u'''(t,r), x_l'''(t,r)\right]$$

Also, here it is considered that, the first, second and third order of derivatives are in $ii-gH$ form and those length functions are decreasing. This process can be developed for the n-th order of differentiability.

- $\underbrace{{}_{i}D_{GL}({}_{i}D_{GL}({}_{i}D_{GL}(\cdots {}_{i}D_{GL}(x(t)))))}_{n-th} = {}_{i}\boldsymbol{D}^{n}_{GL}$

  All the derivatives ${}_{i,i}\boldsymbol{D}^{i}_{GL}, i = 1, 2, \ldots, n$ exist.

- $\underbrace{{}_{ii}D_{GL}({}_{ii}D_{GL}({}_{ii}D_{GL}(\cdots {}_{ii}D_{GL}(x(t)))))}_{n-th} = {}_{ii}\boldsymbol{D}^{n}_{GL}$

  All the derivatives ${}_{ii}\boldsymbol{D}^{j}_{GL}, j = 1, 2, \ldots, n$ exist.

In general, by continuing the process the general form of the n-th gH-derivative is defined as,

$$
\begin{aligned}
{}_{i}\boldsymbol{D}^{n}_{GL}x(t) &= \lim_{h\to 0}\frac{1}{h^n}\sum_{k=0}^{n}(\ominus_H)^k\binom{n}{k}x(t+(n-k)h)\\
&= \lim_{h\to 0}\frac{1}{h^n}\sum_{k=0}^{n}(\ominus_H)^k\binom{n}{k}^{\geq 0}x(t+(n-k)h)\\
&\quad\oplus \lim_{h\to 0}\frac{1}{h^n}\sum_{k=0}^{n}(\ominus_H)^{k+1}(-1)\binom{n}{k}^{<0}x(t+(n-k)h) := \lim_{h\to 0}\frac{\Delta^n_h x(t)}{h^n}\,{}_{ii}\boldsymbol{D}^{n}_{GL}x(t)
\end{aligned}
$$

where $x \in C^{\alpha}_{\mathbb{F}}[t_0, T], T > 0$ and means all the fuzzy derivatives are continuous function and $(\ominus_H)^{even} = \oplus$ and $(\ominus_H)^{odd} = \ominus_H$. In this series n can goes to ∞ and the series can have infinite terms with starting from any point. Now, considering the finite terms of the summations and starting with specified point t_0 and $n = \alpha$, $0<\alpha<1$, and considering the point like t_0 as an initial point, and shifting the points, the n-th order differentiability is defined as the following form for $t = t_0 + h$.

$$
\begin{aligned}
{}_{i}\boldsymbol{D}^{\alpha}_{GL}x(t) &= \lim_{h\to 0}h^{-\alpha}\sum_{k=0}^{\left[\frac{t-t_0}{h}\right]}(\ominus_H)^k\binom{\alpha}{k}x(t-kh)\\
&= \lim_{h\to 0}\frac{1}{h^n}\sum_{k=0}^{\left[\frac{t-t_0}{h}\right]}(\ominus_H)^k\binom{\alpha}{k}^{\geq 0}x(t-kh)\\
&\quad\oplus \lim_{h\to 0}\frac{1}{h^n}\sum_{k=0}^{\left[\frac{t-t_0}{h}\right]}(\ominus_H)^{k+1}(-1)\binom{\alpha}{k}^{<0}x(t-kh) := \lim_{h\to 0}\frac{\Delta^n_h x(t)}{h^n}\,{}_{ii}\boldsymbol{D}^{n}_{GL}x(t)
\end{aligned}
$$

where

$$\binom{\alpha}{k} = \frac{\alpha!}{k!(\alpha-k)!} = \frac{\Gamma(\alpha+1)}{\Gamma(k+1)\Gamma(\alpha-k+1)} = \frac{\Gamma(\alpha+1)}{k!\Gamma(\alpha-k+1)}$$

where

$$\Gamma(z) = \int_0^\infty e^{-t} t^{z-1} dt, \quad \Gamma(\alpha+1) = \alpha\Gamma(\alpha),$$

Note Note that the above binomial coefficients are the coefficients in the binomial series,

$$(1-t)^\alpha = \sum_{k=0}^{\infty} (-1)^k \binom{\alpha}{k} x^k$$

Which for real $\alpha > 0$ converges when $|x| \le 1$. However, they don't vanish anymore for $k > \alpha, \alpha \notin \mathbb{N}$. Because in integer case, if $\alpha = n \in \mathbb{N}$, then $\binom{n}{k} = 0$ for $k > n$. The following relations also are derived,

$$\int_0^t (t-s)^{\frac{1}{\alpha}-1} s^\gamma ds = \frac{\Gamma(\frac{1}{\alpha})\Gamma(\gamma+1)}{\Gamma(\frac{1}{\alpha}+\gamma+1)} t^{\frac{k}{\alpha}+\gamma}, \quad \alpha > 0, \ \gamma > -1$$

Then, the Fuzzy Grunwald-Letnikov Derivative is going to be define as,

$$\begin{aligned}
{}_i\boldsymbol{D}^\alpha_{GL} x(t) &= \lim_{h\to 0} h^{-\alpha} \sum_{k=0}^{\left[\frac{t-t_0}{h}\right]} (\ominus_H)^k \frac{\Gamma(\alpha+1)}{k!\Gamma(\alpha-k+1)} x(t-kh) \\
&= \lim_{h\to 0} h^{-\alpha} \sum_{k=0}^{\left[\frac{t-t_0}{h}\right]} (\ominus_H)^k \left[\frac{\Gamma(\alpha+1)}{k!\Gamma(\alpha-k+1)}\right]^{\ge 0} x(t-kh) \\
&\oplus \lim_{h\to 0} h^{-\alpha} \sum_{k=0}^{\left[\frac{t-t_0}{h}\right]} (\ominus_H)^{k+1} (-1) \left[\frac{\Gamma(\alpha+1)}{k!\Gamma(\alpha-k+1)}\right]^{<0} x(t-kh) := \lim_{h\to 0} \frac{\Delta^\alpha_h x(t)}{h^\alpha} {}_{ii}\boldsymbol{D}^\alpha_{GL} x(t)
\end{aligned}$$

3.2.2 Level-Wise Form of Grunwald-Letnikov GL-Derivative

Suppose that the level-wise form of ${}_i\boldsymbol{D}^{\alpha}_{GL}x(t)$ is an interval like,

$${}_i\boldsymbol{D}^{\alpha}_{GL}x(t,r) := \left[{}_i\boldsymbol{D}^{\alpha}_{GL,l}x(t,r), {}_i\boldsymbol{D}^{\alpha}_{GL,u}x(t,r)\right]$$

Based on the definition of the gH-differentiability, we have,

$$\left[{}_i\boldsymbol{D}^{\alpha}_{GL,l}x(t,r), {}_i\boldsymbol{D}^{\alpha}_{GL,u}x(t,r)\right] = \left[\boldsymbol{D}^{\alpha}_{GL}x_l(t,r), \boldsymbol{D}^{\alpha}_{GL}x_u(t,r)\right]$$

This means, the level-wise form of the derivative on x is the derivative of level-wise form of x. So,

- In case, i—differentiability

$${}_i\boldsymbol{D}^{\alpha}_{GL}x(t,r) = \left[\boldsymbol{D}^{\alpha}_{GL}x_l(t,r), \boldsymbol{D}^{\alpha}_{GL}x_u(t,r)\right]$$

- In case, ii—differentiability

$${}_{ii}\boldsymbol{D}^{\alpha}_{GL}x(t,r) = \left[\boldsymbol{D}^{\alpha}_{GL}x_u(t,r), \boldsymbol{D}^{\alpha}_{GL}x_l(t,r)\right]$$

As we know, the function,

$$\frac{\Gamma(\alpha+1)}{\mathrm{k}!\Gamma(\alpha-k+1)} \leq 0, \quad for \quad \alpha < k-1$$

Then the level-wise form is introduced and separated by the sign of the mentioned Gamma fraction,

$$\begin{aligned}\boldsymbol{D}^{\alpha}_{GL}x_l(t,r) = {} & \lim_{h\to 0} h^{-\alpha}\sum_{k=0}^{\alpha}(-1)^k\left[\frac{\Gamma(\alpha+1)}{\mathrm{k}!\Gamma(\alpha-k+1)}\right]^{\geq 0} x_l(t-kh) \\ & + \lim_{h\to 0} h^{-\alpha}\sum_{k=0}^{\alpha}(-1)^k\left[\frac{\Gamma(\alpha+1)}{\mathrm{k}!\Gamma(\alpha-k+1)}\right]^{<0} x_u(t-kh)\end{aligned}$$

$$\begin{aligned}\boldsymbol{D}^{\alpha}_{GL}x_u(t,r) = {} & \lim_{h\to 0} h^{-\alpha}\sum_{k=0}^{\alpha}(-1)^k\left[\frac{\Gamma(\alpha+1)}{\mathrm{k}!\Gamma(\alpha-k+1)}\right]^{\geq 0} x_u(t-kh) \\ & + \lim_{h\to 0} h^{-\alpha}\sum_{k=0}^{\alpha}(-1)^k\left[\frac{\Gamma(\alpha+1)}{\mathrm{k}!\Gamma(\alpha-k+1)}\right]^{<0} x_l(t-kh)\end{aligned}$$

Note Please note that, finding the derivative for fuzzy function is very difficult because it is always a system of equations in accordance with $x_l(t,r)$ and $x_u(t,r)$.

As you see, it is a system of equation and in the block form it is,

$$\mathbf{\Gamma}(\boldsymbol{\alpha},\mathbf{k})=\begin{pmatrix}\left[\frac{\Gamma(\alpha+1)}{\mathrm{k}!\Gamma(\alpha-k+1)}\right]^{\geq 0} & \left[\frac{\Gamma(\alpha+1)}{\mathrm{k}!\Gamma(\alpha-k+1)}\right]^{<0}\\ \left[\frac{\Gamma(\alpha+1)}{\mathrm{k}!\Gamma(\alpha-k+1)}\right]^{<0} & \left[\frac{\Gamma(\alpha+1)}{\mathrm{k}!\Gamma(\alpha-k+1)}\right]^{\geq 0}\end{pmatrix},$$

$$\boldsymbol{D}^{\alpha}_{GL}\boldsymbol{x}(t)=\begin{pmatrix}\boldsymbol{D}^{\alpha}_{GL}x_l(t,r)\\ \boldsymbol{D}^{\alpha}_{GL}x_u(t,r)\end{pmatrix},\ \boldsymbol{x}(\boldsymbol{t}-\boldsymbol{kh})=\begin{pmatrix}x_l(t-kh)\\ x_l(t-kh)\end{pmatrix}$$

The compact form is, $\mathbf{\Gamma}(\boldsymbol{\alpha},\mathbf{k})\boldsymbol{x}(\boldsymbol{t}-\boldsymbol{kh})=\boldsymbol{D}^{\alpha}\boldsymbol{x}(\boldsymbol{t})$ and the sufficient and necessary condition for having a unique solution is, the matrix $\mathbf{\Gamma}$ be a non-singular matrix. The non-singularity of the matrix is very difficult to consider because its sign is constantly changing. The only way to investigate is, numerical methods to find the approximate solution of the system and setting the fuzzy derivative.

Remark As it is mentioned before, the type of differentiability can also be investigated by the length function,

- In case *i*—differentiability,

$$length\left({}_i\boldsymbol{D}^{\alpha}_{GL}x(t,r)\right)=\boldsymbol{D}^{\alpha}_{GL}x_u(t,r)-\boldsymbol{D}^{\alpha}_{GL}x_l(t,r)\geq 0$$

- In case *ii*—differentiability,

$$length\left({}_{ii}\boldsymbol{D}^{\alpha}_{GL}x(t,r)\right)=-length\left({}_i\boldsymbol{D}^{\alpha}_{GL}x(t,r)\right)=\boldsymbol{D}^{\alpha}_{GL}x_l(t,r)-\boldsymbol{D}^{\alpha}_{GL}x_u(t,r)\leq 0$$

Remark Based on the properties of the shift and forward operators the following properties can be shown easily,

$$\begin{aligned}{}_i\boldsymbol{D}^{\alpha+\beta}_{GL}x(t)&={}_i\boldsymbol{D}^{\alpha}_{GL}x(t)\cdot_i\boldsymbol{D}^{\beta}_{GL}x(t)\\ {}_{ii}\boldsymbol{D}^{\alpha+\beta}_{GL}x(t)&={}_{ii}\boldsymbol{D}^{\alpha}_{GL}x(t)\cdot{}_{ii}\boldsymbol{D}^{\beta}_{GL}x(t)\end{aligned}$$

For instance, for the *i*—derivative,

$$\Delta^{\alpha}_h=(E_h-1)^{\alpha},\ \Delta^{\beta}_h=(E_h-1)^{\beta}$$

Also,

$$\Delta_h^\alpha \Delta_h^\beta = (E_h - 1)^\alpha (E_h - 1)^\beta = (E_h - 1)^{\alpha+\beta} = \Delta_h^{\alpha+\beta}$$

The same reasoning can be used for *ii*—differential.

Remark Before solving some examples, it should be noted that the fuzzy GL-derivative does not exist always, and it is indeed conditional. So, we should go through the other fuzzy derivative to avoid the negative Gamma functions in the matrix.

Example—Fuzzy exponential function Consider the fuzzy exponential function $x(t) = e^{c\odot t}$ where $t \in [t_0, T], c \in \mathbb{F}_R$ is a fuzzy number valued function. The GL-derivative is obtained as,

$$\begin{aligned}
{}_i\boldsymbol{D}_{GL}^{\alpha} e^{c\odot t} &= \lim_{h\to 0} h^{-\alpha} \sum_{k=0}^{\alpha} (\ominus_H)^k \frac{\Gamma(\alpha+1)}{k!\Gamma(\alpha-k+1)} e^{c\odot t + (\alpha-k)h} \\
&= \lim_{h\to 0} h^{-\alpha} \sum_{k=0}^{\alpha} (\ominus_H)^k \left[\frac{\Gamma(\alpha+1)}{k!\Gamma(\alpha-k+1)}\right]^{\geq 0} e^{c\odot t + (\alpha-k)h} \\
&\quad \oplus \lim_{h\to 0} h^{-\alpha} \sum_{k=0}^{\alpha} (\ominus_H)^{k+1} (-1) \left[\frac{\Gamma(\alpha+1)}{k!\Gamma(\alpha-k+1)}\right]^{<0} e^{c\odot t + (\alpha-k)h} \\
&= \lim_{h\to 0} h^{-\alpha} e^{c\odot t} \sum_{k=0}^{\alpha} (\ominus_H)^k \left[\frac{\Gamma(\alpha+1)}{k!\Gamma(\alpha-k+1)}\right]^{\geq 0} e^{c\odot(\alpha-k)h} \\
&\quad \oplus \lim_{h\to 0} h^{-\alpha} e^{c\odot t} \sum_{k=0}^{\alpha} (\ominus_H)^{k+1} (-1) \left[\frac{\Gamma(\alpha+1)}{k!\Gamma(\alpha-k+1)}\right]^{<0} e^{c\odot(\alpha-k)h} \\
&= \lim_{h\to 0} h^{-\alpha} e^{c\odot t} \sum_{k=0}^{\alpha} (\ominus_H)^k \left[\frac{\Gamma(\alpha+1)}{k!\Gamma(\alpha-k+1)}\right]^{\geq 0} \left(e^{c\odot h}\right)^{(\alpha-k)} \\
&\quad \oplus \lim_{h\to 0} h^{-\alpha} e^{c\odot t} \sum_{k=0}^{\alpha} (\ominus_H)^{k+1} (-1) \left[\frac{\Gamma(\alpha+1)}{k!\Gamma(\alpha-k+1)}\right]^{<0} \left(e^{c\odot h}\right)^{(\alpha-k)} \\
&= \lim_{h\to 0} h^{-\alpha} e^{c\odot t} \sum_{k=0}^{\alpha} (\ominus_H)^k \left[\frac{\Gamma(\alpha+1)}{k!\Gamma(\alpha-k+1)}\right]^{\geq 0} \left(e^{c\odot h}\right)^{(\alpha-k)} \\
&\quad \oplus \lim_{h\to 0} h^{-\alpha} e^{c\odot t} \sum_{k=0}^{\alpha} (\ominus_H)^{k+1} (-1) \left[\frac{\Gamma(\alpha+1)}{k!\Gamma(\alpha-k+1)}\right]^{<0} \left(e^{c\odot h}\right)^{(\alpha-k)}
\end{aligned}$$

Since $h^{-\alpha} e^{c\odot t}$ is a positive fuzzy number,

$$= \lim_{h\to 0} h^{-\alpha} e^{c\odot t} \left[\sum_{k=0}^{\alpha} (\ominus_H)^k \left[\frac{\Gamma(\alpha+1)}{k!\Gamma(\alpha-k+1)}\right]^{\geq 0} \left(e^{c\odot h}\right)^{(\alpha-k)} \oplus \sum_{k=0}^{\alpha} (\ominus_H)^{k+1} (-1) \left[\frac{\Gamma(\alpha+1)}{k!\Gamma(\alpha-k+1)}\right]^{<0} \left(e^{c\odot h}\right)^{(\alpha-k)}\right]$$

For more investigation and illustrating, the following equation, as H-difference is true subject to the H-difference exists,

Note The condition for continuing the process is, the following H-difference should exist.

$$\sum_{k=0}^{\alpha}(\ominus_H)^k\left[\frac{\Gamma(\alpha+1)}{k!\Gamma(\alpha-k+1)}\right]^{\geq 0}\left(e^{c\odot h}\right)^{(\alpha-k)}$$
$$=\left(e^{c\odot h}-1\right)^{\alpha}\ominus_H\sum_{k=0}^{\alpha}(\ominus_H)^{k+1}(-1)\left[\frac{\Gamma(\alpha+1)}{k!\Gamma(\alpha-k+1)}\right]^{<0}\left(e^{c\odot h}\right)^{(\alpha-k)}$$

Now based on the definition of the H-difference,

$$\left(e^{c\odot h}-1\right)^{\alpha}=\sum_{k=0}^{\alpha}(\ominus_H)^k\left[\frac{\Gamma(\alpha+1)}{k!\Gamma(\alpha-k+1)}\right]^{\geq 0}\left(e^{c\odot h}\right)^{(\alpha-k)}$$
$$\oplus\sum_{k=0}^{\alpha}(\ominus_H)^{k+1}(-1)\left[\frac{\Gamma(\alpha+1)}{k!\Gamma(\alpha-k+1)}\right]^{<0}\left(e^{c\odot h}\right)^{(\alpha-k)}$$

Thus,

$$=\lim_{h\to 0}h^{-\alpha}e^{c\odot t}\sum_{k=0}^{\alpha}(\ominus_H)^k\left[\frac{\Gamma(\alpha+1)}{k!\Gamma(\alpha-k+1)}\right]\left(e^{c\odot h}\right)^{(\alpha-k)}$$
$$=e^{c\odot t}\lim_{h\to 0}h^{-\alpha}\left(e^{c\odot h}-1\right)^{\alpha}$$

Finally,

$${}_iD_{GL}^{\alpha}e^{c\odot t}=e^{c\odot t}\lim_{h\to 0}h^{-\alpha}\left(e^{c\odot h}-1\right)^{\alpha}=e^{c\odot t}\lim_{h\to 0}\left(\frac{e^{c\odot h}-1}{h}\right)^{\alpha}$$

Since,

$$\lim_{h\to 0}\frac{e^{c\odot h}-1}{h}=c\Rightarrow {}_iD_{GL}^{\alpha}e^{c\odot t}=c^{\alpha}\odot e^{c\odot t}$$

In this example the scalar c is a fuzzy number and sometimes the support of this fuzzy number is non-negative and sometimes, some parts of the support are negative. The values of c^{α} for $0<\alpha<1$, are non-negative fuzzy numbers and positive imaginary numbers respectively. In the other hand, the fuzzy exponential function is a non-negative fuzzy function (The positive and negative fuzzy numbers have been explained in the chapter two). In conclusion, the function is always i—differentiable with the condition of existence of the H-difference that it is noted.

In case, the mentioned H-difference does not exist, we must go to the level-wise form of the derivative and finding the approximate fuzzy derivative function. In the level-wise form it is,

$$ {}_i\boldsymbol{D}^{\alpha}_{GL}x(t,r) = \left[{}_i\boldsymbol{D}^{\alpha}_{GL,l}x(t,r), {}_i\boldsymbol{D}^{\alpha}_{GL,u}x(t,r)\right] $$
$$ \left[{}_i\boldsymbol{D}^{\alpha}_{GL,l}x(t,r), {}_i\boldsymbol{D}^{\alpha}_{GL,u}x(t,r)\right] = \left[\boldsymbol{D}^{\alpha}_{GL}x_l(t,r), \boldsymbol{D}^{\alpha}_{GL}x_u(t,r)\right] $$

Such that,

$$ \boldsymbol{D}^{\alpha}_{GL}x_l(t,r) = c^{\alpha}_l(r)e^{c_l(r)t}, \quad \boldsymbol{D}^{\alpha}_{GL}x_u(t,r) = c^{\alpha}_u(r)e^{c_u(r)t} $$

Since, $c^{\alpha}_l(r) \le c^{\alpha}_u(r)$ and $e^{c_l(r)t} \le e^{c_u(r)t}$ then,

$$ \boldsymbol{D}^{\alpha}_{GL}x_l(t,r) \le \boldsymbol{D}^{\alpha}_{GL}x_u(t,r), \quad 0 \le r \le 1 $$

It follows that, naturally, the *i*—differentiability in the level-wise form is always an interval $\left[\boldsymbol{D}^{\alpha}_{GL}x_l(t,r), \boldsymbol{D}^{\alpha}_{GL}x_u(t,r)\right]$ for all $0 \le r \le 1$ and there is no option for *ii*—differentiability.

Example—Power function with fuzzy coefficient.

Let us consider another simple fuzzy function,

$$ x(t) = c \odot t^j, \quad j \in R, \quad t \ge 0, \quad c \in \mathbb{F}_R. $$

This function does have only *i*—derivative because $t^j \ge 0$, and the mentioned problem and condition for the existence of the GL-derivative are not necessary here.

$$ \begin{aligned} {}_i\boldsymbol{D}^{\alpha}_{GL}\left(c \odot t^j\right) &= c \odot \boldsymbol{D}^{\alpha}_{GL}t^j \\ &= c \odot \lim_{h \to 0} h^{-\alpha} \sum_{k=0}^{\alpha} (-1)^k \left[\frac{\Gamma(\alpha+1)}{k!\Gamma(\alpha-k+1)}\right](t-kh)^j \end{aligned} $$

So,

$$ {}_i\boldsymbol{D}^{\alpha}_{GL}\left(c \odot t^j\right) = c \odot \frac{\Gamma(j+1)}{\Gamma(j-\alpha+1)} t^{j-\alpha} $$

Example—Power series with fuzzy coefficients

Let consider the following series with fuzzy number coefficients,

$$ x(t) = \sum_{n=-\infty}^{\infty} a_n \odot t^n, \quad t \ge 0, \quad a_n \in \mathbb{F}_R $$

Using the linearity property of the GL-derivative,

$$\begin{aligned}
{}_i\boldsymbol{D}_{GL}^{\alpha}x(t) &= \sum_{n=-\infty}^{\infty} a_n \odot \boldsymbol{D}_{GL}^{\alpha}t^n \\
&= \sum_{n=-\infty}^{\infty} a_n \odot \lim_{h\to 0} h^{-\alpha} \sum_{k=0}^{\alpha}(-1)^k \left[\frac{\Gamma(\alpha+1)}{k!\Gamma(\alpha-k+1)}\right](t-kh)^n \\
&= c \odot \sum_{n=-\infty}^{\infty} a_n \odot \frac{\Gamma(j+1)}{\Gamma(j-\alpha+1)} t^{n-\alpha}
\end{aligned}$$

The level-wise form can be explained very easily and,

$$\boldsymbol{D}_{GL}^{\alpha}x_l(t,r) = c_l(r)\cdot \sum_{n=-\infty}^{\infty} a_n \frac{\Gamma(j+1)}{\Gamma(j-\alpha+1)} t^{n-\alpha},$$

$$\boldsymbol{D}_{GL}^{\alpha}x_u(t,r) = c_u(r)\cdot \sum_{n=-\infty}^{\infty} a_n \frac{\Gamma(j+1)}{\Gamma(j-\alpha+1)} t^{n-\alpha},$$

Since, $c_l^{\alpha}(r) \le c_u^{\alpha}(r)$ then,

$$\boldsymbol{D}_{GL}^{\alpha}x_l(t,r) \le \boldsymbol{D}_{GL}^{\alpha}x_u(t,r),\ 0\le r\le 1$$

It follows that, naturally, the *i*—differentiability in the level-wise form is always an interval $\left[\boldsymbol{D}_{GL}^{\alpha}\,{}^{\alpha}x_l(t,r), \boldsymbol{D}_{GL}^{\alpha}x_u(t,r)\right]$ for all $0\le r\le 1$ and there is no option for *ii*—differentiability.

3.2.3 Remark—Fuzzy Fractional Integral Operator

As we know, the integral operator is an anti-operator of differential and vice versa. To this end, it is enough to use $-\alpha$ instead of α in the definition of the fuzzy GL-derivative.

$$\begin{aligned}
{}_i\boldsymbol{D}_{GL}^{-\alpha}x(t) &= \lim_{h\to 0} h^{\alpha} \sum_{k=0}^{\left[\frac{t-t_0}{h}\right]} (\ominus_H)^k \frac{\Gamma(-\alpha+1)}{k!\Gamma(-\alpha-k+1)} x(t-kh) \\
&= \lim_{h\to 0} h^{\alpha} \sum_{k=0}^{\left[\frac{t-t_0}{h}\right]} (\ominus_H)^k \left[\frac{\Gamma(-\alpha+1)}{k!\Gamma(-\alpha-k+1)}\right]^{\ge 0} x(t-kh) \\
&\quad \oplus \lim_{h\to 0} h^{\alpha} \sum_{k=0}^{\left[\frac{t-t_0}{h}\right]} (\ominus_H)^{k+1}(-1)\left[\frac{\Gamma(-\alpha+1)}{k!\Gamma(-\alpha-k+1)}\right]^{<0} x(t-kh)
\end{aligned}$$

3.3 Fuzzy Riemann-Liouville Derivative—Fuzzy RL Derivative

We will begin with a visual interpretation of fractional integrals as a fraction of the total integral of the two types, the left, and right integrals, or the first or first type of integral, and the second type, respectively. In the following Fig. 3.1 one triangle is marked to the left of the line $x_1 = x_2$ and another triangle is marked to the right of the line $x_1 = x_2$. The areas of these triangles are fractions of the full area of the square $(b - a) \times (b - a)$.

Consider the area of the triangle on the left of the line $t_1 = t_2$ in a plane. The area,

$$\int_{t_1=a}^{t_1=t} dt_1 \int_{t_2=t_1}^{t_2=t} dt_2 = \int_{t_1=a}^{t_1=t} (t - t_1)dt_1$$

In three space,

$$\int_{t_1=a}^{t_1=t} dt_1 \int_{t_2=t_1}^{t_2=t} dt_2 \int_{t_3=t_2}^{t_3=t} dt_3 = \int_{t_1=a}^{t_1=t} dt_1 \int_{t_2=t_1}^{t_2=t} dt_2(t - t_2) = \int_{t_1=a}^{t_1=t} \frac{(t - t_1)^2}{2!} dt_1$$

By continuing this process, for the n—space we will find,

$$\int_{t_1=a}^{t_1=t} \frac{(t - t_1)^{n-1}}{(n - 1)!} dt_1$$

If you notice, the corresponding function is a constant function $x(t) = 1$. Now, by considering any fuzzy function like $x(t) \in \mathbb{F}_R$ we will find the following integral in the n-space.

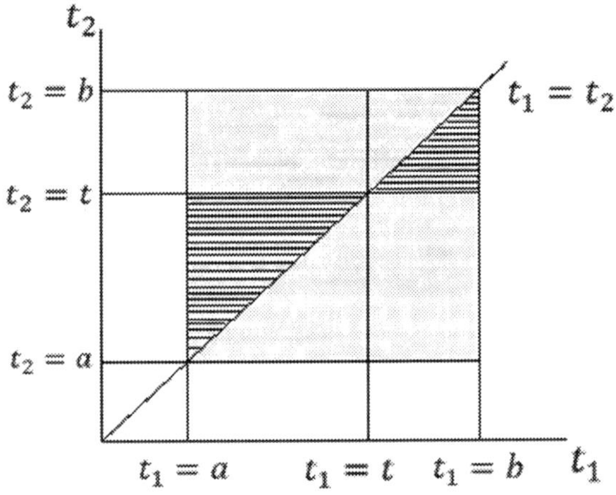

Fig. 3.1 Integration triangle areas

$$\int_{t_1=a}^{t_1=t} \frac{(t-t_1)^{n-1}}{(n-1)!} \odot x(t_1)dt_1 = \int_{t=a}^{t=s} \frac{(s-t)^{n-1}}{(n-1)!} \odot x(t)dt$$

$$= \int_{t=a}^{t=s} \frac{(s-t)^{n-1}}{\Gamma(n)} \odot x(t)dt$$

As it is mentioned early, the integration operator is an anti-operator of the derivative operator, so it can be shown by,

$$\boldsymbol{D}_{\boldsymbol{RL}}^{-n} x(s) = \frac{1}{\Gamma(n)} \odot \int_{t=a}^{t=s} (s-t)^{n-1} \odot x(t)dt, \ s \in [t_0, T]$$

So, the left-sided or first kind Riemann-Liouville fractional integral of order α is denoted by $\boldsymbol{D}_{\boldsymbol{RL}}^{-\alpha} x(s)$ and

$$\boldsymbol{D}_{\boldsymbol{RL}}^{-\alpha} x(s) = \frac{1}{\Gamma(\alpha)} \odot \int_{a}^{s} (s-t)^{\alpha-1} \odot x(t)dt, \ 0 < \alpha < 1$$

Now, consider the area of the triangle on the right of the line $t_1 = t_2$ in the figure. The right area is given by,

$$\int_{t_1=t}^{t_1=b} dt_1 \int_{t_2=t}^{t_2=t_1} dt_2 = \int_{t_1=t}^{t_1=b} (t_1 - t)dt_1$$

In 3-space,

$$\int_{t_1=t}^{t_1=b} dt_1 \int_{t_2=t}^{t_2=t_1} dt_2 \int_{t_3=t}^{t_3=t_2} dt_3 = \int_{t_1=a}^{t_1=t} dt_1 \int_{t_2=t_1}^{t_2=t} (t_2 - t)dt_2 = \int_{t_1=t}^{t_1=b} \frac{(t_1-t)^2}{2!} dt_1$$

By continuing this process, for the n-space we will find,

$$\int_{t_1=t}^{t_1=b} \frac{(t_1-t)^{n-1}}{(n-1)!} dt_1$$

Now, by considering any fuzzy function like $x(t) \in \mathbb{F}_R$ we will find the following integral in the n-space.

$$\int_{t_1=t}^{t_1=b} \frac{(t_1-t)^{n-1}}{(n-1)!} \odot x(t_1)dt_1 := \int_s^b \frac{(t-s)^{n-1}}{(n-1)!} \odot x(t)dt$$
$$= \int_s^b \frac{(t-s)^{n-1}}{\Gamma(n)} \odot x(t)dt$$

Since the integration operator is an anti-operator of the derivative operator, so it can be shown by,

$$\boldsymbol{D}_{\boldsymbol{RL}}^{-n}x(s) = \frac{1}{\Gamma(n)} \odot \int_s^b (t-s)^{n-1} \odot x(t)dt$$

So, the right-sided or second kind Riemann-Liouville fractional integral of order α is denoted by $\boldsymbol{D}_{\boldsymbol{RL}}^{-\alpha}x(s)$ and

$$\boldsymbol{D}_{\boldsymbol{RL}}^{-\alpha}x(s) = \frac{1}{\Gamma(\alpha)} \odot \int_s^b (t-s)^{\alpha-1} \odot x(t)dt, \quad s \in [t_0, T]$$

In conclusion, we have,

- For the **left area**, first kind Riemann-Liouville fractional integral of order α

$$\boldsymbol{I}_{\boldsymbol{RL}}^{\alpha}x(s) := \boldsymbol{D}_{\boldsymbol{RL}}^{-\alpha}x(s) = \frac{1}{\Gamma(\alpha)} \odot \int_a^s (s-t)^{\alpha-1} \odot x(t)dt$$

- For the **right area**, second kind Riemann-Liouville fractional integral of order α

$$\boldsymbol{I}_{\boldsymbol{RL}}^{\alpha}x(s) := \boldsymbol{D}_{\boldsymbol{RL}}^{-\alpha}x(s) = \frac{1}{\Gamma(\alpha)} \odot \int_s^b (t-s)^{\alpha-1} \odot x(t)dt$$

Note Since the only difference between these two operators is in the initial points and, our problems to consider are time-dependent problems with the left initial points. So, the left-sided fractional integral and derivative will be our default operators. Therefore, we will use these operatives as practical operators in our discussions.

$$\boldsymbol{I}_{RL}^{\alpha}x(s) := \boldsymbol{D}_{RL}^{-\alpha}x(s) = \frac{1}{\Gamma(\alpha)} \odot \int_{t_0}^{s} (s-t)^{\alpha-1} \odot x(t)dt, \ s \in [t_0, T],$$

For any $0 < \alpha < 1$.

Note—Combination Property

- $$\boldsymbol{D}_{RL}^{-\alpha}\boldsymbol{D}_{RL}^{-\beta}x(t) = \boldsymbol{D}_{RL}^{-\beta}\boldsymbol{D}_{RL}^{-\alpha}x(t) = \boldsymbol{D}_{RL}^{-(\alpha+\beta)}x(t), \ 0<\alpha<1, 0<\beta<1$$
  $$\boldsymbol{D}_{RL}^{-\alpha}x(s) = \frac{1}{\Gamma(\alpha)} \odot \int_{t_0}^{s} (s-t)^{\alpha-1} \odot x(t)dt, \ 0<\alpha<1$$
  $$\boldsymbol{D}_{RL}^{-\beta}x(s) = \frac{1}{\Gamma(\beta)} \odot \int_{t_0}^{s} (s-t)^{\beta-1} \odot x(t)dt, \ 0<\beta<1$$

Now,

$$\boldsymbol{D}_{RL}^{-\alpha}\boldsymbol{D}_{RL}^{-\beta}x(s) = \frac{1}{\Gamma(\alpha)} \odot \int_{t_0}^{s} (s-t)^{\alpha-1} \odot \left[\frac{1}{\Gamma(\beta)} \odot \int_{u}^{s} (t-u)^{\beta-1} \odot x(u)du\right] dt$$

Since the functions $s-t, t-u$, $\Gamma(\alpha)$, $\Gamma(\beta)$ are positive then the same procedure for the real cases is happened here.

$$\boldsymbol{D}_{RL}^{-\alpha}\boldsymbol{D}_{RL}^{-\beta}x(s) = \frac{1}{\Gamma(\alpha+\beta)} \odot \int_{t_0}^{s} x(u)du \odot \left[\int_{u}^{s} (s-t)^{\alpha-1}(t-u)^{\beta-1}dt\right]$$

By changing some variables like, suppose, $\frac{t-u}{s-u} = v$, where s, u are fixed. Then $\frac{1}{s-u}dt = dv$.

$$(t-u)^{\beta-1} = v^{\beta-1}(s-u)^{\beta-1},$$
$$1 - v = \frac{s-t}{s-u} \Rightarrow (s-t)^{\alpha-1} = (1-v)^{\alpha-1}(s-u)^{\alpha-1}$$

Then

$$\begin{aligned}(s-t)^{\alpha-1}(t-u)^{\beta-1} &= (1-v)^{\alpha-1}(s-u)^{\alpha-1}v^{\beta-1}(s-u)^{\beta-1} \\ &= (s-u)^{\alpha+\beta-2}(1-v)^{\alpha-1}v^{\beta-1}\end{aligned}$$

By substituting,

$$D_{RL}^{-\alpha}D_{RL}^{-\beta}x(s)=\frac{1}{\Gamma(\alpha+\beta)}\odot\int_{t_0}^{s}x(u)du\odot\left[\int_{0}^{1}(s-u)^{\alpha+\beta-1}(1-v)^{\alpha-1}v^{\beta-1}dv\right]$$

$$=\frac{1}{\Gamma(\alpha+\beta)}\odot\int_{t_0}^{s}(s-u)^{\alpha+\beta-1}\odot x(u)du\left[\int_{0}^{1}(1-v)^{\alpha-1}v^{\beta-1}dv\right]$$

Since,

$$\int_{0}^{1}(1-v)^{\alpha-1}v^{\beta-1}dv=1$$

Then

$$D_{RL}^{-\alpha}D_{RL}^{-\beta}x(s)=\frac{1}{\Gamma(\alpha+\beta)}\odot\int_{t_0}^{s}(s-u)^{\alpha+\beta-1}\odot x(u)du=D_{RL}^{-(\alpha+\beta)}x(s)$$

The same procedure can be used to prove $D_{RL}^{-\beta}D_{RL}^{-\alpha}x(t)$.

3.3.1 Level-Wise Form of Fuzzy Riemann-Liouville Integral Operators

Based on the definition of the integral and its level-wise form, we know that, the level-wise form of integral is, the integrals of end points of the interval in the level-wise form of integrand function.

- **The RL fractional integral operator**

$$I_{RL}^{\alpha}x(s,r)=\left[I_{RL}^{\alpha}x_l(s,r),I_{RL}^{\alpha}x_u(s,r)\right],\ s\in[t_0,T]$$

Where

$$I_{RL}^{\alpha}x_l(s,r)=\frac{1}{\Gamma(\alpha)}\int_{t_0}^{s}(s-t)^{\alpha-1}x_l(t,r)dt$$
$$I_{RL}^{\alpha}x_u(s,r)=\frac{1}{\Gamma(\alpha)}\int_{t_0}^{s}(s-t)^{\alpha-1}x_u(t,r)dt$$

Because, $\Gamma(\alpha)>0$ and $(s-t)^{\alpha-1}>0$.

3.3.2 The Fuzzy Riemann-Liouville Derivative Operators

In the definition of RL fractional integral,

$$\boldsymbol{D}_{\boldsymbol{RL}}^{-\alpha}x(s) = \frac{1}{\Gamma(\alpha)} \odot \int\limits_{t_0}^{s} (s-t)^{\alpha-1} \odot x(t)dt, \quad 0<\alpha<1$$

Suppose that instead of α we have $m-\alpha$, then

$$\boldsymbol{D}_{\boldsymbol{RL}}^{-(m-\alpha)}x(s) = \frac{1}{\Gamma(m-\alpha)} \odot \int\limits_{t_0}^{s} (s-t)^{m-\alpha-1} \odot x(t)dt, \quad m-1<\alpha<m$$

Now, $\boldsymbol{D}_{\boldsymbol{RL}}^{m}$ times derivative of a fractional integral operator $\boldsymbol{D}_{\boldsymbol{RL}}^{-(m-\alpha)}$ can be considered as,

$$\boldsymbol{D}_{\boldsymbol{RL}}^{m}\boldsymbol{D}_{\boldsymbol{RL}}^{-(m-\alpha)} = \boldsymbol{D}_{\boldsymbol{RL}}^{\alpha}$$

which is the Riemann-Liouville left-sided (we consider RL-fractional derivative) with order α, and

$$\boldsymbol{D}_{\boldsymbol{RL}}^{-m}\boldsymbol{D}_{\boldsymbol{RL}}^{\alpha} = \boldsymbol{D}_{\boldsymbol{RL}}^{-(m-\alpha)}$$

Is fractional integral operator.

Definition—Riemann-Liouville fractional derivative—RL

Let us consider $x(t)$ is a fuzzy number valued function,

$$\boldsymbol{D}_{\boldsymbol{RL}}^{\alpha}x(s) = \begin{cases} \frac{1}{\Gamma(m-\alpha)} \odot \left(\frac{d}{ds}\right)^{m} \int\limits_{t_0}^{s} (s-t)^{m-\alpha-1} \odot x(t)dt, & m-1<\alpha<m \\ \left(\frac{d}{ds}\right)^{m-1} x(s), & \alpha = m-1 \end{cases}$$

For $s \in [t_0, T]$. Thus, this fractional derivative is defined in terms of a fractional integral or in terms of an integral. This is an important aspect of fractional derivatives. The integral $f(s)$ is defined in $[a,s]$ so $\left(\frac{d}{ds}\right)^{m} f(s)$ is a function in terms of s, and for the derivative $\boldsymbol{D}_{\boldsymbol{RL}}^{\alpha}x(t)$ at the point t, an interval $[a,s]$ specified.

The derivative $\boldsymbol{D}_{\boldsymbol{RL}}^{\alpha}x(t)$ at the point t_0, means computing it at a fixed point t_0. On the other hand, t_0 is an integration variable for the integral which is used for computing the $\boldsymbol{D}_{\boldsymbol{RL}}^{\alpha}x(t)$ at the point t_0. This phenomenon is also seen in practical situations when a practical situation is described in terms of a fractional derivative compared to an integer-order derivative one gets a better fit to the actual data.

Using the definition of gH-derivative, we can consider two cases, If you consider, $f(s)$ is the following fuzzy function,

$$\int_{t_0}^{s} (s-t)^{m-\alpha-1} \odot x(t)dt = f(s)$$

Then $\left(\frac{d}{ds}\right)^m f(s)$ should be defined as a fuzzy number valued function. It means, all the derivatives $\left(\frac{d}{ds}\right)^j f(s), \quad j = 1, 2, \ldots, m$ exist.

$$\left(\frac{d}{ds}\right)^j f(s) = \lim_{h\to 0} \frac{\left(\frac{d}{ds}\right)^{j-1} f(s) \ominus_{gH} \left(\frac{d}{ds}\right)^{j-1} f(s)}{h}$$

By taking to account the nature of the gH-difference, the derivative is defined in two cases,

- $i - gH$ differentiable

$$\left(\frac{d}{ds}\right)^j f(s) = \lim_{h\to 0} \frac{\left(\frac{d}{ds}\right)^{j-1} f(s+h) \ominus_{H} \left(\frac{d}{ds}\right)^{j-1} f(s)}{h}$$

And in level-wise form,

$$\boldsymbol{D}^{\alpha}_{\boldsymbol{RL}_{i-gH}} x(t,r) = \left[\boldsymbol{D}^{\alpha}_{\boldsymbol{RL}} x_l(t,r), \boldsymbol{D}^{\alpha}_{\boldsymbol{RL}} x_u(t,r)\right]$$

- $ii - gH$ differentiable

$$\left(\frac{d}{ds}\right)^j f(s) = \lim_{h\to 0} \frac{\left(\frac{d}{ds}\right)^{j-1} f(s) \ominus_{H} \left(\frac{d}{ds}\right)^{j-1} f(s+h)}{h}$$

And in level-wise form,

$$\boldsymbol{D}^{\alpha}_{\boldsymbol{RL}_{ii-gH}} x(s,r) = \left[\boldsymbol{D}^{\alpha}_{\boldsymbol{RL}} x_u(s,r), \boldsymbol{D}^{\alpha}_{\boldsymbol{RL}} x_l(s,r)\right]$$

Where in two cases,

$$\boldsymbol{D}^{\alpha}_{\boldsymbol{RL}} x_l(s,r) = \begin{cases} \frac{1}{\Gamma(m-\alpha)} \left(\frac{d}{ds}\right)^m \int_{t_0}^{s} (s-t)^{m-\alpha-1} x_l(t,r)dt, & m-1<\alpha<m \\ \left(\frac{d}{ds}\right)^{m-1} x_l(s,r), & \alpha = m-1 \end{cases}$$

And

$$\boldsymbol{D}_{RL}^{\alpha}x_u(s,r) = \begin{cases} \frac{1}{\Gamma(m-\alpha)}\left(\frac{d}{ds}\right)^m \int_{t_0}^{s}(s-t)^{m-\alpha-1}x_u(t,r)dt, & m-1<\alpha<m \\ \left(\frac{d}{ds}\right)^{m-1}x_u(s,r), & \alpha = m-1 \end{cases}$$

Example Assume a fuzzy number $a \in \mathbb{F}_R$, with the membership function,

$$a(r) = (a_l(r), a_u(r)),\ 0 \le r \le 1$$

In this example, the RL derivative is always $i-gH$ differentiable. Because the function $(s-t)^{m-\alpha-1} \ge 0$ and nothing threat the conditions of fuzzy numbers.

$$\boldsymbol{D}_{RL_{i-gH}}^{\alpha}x(s) = \begin{cases} \frac{1}{\Gamma(m-\alpha)} \odot \left(\frac{d}{ds}\right)^m \int_{t_0}^{s}(s-t)^{m-\alpha-1}\odot a dt, & m-1<\alpha<m \\ \left(\frac{d}{ds}\right)^{m-1}a, & \alpha = m-1 \end{cases}$$

$$\boldsymbol{D}_{RL_{i-gH}}^{\alpha}x(s) = \begin{cases} \frac{1}{\Gamma(m-\alpha)}\left(\frac{d}{ds}\right)^m \left[-\frac{a\odot(s-t)^{m-\alpha}}{m-\alpha}\right]_a^s, & m-1<\alpha<m \\ \left(\frac{d}{ds}\right)^m \int_a^s a dt = 0, & \alpha = m-1 \end{cases}$$

$$\begin{aligned} \boldsymbol{D}_{RL_{i-gH}}^{\alpha}x(s) &= \frac{1}{\Gamma(m-\alpha)} \odot \left(\frac{d}{ds}\right)^m \left[\frac{a\odot(s-a)^{m-\alpha}}{m-\alpha}\right] \\ &= \frac{a}{\Gamma(m-\alpha)} \odot \left(\frac{d}{ds}\right)^m \left[\frac{(s-a)^{m-\alpha}}{m-\alpha}\right] = \frac{a}{\Gamma(m-\alpha)} \odot \left(\frac{d}{ds}\right)^{m-1}(s-a)^{m-\alpha-1} \\ &= \frac{a}{\Gamma(m-\alpha)} \odot (s-a)^{-\alpha} \end{aligned}$$

The RL gH-derivative of this fuzzy number in level-wise form is,

$$\boldsymbol{D}_{RL_{i-gH}}^{\alpha}x(s,r) = \left[\boldsymbol{D}_{RL}^{\alpha}x_l(s,r), \boldsymbol{D}_{RL}^{\alpha}x_u(s,r)\right],\ s \in [t_0, T]$$

Where

$$\boldsymbol{D}_{RL}^{\alpha}x_l(s,r) = \frac{a_l(r)}{\Gamma(m-\alpha)}(s-a)^{-\alpha},\ \boldsymbol{D}_{RL}^{\alpha}x_u(s,r) = \frac{a_u(r)}{\Gamma(m-\alpha)}(s-a)^{-\alpha}$$

Thus, the RL gH-derivative of a fuzzy number is not a zero fuzzy number and it is indeed another fuzzy number. Since $\Gamma(m-\alpha) > 0$, $(s-a)^{-\alpha} > 0$ then the derivative satisfies the conditions of a fuzzy number.

Remark Please note that the RL gH-derivative of a fuzzy number valued function is always a fuzzy number valued function and then it exists.

Remark There is a close relationship between the RL derivative and GL derivative. Indeed, it is possible to see that whenever $x \in C^m[t_0, T]$, with $m = \alpha$, then

$$\boldsymbol{D}^{\alpha}_{\boldsymbol{RL}_{gH}} x(t) = \boldsymbol{D}^{\alpha}_{\boldsymbol{GL}} x(t),\ t \in (t_0, T]$$

Example—Fuzzy exponential function

Consider the fuzzy function as $x(t) = e^{c\odot t}$, $t \in (t_0, T]$, $c \in \mathbb{F}_R$, the RL gH-derivative is obtained as,

$$\boldsymbol{D}^{\alpha}_{\boldsymbol{RL}_{gH}} e^{c\odot s} = \begin{cases} \frac{1}{\Gamma(m-\alpha)} \odot \left(\frac{d}{ds}\right)^m \int\limits_{t_0}^{s} (s-t)^{m-\alpha-1} \odot e^{c\odot t} dt, & m-1<\alpha<m \\ \left(\frac{d}{ds}\right)^{m-1} e^{c\odot s}, & \alpha = m-1 \end{cases}$$

And $c(r) = (c_l(r), c_u(r)),\ 0 \leq r \leq 1$.
Using the remark, whenever $x \in C^m[t_0, T],\ t_0 \geq 0$, with $m = \alpha$, then

$$\boldsymbol{D}^{\alpha}_{\boldsymbol{RL}_{gH}} e^{c\odot s} = \boldsymbol{D}^{\alpha}_{\boldsymbol{GL}} e^{c\odot s} = c^{\alpha} \odot e^{c\odot s},\ s \in (t_0, T]$$

Otherwise, it can be handled as the following when $m-1<\alpha<m$,

$$\boldsymbol{D}^{\alpha}_{\boldsymbol{RL}_{gH}} e^{c\odot s} = \frac{1}{\Gamma(m-\alpha)} \odot \left(\frac{d}{ds}\right)^m \int\limits_{t_0}^{s} (s-t)^{m-\alpha-1} \odot e^{c\odot t} dt,\ s \in [t_0, T]$$

let $v = s - t$, and then $dt = -dv$ and $e^{c\odot t} = e^{c\odot(s-v)}$, $s - v \geq 0$.

$$\boldsymbol{D}^{\alpha}_{\boldsymbol{RL}_{gH}} e^{c\odot s} = \frac{1}{\Gamma(m-\alpha)} \odot \left(\frac{d}{ds}\right)^m \int\limits_{s-t_0}^{0} v^{m-\alpha-1} \odot (-1) e^{c\odot(s-v)} dv$$

Since,

$$I := \int\limits_{s-t_0}^{0} v^{m-\alpha-1} \odot (-1) e^{c\odot(s-v)} dv = \int\limits_{0}^{s-t_0} v^{m-\alpha-1} \odot e^{c\odot(s-v)} dv$$

Because in level-wise form,

$$I[r] = [I_l(r), I_u(r)]$$

where

$$I_l(r) = \int_{s-t_0}^{0} v^{m-\alpha-1}(-1)e^{c_l(r)(s-v)}dv = \int_{0}^{s-t_0} v^{m-\alpha-1}e^{c_l(r)(s-v)}dv$$

$$I_u(r) = \int_{s-t_0}^{0} v^{m-\alpha-1}(-1)e^{c_u(r)(s-v)}dv = \int_{0}^{s-t_0} v^{m-\alpha-1}e^{c_u(r)(s-v)}dv$$

For any $r \in [0, 1]$. Finally, we have,

$$\boldsymbol{D}^{\alpha}_{\boldsymbol{RL_{gH}}} e^{c\odot s} = \frac{1}{\Gamma(m-\alpha)} \odot \left(\frac{d}{ds}\right)^m \int_{0}^{s-t_0} v^{m-\alpha-1} \odot e^{c\odot(s-v)}dv$$

The level-wise form of the derivative is,

$$\begin{aligned}
\boldsymbol{D}^{\alpha}_{\boldsymbol{RL_{gH}}} e^{c_l(r)s} &= \frac{1}{\Gamma(m-\alpha)} \left(\frac{d}{ds}\right)^m \int_{0}^{s-t_0} v^{m-\alpha-1}e^{c_l(r)(s-v)}dv \\
&= \frac{1}{\Gamma(m-\alpha)} \left(\frac{d}{ds}\right)^m e^{c_l(r)s} \int_{0}^{s-t_0} v^{m-\alpha-1}e^{-c_l(r)v}dv \\
&= \left(\frac{d}{ds}\right)^m e^{c_l(r)s} \frac{1}{\Gamma(m-\alpha)} \int_{0}^{s-t_0} v^{m-\alpha-1}e^{-c_l(r)v}dv
\end{aligned}$$

The integral $\int_0^{s-t_0} v^{m-\alpha-1}e^{-c_l(r)v}dv$ is an incomplete Gamma density function. Then

$$\boldsymbol{D}^{\alpha}_{\boldsymbol{RL_{gH}}} e^{c_l(r)s} = \left(\frac{d}{ds}\right)^m e^{c_l(r)s} \frac{1}{\Gamma(m-\alpha)} \gamma(v, t)$$

The same process for the upper function,

$$\boldsymbol{D}^{\alpha}_{\boldsymbol{RL_{gH}}} e^{c_u(r)s} = \left(\frac{d}{ds}\right)^m e^{c_u(r)s} \frac{1}{\Gamma(m-\alpha)} \gamma(v, t)$$

It is not clear that,

$$\boldsymbol{D}^{\alpha}_{\boldsymbol{RL_{gH}}} e^{c_l(r)s} \leq \boldsymbol{D}^{\alpha}_{\boldsymbol{RL_{gH}}} e^{c_u(r)s}$$

3.3.3 RL—Fractional Derivative for m = 1

$$D^{\alpha}_{RL_{gH}} x(t) = \frac{1}{\Gamma(1-\alpha)} \odot \frac{d}{ds} \int_{t_0}^{s} \frac{x(t)}{(s-t)^{\alpha}} dt, \quad 0 < \alpha < 1$$

or

$$D^{\alpha}_{RL_{gH}} x(t) = \frac{1}{\Gamma(1-\alpha)} \odot \left(\int_{t_0}^{s} \frac{x(t)}{(s-t)^{\alpha}} dt \right)'_{gH}, \quad 0 < \alpha < 1$$

Assume,

$$\int_{t_0}^{s} \frac{x(t)}{(s-t)^{\alpha}} dt = f(s) \in \mathbb{F}_R, \ x(t) \in \mathbb{F}_R$$

So,

$$\frac{d}{ds} \int_{t_0}^{s} \frac{x(t)}{(s-t)^{\alpha}} dt = f'_{gH}(s), \quad s \in [t_0, T]$$

Remark The function $f(t)$ is gH-differentiable if and only if $f_l'(t,r)$ and $f_u'(t,r)$ are differentiable with respect to t for all $0 \le r \le 1$ and

$$f'_{gH}(t) = \left[\min\left\{ f_l'(t,r), f_u'(t,r) \right\}, \max\left\{ f_l'(t,r), f_u'(t,r) \right\} \right]$$

Theorem The necessary and sufficient condition for RL gH differentiability of $x(t)$ is gH differentiability of $f(s)$. It is,

$$D^{\alpha}_{RL_{gH}} x(t,r) = \left[\min\left\{ D^{\alpha}_{RL} x_l(t,r), D^{\alpha}_{RL} x_u(t,r) \right\}, \max\left\{ D^{\alpha}_{RL} x_l(t,r), D^{\alpha}_{RL} x_u(t,r) \right\} \right]$$

The proof is very easy, since $s - t > 0$,

$$f_l(s,r) = \int_{t_0}^{s} \frac{x_l(t,r)}{(s-t)^{\alpha}} dt, \ f_u(s,r) = \int_{t_0}^{s} \frac{x_u(t,r)}{(s-t)^{\alpha}} dt$$

$$\min\left\{ \tfrac{1}{\Gamma(1-\alpha)} f_l'(t,r), \tfrac{1}{\Gamma(1-\alpha)} f_u'(t,r) \right\} \\ \le \max\left\{ \tfrac{1}{\Gamma(1-\alpha)} f_l'(t,r), \tfrac{1}{\Gamma(1-\alpha)} f_u'(t,r) \right\}$$

Then

$$\min\left\{\boldsymbol{D}_{\boldsymbol{RL}}^{\alpha} x_l(t,r), \boldsymbol{D}_{\boldsymbol{RL}}^{\alpha} x_u(t,r)\right\} \leq \max\left\{\boldsymbol{D}_{\boldsymbol{RL}}^{\alpha} x_l(t,r), \boldsymbol{D}_{\boldsymbol{RL}}^{\alpha} x_u(t,r)\right\}$$

It means that the following interval defines an interval for all $r \in [0,1]$,

$$\boldsymbol{D}_{\boldsymbol{RL}_{gH}}^{\alpha} x(t,r) = \left[\boldsymbol{D}_{\boldsymbol{RL}}^{\alpha} x_l(t,r), \boldsymbol{D}_{\boldsymbol{RL}}^{\alpha} x_u(t,r)\right]$$

3.4 Fuzzy Caputo Fractional Derivative

In this section another form of the fractional derivative is expressed. First the order $m-1<\alpha<m$ and then $0<\alpha<1$ are displayed.

Definition—Caputo fractional derivative of order α

In the definition of RL fractional derivative, suppose the integer order of the derivative is an operator inside of the integral and operating on operand function $x(t) \in \mathbb{F}_R, t \in [t_0, T]$.

$$\boldsymbol{D}_{\boldsymbol{C}_{gH}}^{\alpha} x(s) = \begin{cases} \frac{1}{\Gamma(m-\alpha)} \odot \int\limits_{t_0}^{s} (s-t)^{m-\alpha-1} \odot x_{gH}^{(m)}(t)dt, & m-1<\alpha<m \\ \left(\frac{d}{ds}\right)^{m-1} x(s), & \alpha = m-1 \end{cases}$$

Note The Caputo derivative of a fuzzy number is zero. But in case RL derivative it wasn't.

In the Caputo differential operator, there is a derivative of order m on fuzzy number valued function, $x^{(m)}(t)$. To existence of this derivative for any m, it should be a fuzzy number at any point t. To this end, the gH-difference in the definition of the derivative should be well defined.

$$x^{(m)}(t) = \lim_{h \to 0} \frac{x^{(m-1)}(t+h) \ominus_{gH} x^{(m-1)}(t)}{h}$$

The gH-difference is defined in two cases and

- $i-gH$ differentiable

$$x_{i-gH}^{(m)}(t) = \lim_{h \to 0} \frac{x^{(m-1)}(t+h) \ominus_{H} x^{(m-1)}(t)}{h}$$

Its level-wise form,

$$\boldsymbol{D}^{\alpha}_{C_{i-gH}}x(s,r) = \left[\boldsymbol{D}^{\alpha}_{C}x_l(s,r), \boldsymbol{D}^{\alpha}_{C}x_u(s,r)\right]$$

- $ii - gH$ differentiable

$$x^{(m)}_{ii-gH}(t) = \lim_{h\to 0}\frac{x^{(m-1)}(t)\ominus_H x^{(m-1)}(t+h)}{h}$$

Its level-wise form,

$$\boldsymbol{D}^{\alpha}_{C_{ii-gH}}x(s,r) = \left[\boldsymbol{D}^{\alpha}_{C}x_u(s,r), \boldsymbol{D}^{\alpha}_{C}x_l(s,r)\right]$$

Which are defined as the following form for $m-1<\alpha<m$,

$$\boldsymbol{D}^{\alpha}_{C}x_l(s,r) = \frac{1}{\Gamma(m-\alpha)}\int_{t_0}^{s}(s-t)^{m-\alpha-1}x_l^{(m)}(t,r)dt,$$
$$\boldsymbol{D}^{\alpha}_{C}x_u(s,r) = \frac{1}{\Gamma(m-\alpha)}\int_{t_0}^{s}(s-t)^{m-\alpha-1}x_u^{(m)}(t,r)dt,$$

For $s \in [t_0, T]$.

3.4.1 Caputo—Fuzzy Fractional Derivative for m = 1

Suppose the order of fractional derivative is in $0<\alpha<1, m=1$. Now we are going to cover some properties of Caputo derivative with this fraction order. The definition is as,

$$\boldsymbol{D}^{\alpha}_{C_{gH}}x(s) = \frac{1}{\Gamma(1-\alpha)}\odot\int_{t_0}^{s}(s-t)^{-\alpha}\odot x'_{gH}(t)dt$$
$$:= \frac{1}{\Gamma(1-\alpha)}\odot\int_{t_0}^{s}\frac{x'_{gH}(t)}{(s-t)^{\alpha}}dt$$

Remark The function $x(t)$ is gH-differentiable if and only if $x'_l(t,r)$ and $x'_y(t,r)$ are differentiable with respect to t for all $0\leq r\leq 1$ and

$$x'_{gH}(t,r) = \left[\min\{x'_l(t,r), x'_u(t,r)\}, \max\{x'_l(t,r), x'_u(t,r)\}\right]$$

3.4.2 Caputo gH Differentiability

The fuzzy number valued function $x(t)$ is Caputo gH differentiable if and only if $x_l'(t,r)$ and $x_u'(t,r)$ are differentiable with respect to t for all $0 \le r \le 1$ and

$$\boldsymbol{D}^{\alpha}_{C_{gH}}x(t,r) = \left[\min\left\{\boldsymbol{D}^{\alpha}_{C}x_l(t,r), \boldsymbol{D}^{\alpha}_{C}x_u(t,r)\right\}, \max\left\{\boldsymbol{D}^{\alpha}_{C}x_l(t,r), \boldsymbol{D}^{\alpha}_{C}x_u(t,r)\right\}\right]$$

where

$$\boldsymbol{D}^{\alpha}_{C}x_l(s,r) = \frac{1}{\Gamma(1-\alpha)}\int_{t_0}^{s}\frac{x_l'(t,r)}{(s-t)^{\alpha}}dt,$$

$$\boldsymbol{D}^{\alpha}_{C}x_u(t,r) = \frac{1}{\Gamma(1-\alpha)}\int_{t_0}^{s}\frac{x_u'(t,r)}{(s-t)^{\alpha}}dt$$

The proof is straight,

$$\boldsymbol{D}^{\alpha}_{C_{gH}}x(s,r) = \frac{1}{\Gamma(1-\alpha)} \odot \int_{t_0}^{s}\frac{x_{gH}'(t,r)}{(s-t)^{\alpha}}dt$$

We have

$$x_{gH}'(s,r) = \left[\min\left\{x_l'(s,r), x_u'(s,r)\right\}, \max\left\{x_l'(s,r), x_u'(s,r)\right\}\right]$$

Then

$$x_{gH}'(s,r) = \begin{bmatrix} \min\left\{\frac{1}{\Gamma(1-\alpha)}\int_{t_0}^{s}\frac{x_l'(t,r)}{(s-t)^{\alpha}}dt, \frac{1}{\Gamma(1-\alpha)}\int_{t_0}^{s}\frac{x_u'(t,r)}{(s-t)^{\alpha}}dt\right\}, \\ \max\left\{\frac{1}{\Gamma(1-\alpha)}\int_{t_0}^{s}\frac{x_l'(t,r)}{(s-t)^{\alpha}}dt, \frac{1}{\Gamma(1-\alpha)}\int_{t_0}^{s}\frac{x_u'(t,r)}{(s-t)^{\alpha}}dt\right\} \end{bmatrix}$$

This is exactly,

$$\boldsymbol{D}^{\alpha}_{C_{gH}}x(t,r) = \left[\min\left\{\boldsymbol{D}^{\alpha}_{C}x_l(t,r), \boldsymbol{D}^{\alpha}_{C}x_u(t,r)\right\}, \max\left\{\boldsymbol{D}^{\alpha}_{C}x_l(t,r), \boldsymbol{D}^{\alpha}_{C}x_u(t,r)\right\}\right]$$

Also, in case $0<\alpha<1$, we have two cases of differentiability,

- $i-gH$ differentiable

$$x_{i-gH}'(t) = \lim_{h\to 0}\frac{x(t+h)\ominus_H x(t)}{h}$$

Its level-wise form,

$$\boldsymbol{D}^{\alpha}_{C_{i-gH}}x(s,r) = \left[\boldsymbol{D}^{\alpha}_{C}x_l(s,r), \boldsymbol{D}^{\alpha}_{C}x_u(s,r)\right]$$

- $ii - gH$ differentiable

$$x'_{ii-gH}(t) = \lim_{h\to 0}\frac{x(t)\ominus_H x(t+h)}{h}$$

Its level-wise form,

$$\boldsymbol{D}^{\alpha}_{C_{ii-gH}}x(s,r) = \left[\boldsymbol{D}^{\alpha}_{C}x_u(s,r), \boldsymbol{D}^{\alpha}_{C}x_l(s,r)\right]$$

Example Consider the same fuzzy exponential function $x(t) = e^{c\odot t},\ t \in (t_0, T],\ c \in \mathbb{F}_R,\ c(r) = (c_l(r), c_u(r))$.

$$\boldsymbol{D}^{\alpha}_{C_{gH}}e^{c\odot s} = \frac{1}{\Gamma(1-\alpha)}\odot\int_{t_0}^{s}\frac{c\odot e^{c\odot t}}{(s-t)^{\alpha}}dt$$

We know that

$$x_l(t,r) = e^{c_l(r)t},\ x_u(t,r) = e^{c_u(r)t}$$
$$x'_l(t,r) = c_l(r)e^{c_l(r)t},\ x'_u(t,r) = c_u(r)e^{c_u(r)t}$$

Because $t > 0$ and the exponential function is increasing. Thus,

- $$x'_{gH}(t,r) = \left[c_l(r)e^{c_l(r)t}, c_u(r)e^{c_u(r)t}\right]$$

Finally, the i—differential is found as,

$$\boldsymbol{D}^{\alpha}_{C_{i-gH}}x(s,r) = \left[\boldsymbol{D}^{\alpha}_{C}x_l(s,r), \boldsymbol{D}^{\alpha}_{C}x_u(s,r)\right]$$

Where,

$$\boldsymbol{D}^{\alpha}_{C}x_l(s,r) = \frac{1}{\Gamma(1-\alpha)}\int_{t_0}^{s}\frac{c_l(r)e^{c_l(r)t}}{(s-t)^{\alpha}}dt = \frac{c_l(r)}{\Gamma(1-\alpha)}\int_{t_0}^{s}\frac{e^{c_l(r)t}}{(s-t)^{\alpha}}dt$$

let $v = s - t$, and then $dt = -dv$ and $e^{c\odot t} = e^{c\odot(s-v)},\ s - v \geq 0$. By these conditions, the integral equation, $\int_{t_0}^{s}(s-t)^{-\alpha}\odot e^{c\odot t}dt$ is always $i - gH$ differentiable, and

$$D_C^\alpha x_l(s,r) = \frac{c_l(r)}{\Gamma(1-\alpha)} \int_0^{s-t_0} \frac{e^{c_l(r)(s-v)}}{v^\alpha} dv = \frac{c_l(r)e^{c_l(r)s}}{\Gamma(1-\alpha)} \int_0^{s-t_0} \frac{e^{-c_l(r)v}}{v^\alpha} dv$$

The integral equation,

$$\frac{1}{\Gamma(1-\alpha)} \int_0^{s-t_0} \frac{e^{-c_l(r)v}}{v^\alpha} dv = \frac{1}{\Gamma(1-\alpha)} \int_0^{s-t_0} v^{-\alpha} e^{-c_l(r)v} dv$$

where $\int_0^{s-t_0} v^{-\alpha} e^{-c_l(r)v} dv$ is an incomplete $\gamma(v,t)$ function thus,

$$D_C^\alpha x_l(s,r) = \frac{c_l(r)e^{c_l(r)s}}{\Gamma(1-\alpha)} \gamma(v,t)$$

Using the same process,

$$D_C^\alpha x_u(s,r) = \frac{c_l(r)e^{c_l(r)s}}{\Gamma(1-\alpha)} \gamma(v,t), \ 0<\alpha<1$$

Now if for any $r \in [0,1]$, the conditions of the interval are satisfied then the Caputo derivative for this exponential function exists. This is exactly depending on the sign of $\gamma(v,t)$ and it can be variational. Then it is not easy to claim that,

$$D_C^\alpha x_l(s,r) \le D_C^\alpha x_u(s,r), \quad 0 \le r \le 1$$

3.5 Fuzzy Riemann-Liouville Generalized Fractional Derivative

In this section, the concept is going to be covered for order of $0<\alpha<1$ on absolutely continuous fuzzy number valued functions. First, the fuzzy Riemann-Liouville generalized fractional integral is defined.

3.5.1 Fuzzy Riemann-Liouville Generalized Fractional Integral

To define, assume that $x \in \mathbb{F}_R$ and integrable on $[t_0, T]$, $p > 0$.

$$\boldsymbol{I}_{\boldsymbol{RL}}^{\alpha,p}x(t) = \frac{p^{1-\alpha}}{\Gamma(\alpha)} \odot \int_{t_0}^{t} s^{p-1}(t^p - s^p)^{\alpha-1} \odot x(s)ds$$

In level-wise form,

$$\boldsymbol{I}_{\boldsymbol{RL}}^{\alpha,p}x(t,r) = \left[\min\left\{\boldsymbol{I}_{\boldsymbol{RL}}^{\alpha,p}x_l(t,r), \boldsymbol{I}_{\boldsymbol{RL}}^{\alpha,p}x_u(t,r)\right\}, \max\left\{\boldsymbol{I}_{\boldsymbol{RL}}^{\alpha,p}x_l(t,r), \boldsymbol{I}_{\boldsymbol{RL}}^{\alpha,p}x_u(t,r)\right\}\right]$$

where

$$\min\left\{\boldsymbol{I}_{\boldsymbol{RL}}^{\alpha,p}x_l(t,r), \boldsymbol{I}_{\boldsymbol{RL}}^{\alpha,p}x_u(t,r)\right\} = \boldsymbol{I}_{\boldsymbol{RL}}^{\alpha,p}x_l(t,r) = \frac{p^{1-\alpha}}{\Gamma(\alpha)} \int_{t_0}^{t} s^{p-1}(t^p - s^p)^{\alpha-1} x_l(s,r)ds$$

$$\max\left\{\boldsymbol{I}_{\boldsymbol{RL}}^{\alpha,p}x_l(t,r), \boldsymbol{I}_{\boldsymbol{RL}}^{\alpha,p}x_u(t,r)\right\}\boldsymbol{I}_{\boldsymbol{RL}}^{\alpha,p}x_u(t,r) = \frac{p^{1-\alpha}}{\Gamma(\alpha)} \int_{t_0}^{t} s^{p-1}(t^p - s^p)^{\alpha-1} x_u(s,r)ds$$

Properties

For two fuzzy number valued functions x, y and two fractional orders $\alpha, \beta \in (0,1)$,

- $$\boldsymbol{I}_{\boldsymbol{RL}}^{\alpha,p}(x \oplus y)(t) = \boldsymbol{I}_{\boldsymbol{RL}}^{\alpha,p}x(t) \oplus \boldsymbol{I}_{\boldsymbol{RL}}^{\alpha,p}y(t)$$

- $$length\left(\boldsymbol{I}_{\boldsymbol{RL}}^{\alpha,p}x(t,r)\right) = \boldsymbol{I}_{\boldsymbol{RL}}^{\alpha,p}length(x(t,r))$$

- $$\boldsymbol{I}_{\boldsymbol{RL}}^{\alpha,p}\boldsymbol{I}_{\boldsymbol{RL}}^{\beta,p}x(t) = \boldsymbol{I}_{\boldsymbol{RL}}^{(\alpha+\beta),p}$$

The proofs of two first items are clear and the third can be proved as,

$$\begin{aligned}\boldsymbol{I}_{\boldsymbol{RL}}^{\alpha,p}\boldsymbol{I}_{\boldsymbol{RL}}^{\beta,p}x(t) &= \frac{p^{-(\alpha+\beta)+2}}{\Gamma(\alpha)\Gamma(\beta)} \odot \int_{t_0}^{t} s^{p-1}(t^p - s^p)^{\alpha-1} \odot \int_{u}^{s} u^{p-1}(t^p - u^p)^{\beta-1} \odot x(u)duds \\ &= \frac{p^{-(\alpha+\beta)+2}}{\Gamma(\alpha)\Gamma(\beta)} \odot \int_{t_0}^{t} u^{p-1} \odot x(u) \odot \int_{u}^{s} s^{p-1}(u^p - t^p)^{\beta-1}(s^p - t^p)^{\alpha-1} dsdu\end{aligned}$$

By changing some variables like, suppose, $\frac{u^p - t^p}{s^p - t^p} = v$, where s, u are fixed. Then

$$\int_u^s s^{p-1}(u^p - t^p)^{\beta-1}(s^p - t^p)^{\alpha-1} ds = \frac{(s^p - u^p)^{\alpha+\beta-1}}{p} \int_0^1 (1-v)^{\alpha-1} v^{\beta-1} dv$$

$$= \frac{(s^p - u^p)^{\alpha+\beta-1}}{p} \frac{\Gamma(\alpha)\Gamma(\beta)}{\Gamma(\alpha+\beta)}$$

By substituting,

$$\boldsymbol{I}_{RL}^{\alpha,p}\boldsymbol{I}_{RL}^{\beta,p}x(t) = \frac{p^{-(\alpha+\beta)+1}}{\Gamma(\alpha+\beta)} \odot \int_{t_0}^{t} u^{p-1}(s^p - u^p)^{\alpha+\beta-1} \odot x(u)du$$

$$= \boldsymbol{I}_{RL}^{(\alpha+\beta),p}x(t)$$

Now using the Fuzzy Riemann-Liouville generalized fractional integral $\boldsymbol{I}_{RL}^{\alpha,p}$ we are going to define its fractional derivative based on gH differentiability.

3.5.2 Riemann Liouville–Katugampola gH—Fractional Derivative

With the same assumptions in the integral operator, the $\boldsymbol{D}_{RL}^{\alpha,p}$ is defined as the following form,

$$\boldsymbol{D}_{RL_{gH}}^{\alpha,p}x(t) = t^{1-p}\left(\boldsymbol{I}_{RL}^{(1-\alpha),p}x\right)'_{gH}(t),\ t \in [t_0, T],\ 0<\alpha<1$$

where

$$\boldsymbol{I}_{RL}^{(1-\alpha),p}x(t) = \frac{p^{\alpha}}{\Gamma(1-\alpha)} \odot \int_{t_0}^{t} s^{p-1}(t^p - s^p)^{-\alpha} \odot x(s)ds$$

$$\left(\boldsymbol{I}_{RL}^{(1-\alpha),p}x\right)'_{gH}(t) = \lim_{h\to 0} \frac{\boldsymbol{I}_{RL}^{(1-\alpha),p}x(t+h) \ominus_{gH} \boldsymbol{I}_{RL}^{(1-\alpha),p}x(t)}{h}$$

It is supposed that the gH difference exists. Moreover, based on the nature of gH difference we have two cases,

- $i - gH$ differentiability

$$\left(\boldsymbol{I}_{\boldsymbol{RL}}^{(1-\alpha),p}x\right)'_{gH}(t) = \lim_{h\to 0}\frac{\boldsymbol{I}_{\boldsymbol{RL}}^{(1-\alpha),p}x(t+h)\ominus_H \boldsymbol{I}_{\boldsymbol{RL}}^{(1-\alpha),p}x(t)}{h}$$
$$\boldsymbol{D}_{\boldsymbol{RL}_{i-gH}}^{\alpha,p}x(t) = t^{1-p}\left(\boldsymbol{I}_{\boldsymbol{RL}}^{(1-\alpha),p}x\right)'_{gH}(t), \quad t\in[t_0,T], \quad 0<\alpha<1$$

Where

$$\boldsymbol{I}_{\boldsymbol{RL}}^{(1-\alpha),p}x(t) = \frac{p^{\alpha}}{\Gamma(1-\alpha)}\odot\int_{t_0}^{t} s^{p-1}(t^p - s^p)^{-\alpha}\odot x(s)ds$$
$$\boldsymbol{D}_{\boldsymbol{RL}_{i-gH}}^{\alpha,p}x(t,r) = \left[\boldsymbol{D}_{\boldsymbol{RL}_{gH}}^{\alpha,p}x_l(t,r), \boldsymbol{D}_{\boldsymbol{RL}_{gH}}^{\alpha,p}x_u(t,r)\right]$$

- $ii - gH$ differentiability

$$\left(\boldsymbol{I}_{\boldsymbol{RL}}^{(1-\alpha),p}x\right)_{ii-gH}{}'(t) = \lim_{h\to 0}\frac{\boldsymbol{I}_{\boldsymbol{RL}}^{(1-\alpha),p}x(t)\ominus_H \boldsymbol{I}_{\boldsymbol{RL}}^{(1-\alpha),p}x(t+h)}{h}$$
$$\boldsymbol{D}_{\boldsymbol{RL}_{ii-gH}}^{\alpha,p}x(t) = t^{1-p}\left(\boldsymbol{I}_{\boldsymbol{RL}}^{(1-\alpha),p}x\right)_{ii-gH}{}'(t), \quad t\in[t_0,T], \quad 0<\alpha<1$$

where

$$\boldsymbol{I}_{\boldsymbol{RL}}^{(1-\alpha),p}x(t) = \frac{p^{\alpha}}{\Gamma(1-\alpha)}\odot\int_{t_0}^{t} s^{p-1}(t^p - s^p)^{-\alpha}\odot x(s)ds$$
$$\boldsymbol{D}_{\boldsymbol{RL}_{ii-gH}}^{\alpha,p}x(t,r) = \left[\boldsymbol{D}_{\boldsymbol{RL}_{gH}}^{\alpha,p}x_u(t,r), \boldsymbol{D}_{\boldsymbol{RL}_{gH}}^{\alpha,p}x_l(t,r)\right]$$

Note If the length function of $x(t,r)$ is increasing (decreasing) for all r, then we have $i - gH$ differentiability ($ii - gH$ differentiability).

Example Let consider $p = 1, \ \alpha = \frac{1}{2}$, and $x : (0,1] \to \mathbb{F}_R$ and is given by

$$x(t) = \left(\sqrt{t} - 1, 0, 1 - \sqrt{t}\right), \ t \in (0,1]$$

The derivative of length of $x(t)$ is,

$$\frac{d}{dt}(length(x(t,r))) = \frac{r-1}{\sqrt{t}} \leq 0, \ r \in [0,1]$$

There for the length of $x(t)$ is decreasing. See the following Fig. 3.2.
And for $t \in (0, 1]$,

$$I_{RL}^{\alpha}x(t) = \frac{1}{\Gamma\left(\frac{1}{2}\right)} \odot \int_{0}^{t} (t-s)^{\frac{1}{2}} \odot x(s)ds = \frac{1}{\sqrt{\pi}} \odot \left(\frac{\pi}{2}t - 2\sqrt{t}, 0, 2\sqrt{t} - \frac{\pi}{2}t\right),$$

And

$$\frac{d}{dt} I_{RL}^{\alpha} x(t,r) = \frac{2(r-1)}{\sqrt{\pi}} \left(\frac{\pi}{2} - \frac{1}{\sqrt{t}}\right) := \begin{cases} \geq 0, \ t \in \left(0, \frac{4}{\pi^2}\right], \textit{increasing length} \\ <0, \ t \in \left[\frac{4}{\pi^2}, 1\right], \textit{decreasing length} \end{cases}$$

It is shown in the following Fig. 3.3.

Remark For $x(t) \in \mathbb{F}_R$ and $0<\alpha<\beta\langle 1, p\rangle 0$ we have,

$$\begin{gathered} \left(D_{RL_{gH}}^{\alpha,p} I_{RL}^{\alpha,p} x\right)(t) = x(t) \\ \left(D_{RL_{gH}}^{\alpha,p} I_{RL}^{\beta,p} x\right)(t) = I_{RL}^{(\beta-\alpha),p} x(t) \end{gathered}$$

To show the first item,

$$\begin{aligned} D_{RL_{gH}}^{\alpha,p} I_{RL}^{\alpha,p} x(t) &= t^{1-p} \left(I_{RL}^{(1-\alpha),p} I_{RL}^{\alpha,p} x\right)'_{gH}(t) \\ &= \frac{p^{\alpha} t^{1-p}}{\Gamma(1-\alpha)} \odot \left(\int_{t_0}^{t} s^{p-1} (t^p - s^p)^{-\alpha} \odot I_{RL}^{\alpha,p} x(s) ds\right)'_{gH} \end{aligned}$$

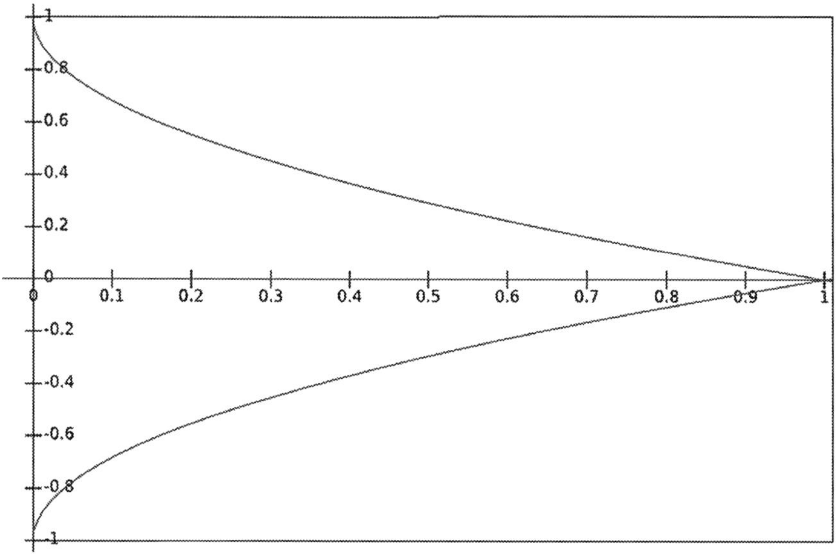

Fig. 3.2 Decreasing length of $x(t)$

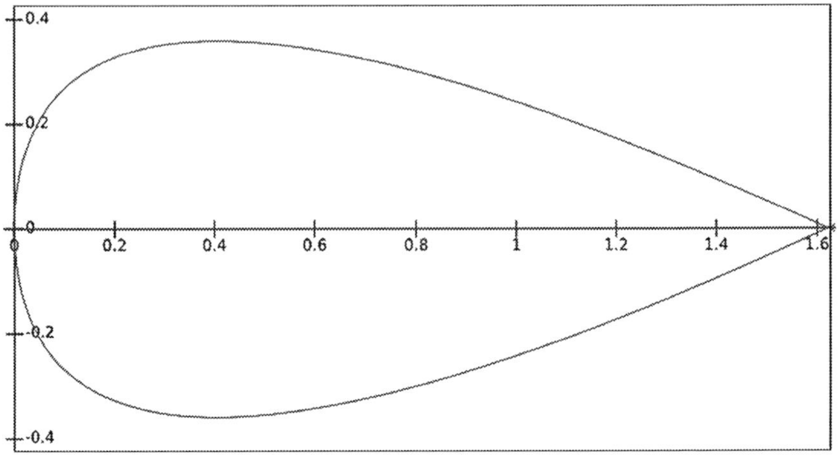

Fig. 3.3 $\frac{d}{dt}\boldsymbol{I}^{\alpha}_{\boldsymbol{RL}}x(t,r)$ length

where

$$\boldsymbol{I}^{\alpha,p}_{\boldsymbol{RL}}x(s) = \frac{p^{1-\alpha}}{\Gamma(\alpha)} \odot \int_{t_0}^{s} u^{p-1}(t^p - u^p)^{\alpha-1} \odot x(u)du$$

By substituting and by Dirichlet formula, the known formula for the Beta function, and setting, $v = \frac{u^p - s^p}{t^p - s^p}$,

$$\begin{aligned}
\boldsymbol{D}^{\alpha,p}_{\boldsymbol{RL}_{gH}}\boldsymbol{I}^{\alpha,p}_{\boldsymbol{RL}}x(t) &= \frac{p^{\alpha}}{\Gamma(1-\alpha)} t^{1-p} \odot \left(\int_{t_0}^{t} \frac{s^{p-1}}{(t^p - s^p)^{\alpha}} \odot \left(\frac{p^{1-\alpha}}{\Gamma(\alpha)} \int_{t_0}^{s} \frac{u^{p-1}}{(S^p - u^p)^{1-\alpha}} x(u)du \right) ds \right)'_{gH} \\
&= \frac{p}{\Gamma(\alpha)\Gamma(1-\alpha)} t^{1-p} \odot \left(\int_{t_0}^{t} x(s)s^{p-1} \odot \int_{s}^{t} u^{p-1} \frac{(u^p - s^p)^{\alpha-1}}{(t^p - u^p)^{\alpha}} duds \right)'_{gH} \\
&= \frac{p}{\Gamma(\alpha)\Gamma(1-\alpha)} t^{1-p} \odot \left(\int_{t_0}^{t} x(s)s^{p-1}ds \odot \underbrace{\int_{0}^{1} \frac{(1-v)^{-\alpha}v^{\alpha-1}}{p} du}_{1} \right)'_{gH} \\
&= \frac{B(1-\alpha,\alpha)}{\Gamma(\alpha)\Gamma(1-\alpha)} t^{1-p} \odot \left(\int_{t_0}^{t} x(s)s^{p-1}ds \right)'_{gH} = x(t)
\end{aligned}$$

The second property,

$$D^{\alpha,p}_{RL_{gH}} I^{\beta,p}_{RL} x(t) = \frac{p^{\alpha-\beta}}{\Gamma(\beta)\Gamma(1-\alpha)} t^{1-p} \odot \left(\int_{t_0}^{t} \frac{s^{p-1}}{(t^p - s^p)^{\alpha-\beta}} \odot x(s)ds \odot \int_0^1 \frac{(1-v)^{-\alpha} v^{\alpha-1}}{p} du \right)'_{gH}$$

$$= \frac{p^{\alpha-\beta}}{\Gamma(\beta-\alpha+1)} \odot \int_{t_0}^{t} \frac{s^{p-1}}{(t^p - s^p)^{\alpha-\beta}} \odot x(s)ds = I^{(\beta-\alpha),p}_{RL} x(t)$$

Remark If in the interval $(t_0, T]$ the type of differentiability does not change, it means, it is either $i - gH$ differentiable or $ii - gH$ differentiable. Then the following relation as a relation of fractional integral and derivative operators is stablished.

$$I^{\alpha,p}_{RL} D^{\alpha,p}_{RL_{gH}} x(t) = x(t) \ominus_{gH} \frac{\left(t^p - t_0^p\right)^{\alpha-1} p^{1-\alpha}}{\Gamma(\alpha)} \odot I^{(1-\alpha),p}_{RL} x(t_0), \ t \in (t_0, T]$$

Where $I^{(1-\alpha),p}_{RL} x(t_0)$ exists and

$$\lim_{t \to t_{0+}} I^{(1-\alpha),p}_{RL} x(t) = I^{(1-\alpha),p}_{RL} x(t_0)$$

To show the assertion, we consider two cases,

- $i - gH$ differentiability

$$I^{\alpha,p}_{RL} D^{\alpha,p}_{RL_{i-gH}} x(t,r) = \left[I^{\alpha,p}_{RL} D^{\alpha,p}_{RL_{gH}} x_l(t,r), I^{\alpha,p}_{RL} D^{\alpha,p}_{RL_{gH}} x_u(t,r) \right]$$

where

$$I^{\alpha,p}_{RL} D^{\alpha,p}_{RL_{gH}} x_l(t,r) = x_l(t,r) - \frac{\left(t^p - t_0^p\right)^{\alpha-1} p^{1-\alpha}}{\Gamma(\alpha)} I^{(1-\alpha),p}_{RL} x_l(t_0,r)$$
$$I^{\alpha,p}_{RL} D^{\alpha,p}_{RL_{gH}} x_u(t,r) = x_u(t,r) - \frac{\left(t^p - t_0^p\right)^{\alpha-1} p^{1-\alpha}}{\Gamma(\alpha)} I^{(1-\alpha),p}_{RL} x_u(t_0,r)$$

Now, since $x_l(t,r)$ is increasing then $I^{\alpha,p}_{RL} D^{\alpha,p}_{RL_{gH}} x_l(t,r)$ is also increasing and $I^{\alpha,p}_{RL} D^{\alpha,p}_{RL_{gH}} x_u(t,r)$ is decreasing because $x_u(t,r)$ is decreasing.

3.6 Caputo–Katugampola gH–Fractional Derivative

If the RL derivative $D^{\alpha,p}_{RL_{gH}} x(t)$ exists in $[t_o, T]$ and $0 < \alpha < 1$,

$$D^{\alpha,p}_{CK_{gH}} x(t) = D^{\alpha,p}_{RL_{gH}} \left(x(t) \ominus_{gH} x(t_0) \right)$$

In the sequel, some relations between fuzzy type Riemann–Liouville–Katugampola generalized fractional derivative and fuzzy type Caputo-Katugampola fractional derivative are shown.

Remark Assume $x(t)$ is absolutely continuous fuzzy number valued function which does have increasing or decreasing length i.e. it is $i-gH$ differentiable or $ii-gH$ differentiable, then

$$\begin{aligned} &\boldsymbol{D}_{\boldsymbol{CK}_{gH}}^{\alpha,p} x(t) \\ &= \frac{p^{\alpha}}{\Gamma(1-\alpha)} \int_{t_0}^{t} (t^p - s^p)^{-\alpha} \odot x'_{gH}(s) ds, \; t \in [t_0, T], \; 0<\alpha<1 \end{aligned}$$

To show the relation, we know the fuzzy function $\boldsymbol{I}_{\boldsymbol{RL}}^{(1-\alpha),p} x(t)$ is absolutely continuous because in its relation,

$$\boldsymbol{I}_{\boldsymbol{RL}}^{(1-\alpha),p} x(t) = \frac{p^{\alpha}}{\Gamma(1-\alpha)} \odot \int_{t_0}^{t} s^{p-1} (t^p - s^p)^{-\alpha} \odot x(s) ds$$

The coefficients $s^{p-1}(t^p - s^p)^{-\alpha} > 0, \frac{p^{\alpha}}{\Gamma(1-\alpha)} > 0$ and the fuzzy function $x(s)$ is absolutely continuous thus $\left(\boldsymbol{I}_{\boldsymbol{RL}}^{(1-\alpha),p} x\right)'_{gH}(t)$ exists and finally, $\boldsymbol{D}_{\boldsymbol{RL}_{gH}}^{\alpha,p} x(t)$ exists for $t \in (t_0, T]$,

$$\boldsymbol{D}_{\boldsymbol{RL}_{gH}}^{\alpha,p} x(t) = t^{1-p} \left(\boldsymbol{I}_{\boldsymbol{RL}}^{(1-\alpha),p} x\right)'_{gH}(t).$$

Now, let us consider a constant fuzzy function like $y \in \mathbb{F}_R$ which is $y(t) := x(t_0)$. Then

$$\boldsymbol{I}_{\boldsymbol{RL}}^{(1-\alpha),p} y(t) = \boldsymbol{I}_{\boldsymbol{RL}}^{(1-\alpha),p} x(t_0) = \frac{\left(t^p - t_0^p\right)^{1-\alpha} p^{\alpha-1}}{\Gamma(2-\alpha)} \odot x(t_0)$$

And if $\alpha :\to 1+\alpha$,

$$\boldsymbol{D}_{\boldsymbol{RL}_{gH}}^{\alpha,p} y(t) = \boldsymbol{I}_{\boldsymbol{RL}}^{-\alpha,p} y(t) = \frac{\left(t^p - t_0^p\right)^{-\alpha} p^{\alpha}}{\Gamma(1-\alpha)} \odot x(t_0)$$

This is the reason for gH differentiability of $\boldsymbol{I}_{\boldsymbol{RL}}^{(1-\alpha),p} x$ (in two cases) on $(t_0, T]$ and it follows,

$$\begin{aligned}\boldsymbol{D}_{\boldsymbol{CK}_{gH}}^{\alpha,p}x(t) &= \boldsymbol{D}_{\boldsymbol{RL}_{gH}}^{\alpha,p}\left(x(t)\ominus_{gH}x(t_0)\right) = \boldsymbol{D}_{\boldsymbol{RL}_{gH}}^{\alpha,p}\left(x(t)\ominus_{gH}y(t)\right)\\ &= \boldsymbol{D}_{\boldsymbol{RL}_{gH}}^{\alpha,p}x(t)\ominus_{gH}\boldsymbol{D}_{\boldsymbol{RL}_{gH}}^{\alpha,p}y(t) = \boldsymbol{D}_{\boldsymbol{RL}_{gH}}^{\alpha,p}x(t)\ominus_{gH}\frac{\left(t^p - t_0^p\right)^{-\alpha}p^\alpha}{\Gamma(1-\alpha)}\odot x(t_0)\end{aligned}$$

So,

$$\boldsymbol{D}_{\boldsymbol{CK}_{gH}}^{\alpha,p}x(t) = \boldsymbol{D}_{\boldsymbol{RL}_{gH}}^{\alpha,p}x(t)\ominus_{gH}\frac{\left(t^p - t_0^p\right)^{-\alpha}p^\alpha}{\Gamma(1-\alpha)}\odot x(t_0)$$

Two sides in the level-form and based on the first case of gH difference, we have,

Case $i - gH$ difference,

$$\boldsymbol{D}_{\boldsymbol{RL}_{gH}}^{\alpha,p}x(t,r) = \boldsymbol{D}_{\boldsymbol{CK}_{gH}}^{\alpha,p}x(t,r)\oplus\frac{\left(t^p - t_0^p\right)^{-\alpha}p^\alpha}{\Gamma(1-\alpha)}x(t_0,r)$$

In the level-wise form,

$$\begin{aligned}\left[\boldsymbol{D}_{\boldsymbol{RL}}^{\alpha,p}x_l(t,r),\boldsymbol{D}_{\boldsymbol{RL}}^{\alpha,p}x_u(t,r)\right] &= \left[\boldsymbol{D}_{\boldsymbol{CK}}^{\alpha,p}x_l(t,r),\boldsymbol{D}_{\boldsymbol{CK}}^{\alpha,p}x_u(t,r)\right]\\ &\quad+\frac{\left(t^p - t_0^p\right)^{-\alpha}p^\alpha}{\Gamma(1-\alpha)}\left[x_l(t_0,r),x_u(t_0,r)\right]\end{aligned}$$

There for any $r\in[0,1]$ and $t\in(t_0,T]$,

$$\begin{aligned}\boldsymbol{D}_{\boldsymbol{RL}_{gH}}^{\alpha,p}x(t) &= \boldsymbol{D}_{\boldsymbol{CK}_{gH}}^{\alpha,p}x(t)\oplus\frac{\left(t^p - t_0^p\right)^{-\alpha}p^\alpha}{\Gamma(1-\alpha)}\odot x(t_0)\\ &= \left(\boldsymbol{I}_{\boldsymbol{RL}}^{(1-\alpha),p}s^{1-p}\frac{d}{ds}x\right)(t)\oplus\frac{\left(t^p - t_0^p\right)^{-\alpha}p^\alpha}{\Gamma(1-\alpha)}\odot x(t_0)\end{aligned}$$

Substituting in the following,

$$\boldsymbol{D}_{\boldsymbol{CK}_{gH}}^{\alpha,p}x(t) = \boldsymbol{D}_{\boldsymbol{RL}_{gH}}^{\alpha,p}x(t)\ominus_{gH}\frac{\left(t^p - t_0^p\right)^{-\alpha}p^\alpha}{\Gamma(1-\alpha)}\odot x(t_0)$$

We get,

$$\boldsymbol{D}_{\boldsymbol{CK}_{gH}}^{\alpha,p}x(t) = \left(\boldsymbol{I}_{\boldsymbol{RL}}^{(1-\alpha),p}s^{1-p}\frac{d}{ds}x\right)(t)$$

Based on the definition of

$$\boldsymbol{I}_{\boldsymbol{RL}}^{(1-\alpha),p}x(t) = \frac{p^\alpha}{\Gamma(1-\alpha)}\odot\int_{t_0}^{t}s^{p-1}\left(t^p - s^p\right)^{-\alpha}\odot x(s)ds$$

It is concluded,

$$\left(I_{RL}^{(1-\alpha),p} s^{1-p} \frac{d}{ds} x\right)(t) = \frac{p^\alpha}{\Gamma(1-\alpha)} \odot \int_{t_0}^{t} (t^p - s^p)^{-\alpha} \odot x'(s) ds$$

Thus, the proof is completed in case $i - gH$ difference.
Case $ii - gH$ difference,

$$D_{CK_{gH}}^{\alpha,p} x(t,r) = D_{RL_{gH}}^{\alpha,p} x(t,r) \oplus (-1) \frac{\left(t^p - t_0^p\right)^{-\alpha} p^\alpha}{\Gamma(1-\alpha)} \odot x(t_0, r)$$

Because in level-wise form, since the function $x(t)$ is $ii - gH$ differentiable, so

$$\begin{aligned} D_{RL_{gH}}^{\alpha,p} x(t,r) &= \left[D_{RL}^{\alpha,p} x_u(t,r), D_{RL}^{\alpha,p} x_l(t,r)\right], \\ D_{CK_{gH}}^{\alpha,p} x(t,r) &= \left[D_{CK}^{\alpha,p} x_u(t,r), D_{CK}^{\alpha,p} x_l(t,r)\right] \end{aligned}$$

$$\begin{aligned} \left[D_{CK_{gH}}^{\alpha,p} x_u(t,r), D_{CK_{gH}}^{\alpha,p} x_l(t,r)\right] &= \left[D_{RL_{gH}}^{\alpha,p} x_u(t,r), D_{RL_{gH}}^{\alpha,p} x_l(t,r)\right] \\ &\quad + (-1) \frac{\left(t^p - t_0^p\right)^{-\alpha} p^\alpha}{\Gamma(1-\alpha)} [x_l(t_0, r), x_u(t_0, r)] \end{aligned}$$

There for for any $r \in [0, 1]$ and $t \in (t_0, T]$,

$$\begin{aligned} \left[\left(I_{RL}^{(1-\alpha),p} s^{1-p} \frac{d}{ds} x_l\right)(t,r), \left(I_{RL}^{(1-\alpha),p} s^{1-p} \frac{d}{ds} x_u\right)(t,r)\right] &= \left[D_{RL}^{\alpha,p} x_u(t,r), D_{RL}^{\alpha,p} x_l(t,r)\right] \\ &\quad + (-1) \frac{\left(t^p - t_0^p\right)^{-\alpha} p^\alpha}{\Gamma(1-\alpha)} [x_l(t_0, r), x_u(t_0, r)] \end{aligned}$$

So, we have

$$\left(I_{RL}^{(1-\alpha),p} s^{1-p} \frac{d}{ds} x\right)(t) = D_{RL_{gH}}^{\alpha,p} x(t) \oplus \frac{\left(t^p - t_0^p\right)^{-\alpha} p^\alpha}{\Gamma(1-\alpha)} (-1) \odot x(t_0)$$

Substituting in the following,

$$D_{CK_{gH}}^{\alpha,p} x(t) = D_{RL_{gH}}^{\alpha,p} x(t) \ominus_{gH} \frac{\left(t^p - t_0^p\right)^{-\alpha} p^\alpha}{\Gamma(1-\alpha)} \odot x(t_0)$$

We get,

$$D_{CK_{gH}}^{\alpha,p} x(t) = \left(I_{RL}^{(1-\alpha),p} s^{1-p} \frac{d}{ds} x\right)(t)$$

Based on the definition of

$$\boldsymbol{I}_{\boldsymbol{RL}}^{(1-\alpha),p}x(t)=\frac{p^{\alpha}}{\Gamma(1-\alpha)}\odot\int\limits_{t_0}^{t}s^{p-1}(t^p-s^p)^{-\alpha}\odot x(s)ds$$

It is concluded,

$$\left(\boldsymbol{I}_{\boldsymbol{RL}}^{(1-\alpha),p}s^{1-p}\frac{d}{ds}x\right)(t)=\frac{p^{\alpha}}{\Gamma(1-\alpha)}\odot\int\limits_{t_0}^{t}(t^p-s^p)^{-\alpha}\odot x'(s)ds$$

Thus, the proof is also completed in case $ii-gH$ difference.

Remark The RL integral operator is bounded.

$$\boldsymbol{I}_{\boldsymbol{RL}}^{\alpha,p}x(t)=\frac{p^{1-\alpha}}{\Gamma(\alpha)}\odot\int\limits_{t_0}^{t}s^{p-1}(t^p-s^p)^{\alpha-1}\odot x(s)ds$$

It can be shown by Hausdorff distance.

$$\sup_{t\in[t_0,T]}D_H\left(\boldsymbol{I}_{\boldsymbol{RL}}^{\alpha,p}x(t),0\right)\leq D_H(x(s),0)\frac{p^{1-\alpha}}{\Gamma(\alpha)}\int\limits_{t_0}^{t}s^{p-1}(t^p-s^p)^{\alpha-1}ds$$

$$=\frac{p^{-\alpha}}{\Gamma(\alpha+1)}D_H(x(s),0)\left(T^p-t_0^p\right)^{\alpha}$$

As it is mentioned before, the existence of the Caputo- Katugampola depends on the continuity of $x(t)$ and it has been supposed that is absolutely continuous. Then we claim the following remark.

Remark $\boldsymbol{D}_{\boldsymbol{CK}_{gH}}^{\alpha,p}x(t)=0$ at $t=0$.

Since $\boldsymbol{I}_{\boldsymbol{RL}}^{(1-\alpha),p}$ is bounded then the Caputo-Katugampola derivative is continuous.

$$\boldsymbol{D}_{\boldsymbol{CK}_{gH}}^{\alpha,p}x(t)=\left(\boldsymbol{I}_{\boldsymbol{RL}}^{(1-\alpha),p}s^{1-p}\frac{d}{ds}x\right)(t)$$

To show the assertion, it is enough to show that the upper bound of the derivative goes to zero at the point $t=0$ Then

$$\begin{aligned} D_H\left(\boldsymbol{D}^{\alpha,p}_{\boldsymbol{CK}_{gH}}x(t),0\right) &= D_H\left(\left(\boldsymbol{I}^{(1-\alpha),p}_{\boldsymbol{RL}}s^{1-p}x'\right)(t),0\right) \\ &\le \frac{p^{\alpha-1}}{\Gamma(2-\alpha)}D_H\left(s^{1-p}x',0\right)\left(t^p-t_0^p\right)^{1-\alpha} \\ &\le \frac{p^{\alpha-1}}{\Gamma(2-\alpha)}\sup_{s\in[t_0,T]}D_H\left(s^{1-p}x',0\right)\left(t^p-t_0^p\right)^{1-\alpha} \end{aligned}$$

The supremum $\sup_{s\in[t_0,T]}D_H(s^{1-p}x',0)$ is a real number like k then

$$D_H\left(\boldsymbol{D}^{\alpha,p}_{\boldsymbol{CK}_{gH}}x(t),0\right) \le \frac{p^{\alpha-1}}{\Gamma(2-\alpha)}k\left(t^p-t_0^p\right)^{1-\alpha}$$

It is clear that, at the point $t=0$, the distance goes zero and it completes the proof.

Remark If the function $x(t)$ is $i-gH$ or $ii-gH$ differentiable on $(t_0,T]$, for $0<\alpha<1$ we have,

- $$\boldsymbol{I}^{\alpha,p}_{\boldsymbol{RL}}\boldsymbol{D}^{\alpha,p}_{\boldsymbol{CK}_{gH}}x(t) = x(t)\ominus_{gH}x(t_0)$$

- $$\boldsymbol{D}^{\alpha,p}_{\boldsymbol{CK}_{gH}}\boldsymbol{I}^{\alpha,p}_{\boldsymbol{RL}}x(t) = x(t)$$

The first item,

$$\begin{aligned} \boldsymbol{I}^{\alpha,p}_{\boldsymbol{RL}}\boldsymbol{D}^{\alpha,p}_{\boldsymbol{CK}_{gH}}x(t) &= \boldsymbol{I}^{\alpha,p}_{\boldsymbol{RL}}\left(\boldsymbol{I}^{(1-\alpha),p}_{\boldsymbol{RL}}s^{1-p}\frac{d}{ds}x\right)(t) \\ &= \left(\boldsymbol{I}^{1,p}_{\boldsymbol{RL}}s^{1-p}\frac{d}{ds}x\right)(t) = \int_{t_0}^{t}x'(s)ds = x(t)\ominus_{gH}x(t_0) \end{aligned}$$

Because in the definition,

$$\boldsymbol{I}^{\alpha,p}_{\boldsymbol{RL}}x(t) = \frac{p^{1-\alpha}}{\Gamma(\alpha)}\odot\int_{t_0}^{t}s^{p-1}\left(t^p-s^p\right)^{\alpha-1}\odot x(s)ds$$

If $\alpha=1, p=1$ then,

$$\boldsymbol{I}^{1,1}_{\boldsymbol{RL}}x(t) = \int_{t_0}^{t}x(s)ds$$

To prove the second item, we have the following relations,
For $x(t_0)$ as a constant fuzzy, we have

$$\boldsymbol{I}_{\boldsymbol{RL}}^{(1-\alpha),p}x(t_0)=\frac{p^{\alpha}}{\Gamma(1-\alpha)}\odot\int_{t_0}^{t}s^{p-1}(t^p-s^p)^{-\alpha}\odot x(t_0)ds=\frac{\left(t^p-t_0^p\right)^{1-\alpha}p^{\alpha-1}}{\Gamma(2-\alpha)}\odot x(t_0)$$

And if $\alpha:\rightarrow 1+\alpha$,

$$\boldsymbol{D}_{\boldsymbol{RL}_{gH}}^{\alpha,p}x(t_0)=\boldsymbol{I}_{\boldsymbol{RL}}^{-\alpha,p}y(t)=\frac{\left(t^p-t_0^p\right)^{-\alpha}p^{\alpha}}{\Gamma(1-\alpha)}\odot x(t_0)$$
$$\boldsymbol{D}_{\boldsymbol{CK}_{gH}}^{\alpha,p}x(t)=\boldsymbol{D}_{\boldsymbol{RL}_{gH}}^{\alpha,p}\left(x(t)\ominus_{gH}x(t_0)\right)=\boldsymbol{D}_{\boldsymbol{RL}_{gH}}^{\alpha,p}x(t)\ominus_{gH}\frac{\left(t^p-t_0^p\right)^{-\alpha}p^{\alpha}}{\Gamma(1-\alpha)}\odot x(t_0)$$

Now by substituting $\boldsymbol{I}_{\boldsymbol{RL}}^{\alpha,p}x(t)\rightarrow x(t)$ we get,

$$\boldsymbol{D}_{\boldsymbol{CK}_{gH}}^{\alpha,p}\boldsymbol{I}_{\boldsymbol{RL}}^{\alpha,p}x(t)=\boldsymbol{D}_{\boldsymbol{RL}_{gH}}^{\alpha,p}\boldsymbol{I}_{\boldsymbol{RL}}^{\alpha,p}x(t)\ominus_{gH}\frac{\left(t^p-t_0^p\right)^{-\alpha}p^{\alpha}}{\Gamma(1-\alpha)}\odot\left(\boldsymbol{I}_{\boldsymbol{RL}}^{\alpha,p}x\right)(t_0)$$

If we show $\left(\boldsymbol{I}_{\boldsymbol{RL}}^{\alpha,p}x\right)(t_0)=0$ the proof is completed. To do this, we use the distance.

$$D_H\left(\left(\boldsymbol{I}_{\boldsymbol{RL}}^{\alpha,p}x\right)(t_0),0\right)\leq k\frac{p^{1-\alpha}}{\Gamma(\alpha)}\int_{t_0}^{t}s^{p-1}(t^p-s^p)^{\alpha-1}ds=k\frac{\left(t^p-t_0^p\right)^{\alpha}p^{-\alpha}}{\Gamma(1+\alpha)}$$

Where $\sup_{t\in[t_0,T]}D_H(x(t),0)=k$. Then

$$D_H\left(\left(\boldsymbol{I}_{\boldsymbol{RL}}^{\alpha,p}x\right)(t_0),0\right)\leq k\frac{\left(t^p-t_0^p\right)^{\alpha}p^{-\alpha}}{\Gamma(1+\alpha)}\rightarrow 0$$

3.7 Riemann-Liouville gH-Fractional Derivative of $\boldsymbol{\alpha}\in(1,2)$

To discuss the subject, we need some primary definitions like the second order of gH-differentiability of a fuzzy number valued function $x(t)\in\mathbb{F}_R$ that is absolutely continuously gH-differentiable. It means the lower and upper parametric functions $x_l(t,r),x_u(t,r)$ in the level-wise form are absolutely continuous. This is because this continuity is evaluated by Hausdorff distance. As we know, the definition of second order of gH-differential on $[t_0,T]$ is as follows, such that $x''_{gH}(t)\in\mathbb{F}_R$

$$x''_{gH}(t) = \lim_{h\to 0} \frac{x'(t+h)\ominus_{gH}x'(t)}{h}$$

Remark With the same assumptions for $x(t)$ and more over $x(t), x'_{gH}(t)$ do have monotone length, we have

$$x(t)\ominus_{gH}x(t_0)\ominus_{gH}(t-t_0)x'_{gH}(t_0) = \int_{t_0}^{t}\int_{t_0}^{s} x''_{gH}(u)duds, t_0 \le u \le s \le t \le T$$

To show the relation,

$$\int_{t_0}^{t}\int_{t_0}^{s} x''_{gH}(u)duds = \int_{t_0}^{t} \left(x'(s)\ominus_{gH}x'(t_0)\right)ds$$

Let us know about the length of $x'(s)\ominus_{gH}x'(t_0)$,

$$\begin{aligned} length\left(x'(s)\ominus_{gH}x'(t_0)\right) &= \left(x'_u(s,r) - x'_u(t_0,r)\right) - \left(x'_l(s,r) - x'_l(t_0,r)\right) \\ &= \left(x'_u(s,r) - x'_l(s,r)\right) - \left(x'_u(t_0,r) - x'_l(t_0,r)\right) \\ &= length(x'(s)) - length(x'(t_0)) \end{aligned}$$

Finally, we have,

$$length\left(x'(s)\ominus_{gH}x'(t_0)\right) = length(x'(s)) - length(x'(t_0))$$

- If $length(x'(s)) \ge length(x'(t_0))$ then $length\left(x'(s)\ominus_{gH}x'(t_0)\right) \ge 0$
- If $length(x'(s)) \le length(x'(t_0))$ then $length\left(x'(s)\ominus_{gH}x'(t_0)\right) \le 0$

These two cases mean, if the length of $x'(s)$ is increasing (decreasing) then the length of $x'(s)\ominus_{gH}x'(t_0)$ is increasing (decreasing) too. Then the sign of gH-difference $x'(s)\ominus_{gH}x'(t_0)$ is constant,

$$\begin{aligned} \int_{t_0}^{t}\int_{t_0}^{s} x''_{gH}(u)duds &= \int_{t_0}^{t}\left(x'(s)\ominus_{gH}x'(t_0)\right)ds = \int_{t_0}^{t} x'(s)ds\ominus_{gH}\int_{t_0}^{t} x'(t_0)ds \\ &= x(t)\ominus_{gH}x(t_0)\ominus_{gH}(t-t_0)\odot x'(t_0) \end{aligned}$$

It is completed.

Definition—The fuzzy Riemann-Liouville gH-fractional derivative of order $\alpha \in (1,2)$

$$D_{RL_{gH}}^{\alpha>1}x(t) = \left(I_{RL}^{2-\alpha}x(t)\right)'', \ t \in [t_0, T]$$

Subject to the second derivative in the sense of gH-differentiability exists. In the other words, these limits exist.

$$\left(I_{RL}^{2-\alpha}x(t)\right)'' = \lim_{h\to 0}\frac{\left(I_{RL}^{2-\alpha}x\right)'(t+h)\ominus_{gH}\left(I_{RL}^{2-\alpha}x\right)'(t)}{h}$$
$$\left(I_{RL}^{2-\alpha}x(t)\right)' = \lim_{h\to 0}\frac{I_{RL}^{2-\alpha}x(t+h)\ominus_{gH}I_{RL}^{2-\alpha}x(t)}{h}$$

Where

$$I_{RL}^{2-\alpha}x(t) = \frac{1}{\Gamma(2-\alpha)}\int_{t_0}^{t}(t-s)^{1-\alpha}\odot x(s)ds, \ t \in [t_0, T]$$

- Suppose that the functions $I_{RL}^{2-\alpha}x(t)$ and $\left(I_{RL}^{2-\alpha}x(t)\right)'$ are $i-gH$ differentiable on $[t_0, T]$. The level-wise form of the derivative is as,

$$D_{RL_{gH}}^{\alpha>1}x(t,r) = \left(I_{RL}^{2-\alpha}x(t,r)\right)'', \ r \in [0,1]$$

Then

$$D_{RL_{gH}}^{\alpha>1}x_l(t,r) = \left(I_{RL}^{2-\alpha}x_l(t,r)\right)'' = \frac{1}{\Gamma(2-\alpha)}\int_{t_0}^{t}(t-s)^{1-\alpha}x_l(s,r)ds$$

$$D_{RL_{gH}}^{\alpha>1}x_u(t,r) = \left(I_{RL}^{2-\alpha}x_u(t,r)\right)'' = \frac{1}{\Gamma(2-\alpha)}\int_{t_0}^{t}(t-s)^{1-\alpha}x_u(s,r)ds$$

- Suppose that the functions $I_{RL}^{2-\alpha}x(t)$ and $\left(I_{RL}^{2-\alpha}x(t)\right)'$ are $ii-gH$ differentiable on $[t_0, T]$. The level-wise form of the derivative is as,

$$\left(I_{RL}^{2-\alpha}x(t,r)\right)'' = \left[I_{RL}^{2-\alpha}x_u(t,r), I_{RL}^{2-\alpha}x_l(t,r)\right]$$
$$D_{RL_{gH}}^{\alpha>1}x(t,r) = \left(I_{RL}^{2-\alpha}x(t,r)\right)'', \ r \in [0,1]$$

In this case to define the derivative, the Caputo operative is also $ii-gH$ differentiable, then

$$D^{\alpha>1}_{RL_{gH}}x(t,r) = \left[D^{\alpha>1}_{RL_{gH}}x_u(t,r), D^{\alpha>1}_{RL_{gH}}x_l(t,r)\right]$$

where

$$D^{\alpha>1}_{RL_{gH}}x_l(t,r) = \left(I^{2-\alpha}_{RL}x_l(t,r)\right)'' = \frac{1}{\Gamma(2-\alpha)}\int_{t_0}^{t}(t-s)^{1-\alpha}x_l(s,r)ds$$

$$D^{\alpha>1}_{RL_{gH}}x_u(t,r) = \left(I^{2-\alpha}_{RL}x_u(t,r)\right)'' = \frac{1}{\Gamma(2-\alpha)}\int_{t_0}^{t}(t-s)^{1-\alpha}x_u(s,r)ds$$

In fact, the type of gH-differentiability of Caputo depends on the type of second order differentiability of RL integral operator.

- Assume the fuzzy functions $x'(t)$ is $i-gH$ differentiable and $x(t)$ is $ii-RL_{gH}$ differentiable, $D^{\alpha>1}_{RL_{ii-gH}}$, in this case,

$$x''(t,r) = \left[x_l''(t,r), x_u''(t,r)\right]$$

Then

$$I^{2-\alpha}_{RL}x''(t,r) = \left[I^{2-\alpha}_{RL}x_l''(t,r), I^{2-\alpha}_{RL}x_u''(t,r)\right]$$
$$D^{\alpha>1}_{RL_{ii-gH}}x(t,r) = \left[D^{\alpha>1}_{RL}x_u(t,r), D^{\alpha>1}_{RL}x_l(t,r)\right]$$

where

$$D^{\alpha>1}_{RL}x_l(t,r) = I^{2-\alpha}_{RL}x_u''(t,r),\ D^{\alpha>1}_{RL_{gH}}x_u(t,r) = I^{2-\alpha}_{RL}x_l''(t,r)$$

Basically, this case forms a system of fractional differential equations and it is occasionally is easy to solve. Another case is similar to this case.

The following properties can be proved for any $t \in [t_0, T]$.

Properties

- $$D^{\alpha>1}_{RL_{gH}}I^{\alpha}_{RL}x(t) = x(t)$$

- $$D^{\alpha>1}_{RL_{gH}}I^{\beta}_{RL}x(t) = I^{\beta-\alpha}_{RL}x(t),\ \beta > \alpha$$

For the first one, by using the definition,

$$\boldsymbol{D}^{\alpha>1}_{\boldsymbol{RL}_{gH}} \boldsymbol{I}^{\alpha}_{\boldsymbol{RL}} x(t) = \boldsymbol{D}^{\alpha>1}_{\boldsymbol{RL}_{gH}} \left(\boldsymbol{I}^{\alpha}_{\boldsymbol{RL}} x(t) \right) = \left(\boldsymbol{I}^{2-\alpha}_{\boldsymbol{RL}} \boldsymbol{I}^{\alpha}_{\boldsymbol{RL}} x(t) \right)'' = \left(\boldsymbol{I}^{2}_{\boldsymbol{RL}} x(t) \right)''$$

And

$$\left(\boldsymbol{I}^{2}_{\boldsymbol{RL}} x(t) \right)'' = \left(\int_{t_0}^{t} \int_{t_0}^{s} x(u) du ds \right)''$$

Finally,

$$\boldsymbol{D}^{\alpha>1}_{\boldsymbol{RL}_{gH}} \boldsymbol{I}^{\alpha}_{\boldsymbol{RL}} x(t) = \left(\int_{t_0}^{t} \int_{t_0}^{s} x(u) du ds \right)'' = x(t)$$

The second property,

$$\boldsymbol{D}^{\alpha>1}_{\boldsymbol{RL}_{gH}} \boldsymbol{I}^{\beta}_{\boldsymbol{RL}} x(t) = \left(\boldsymbol{I}^{2-\alpha}_{\boldsymbol{RL}} \boldsymbol{I}^{\beta}_{\boldsymbol{RL}} x(t) \right)'' = \frac{1}{\Gamma(2-\alpha)} \odot \left(\int_{t_0}^{t} (t-s)^{1-\alpha} \odot \boldsymbol{I}^{\beta}_{\boldsymbol{RL}} x(s) ds \right)''$$

$$= \frac{1}{\Gamma(2-\alpha)\Gamma(\beta)} \odot \left(\int_{t_0}^{t} (t-s)^{1-\alpha} \int_{t_0}^{s} (s-u)^{\beta-1} \odot x(u) du ds \right)''$$

If we suppose $v = \frac{u-s}{t-s}$ then

$$(t-s)^{1-\alpha} = (u-s)^{1-\alpha} v^{\alpha-1}, \ (s-u)^{\beta-1} = v^{\beta-1} (s-t)^{\beta-1}$$

$$\boldsymbol{D}^{\alpha>1}_{\boldsymbol{RL}_{gH}} \boldsymbol{I}^{\beta}_{\boldsymbol{RL}} x(t) = \frac{1}{\Gamma(2-\alpha)\Gamma(\beta)} \odot \left(\int_{t_0}^{t} (t-s)^{1+\beta-\alpha} \odot x(s) ds \int_{0}^{1} (1-v)^{2-\alpha-1} v^{\beta-1} dv \right)''$$

On the other hand,

$$\int_{0}^{1} (1-v)^{2-\alpha-1} v^{\beta-1} dv = \frac{\Gamma(2-\alpha)\Gamma(\beta)}{\Gamma(\beta-\alpha+2)}$$

Then

$$D_{RL_{gH}}^{\alpha>1} I_{RL}^{\beta} x(t) = \frac{1}{\Gamma(\beta-\alpha+2)} \odot \left(\int_{t_0}^{t} (t-s)^{1+\beta-\alpha} \odot x(s)ds \right)''$$

Now we are going to find the $\left(\int_{t_0}^{t} (t-s)^{1+\beta-\alpha} \odot x(s)ds \right)''$,

We have,

$$\left(\int_{t_0}^{t} (t-s)^{1+\beta-\alpha} \odot x(s)ds \right)' = \int_{t_0}^{t} (1+\beta-\alpha)(t-s)^{\beta-\alpha} \odot x(s)ds$$

Also,

$$\left(\int_{t_0}^{t} (1+\beta-\alpha)(t-s)^{\beta-\alpha} \odot x(s)ds \right)' = (1+\beta-\alpha)(\beta-\alpha) \int_{t_0}^{t} (t-s)^{\beta-\alpha-1} \odot x(s)ds$$

By substituting in,

$$\begin{aligned} D_{RL_{gH}}^{\alpha>1} I_{RL}^{\beta} x(t) &= \frac{1}{\Gamma(\beta-\alpha+2)} \odot \left(\int_{t_0}^{t} (t-s)^{1+\beta-\alpha} \odot x(s)ds \right)'' \\ &= \frac{(1+\beta-\alpha)(\beta-\alpha)}{\Gamma(\beta-\alpha+2)} \odot \int_{t_0}^{t} (t-s)^{\beta-\alpha-1} \odot x(s)ds \end{aligned}$$

Since,

$$\frac{(1+\beta-\alpha)(\beta-\alpha)}{\Gamma(\beta-\alpha+2)} = \frac{1}{\Gamma(\beta-\alpha)}$$

$$D_{RL_{gH}}^{\alpha>1} I_{RL}^{\beta} x(t) = \frac{1}{\Gamma(\beta-\alpha)} \odot \int_{t_0}^{t} (t-s)^{\beta-\alpha-1} \odot x(s)ds = I_{RL}^{\beta-\alpha} x(t)$$

Example Consider the fuzzy function $x(t) = c \odot e^t$, $c \in \mathbb{F}_R$, $t \in [0, 1]$, $\alpha = \frac{3}{2}$

$$D^{\frac{3}{2}}_{RL_{gH}} x(t) = \left(I^{\frac{1}{2}}_{RL} c \odot e^t \right)'',$$

where

$$I^{\frac{1}{2}}_{RL} c \odot e^t = \frac{c}{\Gamma\left(\frac{1}{2}\right)} \odot \int_0^1 (t-s)^{-\frac{1}{2}} e^s ds, \; t \in [t_0, T]$$

3.7.1 Definition—Fuzzy Caputo Fractional Derivative of Order $\alpha \in (1, 2)$

Considering the same assumptions on the fuzzy number valued function $x(t)$,

$$D^{\alpha>1}_{C_{gH}} x(t) = D^{\alpha>1}_{RL_{gH}} \left(x(t) \ominus_{gH} x(t_0) \ominus_{gH} (t - t_0) \odot x'(t_0) \right), \; t \in [t_0, T]$$

We obtained the following relation,

$$\int_{t_0}^{t} \int_{t_0}^{s} x''_{gH}(u) du ds = x(t) \ominus_{gH} x(t_0) \ominus_{gH} (t - t_0) \odot x'(t_0)$$

By substituting,

$$D^{\alpha>1}_{RL_{gH}} \left(x(t) \ominus_{gH} x(t_0) \ominus_{gH} (t - t_0) \odot x'(t_0) \right) = D^{\alpha>1}_{RL_{gH}} \left(\int_{t_0}^{t} \int_{t_0}^{s} x''_{gH}(u) du ds \right)$$

Indeed,

$$D^{\alpha>1}_{C_{gH}} x(t) = D^{\alpha>1}_{RL_{gH}} \left(\int_{t_0}^{t} \int_{t_0}^{s} x''_{gH}(u) du ds \right) = D^{\alpha>1}_{RL_{gH}} \left(I^{2}_{RL} x''(t) \right) = I^{2-\alpha}_{RL} x''(t)$$

Finally,

$$D^{\alpha>1}_{C_{gH}} x(t) = I^{2-\alpha}_{RL} x''(t) = \frac{1}{\Gamma(2-\alpha)} \int_{t_0}^{t} (t-s)^{1-\alpha} \odot x''(s) ds$$

In level-wise form,

- Assume the fuzzy functions $x(t), x'(t)$ are $i-gH$ differentiable,

$$x''(t,r) = \left[x_l''(t,r), x_u''(t,r)\right]$$

Then

$$\boldsymbol{I}_{\boldsymbol{RL}}^{2-\alpha} x''(t,r) = \left[\boldsymbol{I}_{\boldsymbol{RL}}^{2-\alpha} x_l''(t,r), \boldsymbol{I}_{\boldsymbol{RL}}^{2-\alpha} x_u''(t,r)\right]$$

where

$$\boldsymbol{I}_{\boldsymbol{RL}}^{2-\alpha} x_l''(t,r) = \frac{1}{\Gamma(2-\alpha)} \int_{t_0}^{t} (t-s)^{1-\alpha} x_l''(s) ds$$

$$\boldsymbol{I}_{\boldsymbol{RL}}^{2-\alpha} x_u''(t,r) = \frac{1}{\Gamma(2-\alpha)} \int_{t_0}^{t} (t-s)^{1-\alpha} x_u''(s) ds$$

$$\boldsymbol{D}_{\boldsymbol{C}_{gH}}^{\alpha>1} x(t,r) = \left[\boldsymbol{D}_{\boldsymbol{C}_{gH}}^{\alpha>1} x_l(t,r), \boldsymbol{D}_{\boldsymbol{C}_{gH}}^{\alpha>1} x_u(t,r)\right]$$

where

$$\boldsymbol{D}_{\boldsymbol{C}_{gH}}^{\alpha>1} x_l(t,r) = \boldsymbol{I}_{\boldsymbol{RL}}^{2-\alpha} x_l''(t,r),\ \boldsymbol{D}_{\boldsymbol{C}_{gH}}^{\alpha>1} x_u(t,r) = \boldsymbol{I}_{\boldsymbol{RL}}^{2-\alpha} x_u''(t,r)$$

- Assume the fuzzy functions $x(t), x'(t)$ are $ii-gH$ differentiable,

$$x''(t,r) = \left[x_u''(t,r), x_l''(t,r)\right]$$

Then

$$\boldsymbol{I}_{\boldsymbol{RL}}^{2-\alpha} x''(t,r) = \left[\boldsymbol{I}_{\boldsymbol{RL}}^{2-\alpha} x_u''(t,r), \boldsymbol{I}_{\boldsymbol{RL}}^{2-\alpha} x_l''(t,r)\right]$$
$$\boldsymbol{D}_{\boldsymbol{C}_{gH}}^{\alpha>1} x(t,r) = \left[\boldsymbol{D}_{\boldsymbol{C}_{gH}}^{\alpha>1} x_u(t,r), \boldsymbol{D}_{\boldsymbol{C}_{gH}}^{\alpha>1} x_l(t,r)\right]$$

where

$$\boldsymbol{D}_{\boldsymbol{C}_{gH}}^{\alpha>1} x_l(t,r) = \boldsymbol{I}_{\boldsymbol{RL}}^{2-\alpha} x_l''(t,r),\ \boldsymbol{D}_{\boldsymbol{C}_{gH}}^{\alpha>1} x_u(t,r) = \boldsymbol{I}_{\boldsymbol{RL}}^{2-\alpha} x_u''(t,r)$$

- Assume the fuzzy functions $x'(t)$ is $ii-gH$ differentiable and $x(t)$ is $i-C_{gH}$ differentiable, $D^{\alpha>1}_{C_{i-gH}}$, in this case,

$$x''(t,r) = \left[x''_u(t,r), x''_l(t,r)\right]$$

Then

$$\begin{aligned} I^{2-\alpha}_{RL} x''(t,r) &= \left[I^{2-\alpha}_{RL} x''_u(t,r), I^{2-\alpha}_{RL} x''_l(t,r)\right] \\ D^{\alpha>1}_{C_{i-gH}} x(t,r) &= \left[D^{\alpha>1}_{C} x_l(t,r), D^{\alpha>1}_{C} x_u(t,r)\right] \end{aligned}$$

where

$$D^{\alpha>1}_{C} x_l(t,r) = I^{2-\alpha}_{RL} x''_u(t,r),\ D^{\alpha>1}_{C_{gH}} x_u(t,r) = I^{2-\alpha}_{RL} x''_l(t,r)$$

Basically, this case forms a system of fractional differential equations and it is occasionally is easy to solve. Another case is similar to this case.

Example Consider the fuzzy function $x(t) = c \odot t^2,\ , c \in \mathbb{F}_R,\ t \in [0,1],\ \alpha = \frac{3}{2}$
It is clear the function $x(t), x'(t)$ are $i-gH$ differentiable and

$$D^{\alpha>1}_{C_{gH}}\left(c \odot t^2\right) = I^{2-\alpha}_{RL}(2t \odot c) = \frac{4t\sqrt{t}}{\Gamma\left(\frac{1}{2}\right)} \odot c$$

It can be seen that $D^{\alpha>1}_{C_{gH}}(c \odot t^2)$ is $i-C_{gH}$ differentiable.

Remark

$$I^{\alpha>1}_{RL} D^{\alpha>1}_{C_{gH}} x(t) = x(t) \ominus_{gH} x(t_0) \ominus_{gH} (t-t_0) \odot x'_{gH}(t_0)$$

To show it, we have,

$$D^{\alpha>1}_{C_{gH}} x(t) = D^{\alpha>1}_{RL_{gH}}\left(x(t) \ominus_{gH} x(t_0) \ominus_{gH} (t-t_0) \odot x'_{gH}(t_0)\right)$$

using the RL fractional integration in both sides, we have

$$I^{\alpha>1}_{RL} D^{\alpha>1}_{C_{gH}} x(t) = x(t) \ominus_{gH} x(t_0) \ominus_{gH} (t-t_0) \odot x'_{gH}(t_0)$$

This relation is also can be considered in the level-wise form,

- If $x(t) \ominus_{gH} x(t_0)$ exists in case of $i-gH$, then $x'_{i-gH}(t_0)$ exists and if $x(t)$ is $i-C_{gH}$ differentiable, $D^{\alpha>1}_{C_{i-gH}}$, in this case,

$$I^{\alpha>1}_{RL} D^{\alpha>1}_{C_{i-gH}} x(t) = x(t) \ominus_{i-gH} x(t_0) \ominus_{i-gH} (t-t_0) \odot x'_{i-gH}(t_0)$$

In level-wise form,

$$\begin{aligned} I_{RL}^{\alpha>1} D_{C_{gH}}^{\alpha>1} x_l(t,r) &= x_l(t,r) - x_l(t_0,r) - (t-t_0)x_l'(t_0,r) \\ I_{RL}^{\alpha>1} D_{C_{gH}}^{\alpha>1} x_u(t,r) &= x_u(t,r) - x_u(t_0,r) - (t-t_0)x_u'(t_0,r) \end{aligned}$$

- If $x(t)\ominus_{gH}x(t_0)$ exists in case of $ii-gH$, then $x_{ii-gH}'(t_0)$ exists and if $x(t)$ is $ii-C_{gH}$ differentiable, $D_{C_{ii-gH}}^{\alpha>1}$, again in this case,

$$I_{RL}^{\alpha>1} D_{C_{ii-gH}}^{\alpha>1} x(t) = x(t)\ominus_{ii-gH}x(t_0)\ominus_{i-gH}(t-t_0)\odot x_{ii-gH}'(t_0)$$

In level-wise form,

$$\begin{aligned} I_{RL}^{\alpha>1} D_{C_{gH}}^{\alpha>1} x_l(t,r) &= x_l(t,r) - x_l(t_0,r) - (t-t_0)x_u'(t_0,r) \\ I_{RL}^{\alpha>1} D_{C_{gH}}^{\alpha>1} x_u(t,r) &= x_u(t,r) - x_u(t_0,r) - (t-t_0)x_l'(t_0,r) \end{aligned}$$

- If $x(t)\ominus_{gH}x(t_0)$ exists in case of $i-gH$, then $x_{i-gH}'(t_0)$ exists and if $x(t)$ is $ii-C_{gH}$ differentiable, $D_{C_{ii-gH}}^{\alpha>1}$, again in this case,

$$I_{RL}^{\alpha>1} D_{C_{ii-gH}}^{\alpha>1} x(t) = x(t)\ominus_{i-gH}x(t_0)\ominus_{i-gH}(t-t_0)\odot x_{i-gH}'(t_0)$$

In level-wise form,

$$\begin{aligned} I_{RL}^{\alpha>1} D_{C_{gH}}^{\alpha>1} x_u(t,r) &= x_l(t,r) - x_l(t_0,r) - (t-t_0)x_l'(t_0,r) \\ I_{RL}^{\alpha>1} D_{C_{gH}}^{\alpha>1} x_l(t,r) &= x_u(t,r) - x_u(t_0,r) - (t-t_0)x_u'(t_0,r) \end{aligned}$$

This case will be very difficult to solve, because it is a system of fractional differential equations.

3.8 Generalized Fuzzy ABC Fractional Derivative

The generalized ABC_{gH} derivative of a fuzzy number valued function $x(t)$ on interval $[t_0, T]$ starting at t_0 with kernel $E_{\alpha,\mu}^{\gamma}(\lambda, t)$ where $0<\alpha\langle 1, Re(\mu)\rangle 0, \gamma \in R$, $\lambda = \frac{-\alpha}{1-\alpha}$ is defined in the following form,

$$D_{ABC_{gH}}^{\alpha,\mu,\gamma} x(t) = \frac{B(\alpha)}{1-\alpha} \odot \int_{t_0}^{t} x_{gH}'(\tau) \odot E_{\alpha,\mu}^{\gamma}(\lambda, t-\tau) d\tau$$

where $B(\alpha) > 0$ is a normalizing function and is defined as,

$$B(\alpha) = 1 - \alpha + \frac{\alpha}{\Gamma(\alpha)}$$

With properties $B(0) = B(1) = 1$. The level-wise form of the derivative is defined in the following form with considering the type of gH-differential.

Case 1. If x is $i - gH$ differentiable,

$$\boldsymbol{D}^{\alpha,\mu,\gamma}_{\boldsymbol{ABC}_{gH}} x_l(t,r) = \frac{B(\alpha)}{1-\alpha} \odot \int_{t_0}^{t} x_l'(\tau,r) \odot E^{\gamma}_{\alpha,\mu}(\lambda, t-\tau)d\tau$$

$$\boldsymbol{D}^{\alpha,\mu,\gamma}_{\boldsymbol{ABC}_{gH}} x_u(t,r) = \frac{B(\alpha)}{1-\alpha} \odot \int_{t_0}^{t} x_u'(\tau,r) \odot E^{\gamma}_{\alpha,\mu}(\lambda, t-\tau)d\tau$$

Case 2. If x is $ii - gH$ differentiable,

$$\boldsymbol{D}^{\alpha,\mu,\gamma}_{\boldsymbol{ABC}_{gH}} x_l(t,r) = \frac{B(\alpha)}{1-\alpha} \odot \int_{t_0}^{t} x_u'(\tau,r) \odot E^{\gamma}_{\alpha,\mu}(\lambda, t-\tau)d\tau$$

$$\boldsymbol{D}^{\alpha,\mu,\gamma}_{\boldsymbol{ABC}_{gH}} x_u(t,r) = \frac{B(\alpha)}{1-\alpha} \odot \int_{t_0}^{t} x_l'(\tau,r) \odot E^{\gamma}_{\alpha,\mu}(\lambda, t-\tau)d\tau$$

Note In case 2, the fuzzy ABC differential cannot be defined because the left side and right side do not have the same function. We only claim that how can we have an interval with this derivative in the level-wise form.

Case 1. $x(t)$ is $i - gH$ differentiable,

$$\boldsymbol{D}^{\alpha,\mu,\gamma}_{\boldsymbol{ABC}_{i-gH}} x(t,r) = \left[\boldsymbol{D}^{\alpha,\mu,\gamma}_{\boldsymbol{ABC}} x_l(t,r), \boldsymbol{D}^{\alpha,\mu,\gamma}_{\boldsymbol{ABC}} x_u(t,r)\right]$$

Case 2. $x(t)$ is $ii - gH$ differentiable,

$$\boldsymbol{D}^{\alpha,\mu,\gamma}_{\boldsymbol{ABC}_{ii-gH}} x(t,r) = \left[\boldsymbol{D}^{\alpha,\mu,\gamma}_{\boldsymbol{ABC}} x_u(t,r), \boldsymbol{D}^{\alpha,\mu,\gamma}_{\boldsymbol{ABC}} x_l(t,r)\right]$$

where

$$\boldsymbol{D}^{\alpha,\mu,\gamma}_{\boldsymbol{ABC}} x_l(t,r) = \tfrac{B(\alpha)}{1-\alpha} \int_{t_0}^{t} x_l'(\tau,r) E_\alpha\left(-\tfrac{\alpha}{1-\alpha}(t-\tau)^\alpha\right)d\tau$$

$$\boldsymbol{D}^{\alpha,\mu,\gamma}_{\boldsymbol{ABC}} x_u(t,r) = \tfrac{B(\alpha)}{1-\alpha} \int_{t_0}^{t} x_u'(\tau,r) E_\alpha\left(-\tfrac{\alpha}{1-\alpha}(t-\tau)^\alpha\right)d\tau$$

Its corresponding AB fractional integral operator is defined as,

$$\boldsymbol{I}_{\boldsymbol{AB}}^{\alpha,\mu,\gamma}x(t)=\sum_{i=0}^{\gamma}\binom{\gamma}{i}\frac{\alpha^{i}}{B(\alpha)(1-\alpha)^{i-1}}\boldsymbol{I}^{\alpha,\mu,\gamma}x(t)$$

Subject to,

$$\boldsymbol{I}_{\boldsymbol{AB}}^{\alpha,\mu.\gamma}\boldsymbol{D}_{\boldsymbol{ABC}_{gH}}^{\alpha,\mu.\gamma}x(t)=x(t)\ominus_{gH}x(t_0)$$

Based on the definition of the gH-difference,
Case 1. $x(t)$ is $i-gH$ difference,

$$\begin{aligned}&\boldsymbol{I}_{\boldsymbol{AB}}^{\alpha,\mu.\gamma}\boldsymbol{D}_{\boldsymbol{ABC}_{gH}}^{\alpha,\mu.\gamma}x(t)=x(t)\ominus_{H}x(t_0)\\&x(t)=\boldsymbol{I}_{\boldsymbol{AB}}^{\alpha,\mu.\gamma}\boldsymbol{D}_{\boldsymbol{ABC}_{gH}}^{\alpha,\mu.\gamma}x(t)\oplus x(t_0)\\&x_l(t,r)=x_l(t_0,r)+\boldsymbol{I}_{\boldsymbol{AB}}^{\alpha,\mu.\gamma}\boldsymbol{D}_{\boldsymbol{ABC}_{gH}}^{\alpha,\mu.\gamma}x_l(t,r)\\&x_u(t,r)=x_u(t_0,r)+\boldsymbol{I}_{\boldsymbol{AB}}^{\alpha,\mu.\gamma}\boldsymbol{D}_{\boldsymbol{ABC}_{gH}}^{\alpha,\mu.\gamma}x_u(t,r)\end{aligned}$$

Case 2. $x(t)$ is $ii-gH$ difference,

$$\begin{aligned}&\boldsymbol{I}_{\boldsymbol{AB}}^{\alpha,\mu.\gamma}\boldsymbol{D}_{\boldsymbol{ABC}_{gH}}^{\alpha,\mu.\gamma}x(t)=\ominus_{H}(-1)x(t)\oplus(-1)x(t_0)\\&x(t_0)=x(t)\oplus(-1)\boldsymbol{I}_{\boldsymbol{AB}}^{\alpha,\mu.\gamma}\boldsymbol{D}_{\boldsymbol{ABC}_{gH}}^{\alpha,\mu.\gamma}x(t)\\&x_u(t,r)=x_u(t_0,r)+\boldsymbol{I}_{\boldsymbol{AB}}^{\alpha,\mu.\gamma}\boldsymbol{D}_{\boldsymbol{ABC}_{gH}}^{\alpha,\mu.\gamma}x_l(t,r)\\&x_l(t,r)=x_l(t_0,r)+\boldsymbol{I}_{\boldsymbol{AB}}^{\alpha,\mu.\gamma}\boldsymbol{D}_{\boldsymbol{ABC}_{gH}}^{\alpha,\mu.\gamma}x_u(t,r)\end{aligned}$$

It is seen that in case 2 we have a system of fractional differential equations and analytical solving of such system is so difficult and we should find the numerical solutions. The numerical method for this problem is explained in Chap. 5.

Chapter 4
Fuzzy Fractional Differential Equations

4.1 Introduction

Different materials and processes in many applied sciences like electrical circuits, biology, biomechanics, electrochemistry, electromagnetic processes and, others are widely recognized to be well predicted by using fractional differential operators in accordance with their memory and hereditary properties. For the complex phenomena, the modeling and their results in diverse widespread fields of science and engineering, are also so complicated and for achieving the accurate method the only powerful tool is fractional calculus. Indeed, the fractional calculus is not only a very important and productive topic, it also represents a new point of view that how to construct and apply a certain type of non-local operators to real-world problems.

In general, the majority of the fuzzy fractional differential equations as same as fuzzy differential equations do not have exact solutions. This is why, approximate and numerical procedures are important to be developed. On the other hand, the complicity of many parameters in mathematical modeling of natural phenomena appears as an uncertain fractional model and they play an important role in various disciplines. Hence, it motivates the researchers to investigate effective numerical methods with error analysis to approximate the fuzzy differential equations.

The objectives of this chapter are to consider fractional operators such as differentials and integral operators on fuzzy number valued functions. Then the fuzzy fractional differential equations are introduced and solved and analyzed by some theoretical methods.

T. Allahviranloo, *Fuzzy Fractional Differential Operators and Equations*, Studies in Fuzziness and Soft Computing 397, https://doi.org/10.1007/978-3-030-51272-9_4

4.2 Fuzzy Fractional Differential Equations—Caputo-Katugampola Derivative

In this section, we are going to cover the topic with different types of fuzzy fractional operators. The fuzzy fractional differential equations can be defined by each operator which are defined before. In each type of differentiability, the existence and uniqueness of the solutions are discussed, and some analytical methods are introduced to obtain the solutions. One of the popular methods that is going to be explained is Laplace transforms. Thus, these transforms are introduced and discussed with details in this section as well.

Definition—Fuzzy fractional differential equations
Consider the following fuzzy fractional differential equation with fuzzy initial value.

$$\boldsymbol{D}^{\alpha,p}_{\boldsymbol{CK}_{gH}}x(t)=f(t,x(t)),\quad x(t_0)=x_0,\quad t\in[t_0,T],\quad 0<\alpha<1.$$

where $f:[t_0,T]\times\mathbb{F}_R\rightarrow\mathbb{F}_R$ is a fuzzy number valued function and $x(t)$ is a continuous fuzzy set valued solution and $\boldsymbol{D}^{\alpha,p}_{\boldsymbol{CK}_{gH}}$ is the fractional Caputo- Katugampola fractional operator. The first discussion is about the relation between the solution of fractional differential equations and solution of its corresponding fuzzy fractional differential equation. It means the fuzzy solution is the common solution of two fractional operators, differential and integral. In the following remark, we suppose that, the fuzzy solution is $i-gH$ differentiable or $ii-gH$ differentiable on $(t_0,T]$.

Remark Considering the mentioned above assumptions, the continuous fuzzy function $x(t)$ is the solution of

$$\boldsymbol{D}^{\alpha,p}_{\boldsymbol{CK}_{gH}}x(t)=f(t,x(t)),\quad x(t_0)=x_0,\quad t\in[t_0,T],\quad 0<\alpha<1$$

If and only if, $x(t)$ satisfies the following integral equation,

$$x(t)\ominus_{gH}x(t_0)=\frac{p^{1-\alpha}}{\Gamma(\alpha)}\int\limits_{t_0}^{t}s^{p-1}(t^p-s^p)^{\alpha-1}f(s,x(s))ds,\quad t\in[t_0,T]$$

Before to show the assertion, it is mentioned that, the function $x(t)$ can be $i-gH$ differentiable (increasing length) or $ii-gH$ differentiable (decreasing length) on $(t_0,T]$ but the fractional integral operator $\boldsymbol{I}^{\alpha,p}_{\boldsymbol{RL}}$ always does have increasing length. To prove the remark, first suppose that $x(t)$ is the solution of differential equation, and moreover, suppose $x(t)\ominus_{gH}x(t_0):=y(t)$. The length of $y(t)$ is increasing and it is $i-gH$ differentiable, because if $x(t)$ is $i-gH$ differentiable, then

$$t_0 < t \Rightarrow x_u(t_0, r) - x_l(t_0, r) < x_u(t, r) - x_l(t, r)$$

Then

$$y_l(t, r) = x_l(t, r) - x_l(t_0, r) < x_u(t, r) - x_u(t_0, r) = y_u(t, r)$$

It means $y(t, r)$ is an interval for any $r \in [0, 1]$ and concludes it satisfies the type (1) of gH-difference. As it is shown before,

$$\boldsymbol{I}_{\boldsymbol{RL}}^{\alpha,p} \boldsymbol{D}_{\boldsymbol{CK}_{gH}}^{\alpha,p} x(t) = x(t) \ominus_{gH} x(t_0)$$

And $\boldsymbol{D}_{\boldsymbol{CK}_{gH}}^{\alpha,p} x(t) = f(t, x(t))$ then the following relation is concluded,

$$\boldsymbol{I}_{\boldsymbol{RL}}^{\alpha,p} f(t, x(t)) = x(t) \ominus_{gH} x(t_0)$$

On the other hand, based on the definition of fractional integral

$$\boldsymbol{I}_{\boldsymbol{RL}}^{\alpha,p} x(t) = \frac{p^{1-\alpha}}{\Gamma(\alpha)} \odot \int_{t_0}^{t} s^{p-1} (t^p - s^p)^{\alpha-1} \odot x(s) ds$$

Then

$$\boldsymbol{I}_{\boldsymbol{RL}}^{\alpha,p} f(t, x(t)) = \frac{p^{1-\alpha}}{\Gamma(\alpha)} \odot \int_{t_0}^{t} s^{p-1} (t^p - s^p)^{\alpha-1} \odot f(s, x(s)) ds$$

So, the necessary condition is now proved. Next, the sufficient condition is going to be proved. Suppose we have,

$$x(t) \ominus_{gH} x(t_0) = \frac{p^{1-\alpha}}{\Gamma(\alpha)} \int_{t_0}^{t} s^{p-1} (t^p - s^p)^{\alpha-1} \odot f(s, x(s)) ds$$

Effecting the fractional derivative $\boldsymbol{D}_{\boldsymbol{RL}_{gH}}^{\alpha,p}$ to both sides,

$$\boldsymbol{D}_{\boldsymbol{RL}_{gH}}^{\alpha,p} \left(x(t) \ominus_{gH} x(t_0) \right) = \boldsymbol{D}_{\boldsymbol{RL}_{gH}}^{\alpha,p} \frac{p^{1-\alpha}}{\Gamma(\alpha)} \int_{t_0}^{t} s^{p-1} (t^p - s^p)^{\alpha-1} \odot f(s, x(s)) ds$$

thus

$$\boldsymbol{D}_{\boldsymbol{RL}_{gH}}^{\alpha,p}\left(x(t)\ominus_{gH}x(t_0)\right) = \boldsymbol{D}_{\boldsymbol{RL}_{gH}}^{\alpha,p}\boldsymbol{I}_{\boldsymbol{RL}}^{\alpha,p}f(t,x(t)) = f(t,x(t))$$

Therefor,

$$\boldsymbol{D}_{\boldsymbol{RL}_{gH}}^{\alpha,p}\left(x(t)\ominus_{gH}x(t_0)\right) = f(t,x(t))$$

Since we had,

$$\boldsymbol{D}_{\boldsymbol{CK}_{gH}}^{\alpha,p}x(t) = \boldsymbol{D}_{\boldsymbol{RL}_{gH}}^{\alpha,p}\left(x(t)\ominus_{gH}x(t_0)\right)$$

Thus,

$$\boldsymbol{D}_{\boldsymbol{CK}_{gH}}^{\alpha,p}x(t) = f(t,x(t))$$

And the proof is completed.

Note Based on the definition of the gH-difference in the integral equation,

$$x(t)\ominus_{gH}x(t_0) = \frac{p^{1-\alpha}}{\Gamma(\alpha)}\int_{t_0}^{t} s^{p-1}(t^p - s^p)^{\alpha-1}\odot f(s,x(s))ds, \quad t\in[t_0,T]$$

- In case the gH-difference is $i - gH$ difference,

$$x(t) = x(t_0)\oplus\frac{p^{1-\alpha}}{\Gamma(\alpha)}\int_{t_0}^{t} s^{p-1}(t^p - s^p)^{\alpha-1}\odot f(s,x(s))ds, \quad t\in[t_0,T]$$

- In case the gH-difference is $ii - gH$ difference,

$$x(t) = x(t_0)\ominus_H(-1)\frac{p^{1-\alpha}}{\Gamma(\alpha)}\int_{t_0}^{t} s^{p-1}(t^p - s^p)^{\alpha-1}\odot f(s,x(s))ds, \quad t\in[t_0,T]$$

Example Consider the following fuzzy initial value problem,

$$\boldsymbol{D}_{\boldsymbol{CK}_{gH}}^{\frac{1}{2},1}x(t) = \left(\sqrt{t},\frac{1}{\sqrt{t}},\frac{2}{\sqrt{t}}\right) = f(t), \quad x(0) = (-2,0,1), \quad t\in(0,1],$$

Now using the fractional derivative operator in both sides,

$$\boldsymbol{I}_{\boldsymbol{RL}}^{\frac{1}{2},p}\boldsymbol{D}_{\boldsymbol{CK}_{gH}}^{\frac{1}{2},p}x(t) = x(t)\ominus_{gH}x(t_0)$$

with

$$x(t)\ominus_{gH}x(t_0) = \boldsymbol{I}_{\boldsymbol{RL}}^{\frac{1}{2},p}f(t) = \frac{p^{1-\alpha}}{\Gamma(\alpha)}\int_{t_0}^{t} s^{p-1}(t^p - s^p)^{\alpha-1}\odot f(s,x(s))ds, \quad t \in (0,1]$$

And $p = 1$ we have,

$$\boldsymbol{I}_{\boldsymbol{RL}}^{\frac{1}{2},1}f(t) = \frac{1}{\Gamma\left(\frac{1}{2}\right)}\odot\int_{0}^{t}(t-s)^{-\frac{1}{2}}\odot\left(\sqrt{s},\frac{1}{\sqrt{s}},\frac{2}{\sqrt{s}}\right)ds = \left(\frac{\sqrt{\pi}}{2}t,\sqrt{\pi},2\sqrt{\pi}\right)$$

then

$$x(t)\ominus_{gH}x(t_0) = \left(\frac{\sqrt{\pi}}{2}t,\sqrt{\pi},2\sqrt{\pi}\right)$$

By substituting the initial value,

$$x(t)\ominus_{gH}(-2,0,1) = \left(\frac{\sqrt{\pi}}{2}t,\sqrt{\pi},2\sqrt{\pi}\right)$$

The length of $\left(\frac{\sqrt{\pi}}{2}t,\sqrt{\pi},2\sqrt{\pi}\right)$ is $(1-r)\sqrt{\pi}\left(2-\frac{t}{2}\right), t \in (0,1], r \in [0,1]$. The length function is decreasing because,

$$\frac{d}{dt}\left[(1-r)\sqrt{\pi}\left(2-\frac{t}{2}\right)\right] < 0, \quad r \in [0,1]$$

In case $i-gH$ difference, (Fig. 4.1)

$$x(t) = (-2,0,1)\oplus\left(\frac{\sqrt{\pi}}{2}t,\sqrt{\pi},2\sqrt{\pi}\right) = \left(-2+\frac{\sqrt{\pi}}{2}t,\sqrt{\pi},1+2\sqrt{\pi}\right)$$

The level-wise form of the solution is,

$$x_l(t,r) = \sqrt{\pi} + \left(\sqrt{\pi}+2-\frac{\sqrt{\pi}}{2}t\right)(r-1), \quad x_u(t,r) = \sqrt{\pi} + \left(1+\sqrt{\pi}\right)(1-r)$$

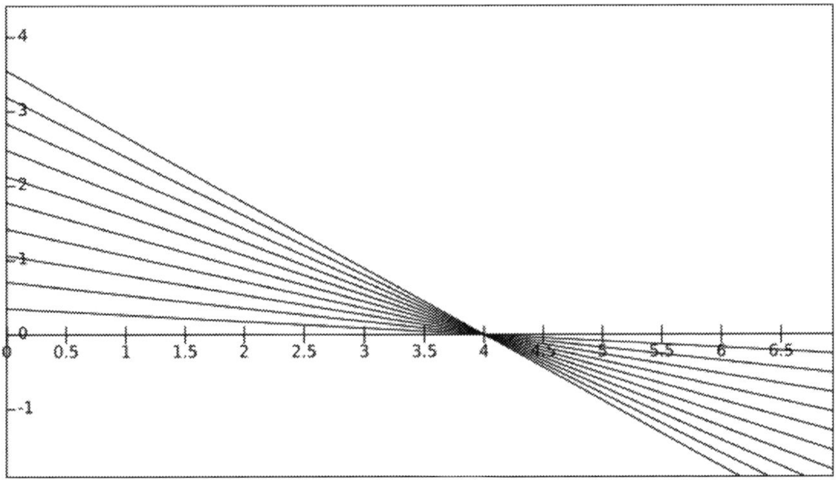

Fig. 4.1 The length of $x(t) \ominus_{gH} x(t_0)$ is decreasing

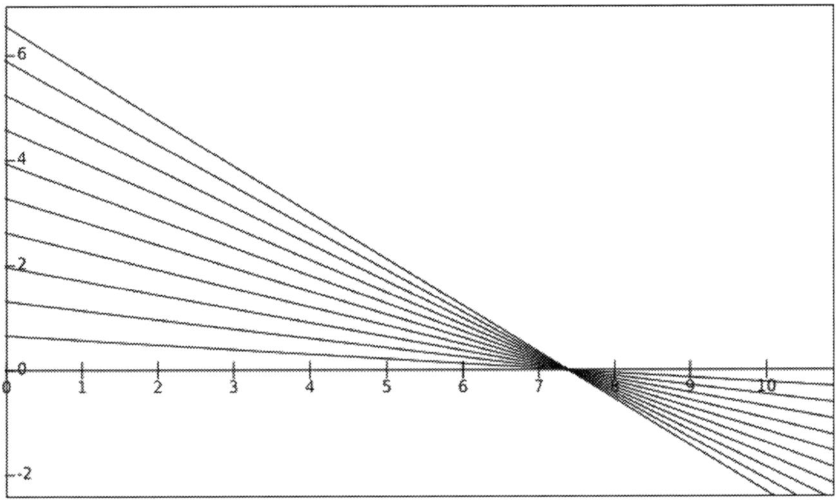

Fig. 4.2 Length of $x(t)$ is decreasing

And

$$length(x(t)) = \left(3 + 2\sqrt{\pi} - \frac{\sqrt{\pi}}{2}t\right)(1 - r), \quad \frac{d}{dt}\left[3 + 2\sqrt{\pi} - \frac{\sqrt{\pi}}{2}t\right] < 0, r \in [0, 1]$$

It is seen from the figures both $x(t)$ and $x(t) \ominus_{gH} x_0$ do have decreasing length and the solution is as, (Fig. 4.2)

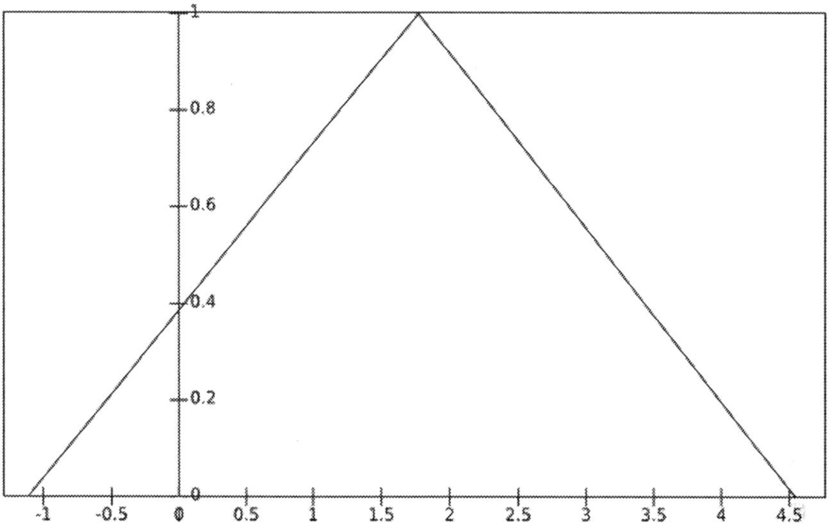

Fig. 4.3 The solution $x(t)$ at $t = 1$

$$x(t) = x_0 \oplus \boldsymbol{I}_{RL}^{\frac{1}{2},1} f(t) = \left(-2 + \frac{\sqrt{\pi}}{2}t, \sqrt{\pi}, 1 + 2\sqrt{\pi}\right)$$

The figure shows that the solution is fuzzy number at each point like $t = 1$. To check the function is the solution of the problem, (Fig. 4.3)

$$\begin{aligned} \boldsymbol{D}_{CK_{gH}}^{\alpha,p} x(t) &= \frac{1}{\Gamma\left(\frac{1}{2}\right)} \int_0^t (t-s)^{-\frac{1}{2}} \odot \left(-2 + \frac{\sqrt{\pi}}{2}s, \sqrt{\pi}, 1 + 2\sqrt{\pi}\right)' ds \\ &= \left(\frac{\frac{2}{3}\sqrt{\pi}t^{\frac{3}{2}} - 4\sqrt{t}}{\sqrt{\pi}}, 2\sqrt{t}, \frac{2(1+2\sqrt{\pi})\sqrt{t}}{\sqrt{\pi}}\right) \neq \left(\sqrt{t}, \frac{1}{\sqrt{t}}, \frac{2}{\sqrt{t}}\right) \end{aligned}$$

Despite the fuzzy function $x(t)$ is a fuzzy number valued function but it is not the solution of fuzzy fractional differential equation. The reason is the length of $x(t) \ominus_{gH} x(t_0)$ is not increasing.

Example Consider the following fuzzy initial value problem,

$$\boldsymbol{D}_{CK_{gH}}^{\frac{1}{2},1} x(t) = \sqrt{\pi}(-t^2, 0, 2t - t^2) = f(t), \quad x(0) = (-2, 0, 2), \quad t \in (0, 1],$$

Now using the fractional derivative operator in both sides,

$$\boldsymbol{I}_{\boldsymbol{RL}}^{\frac{1}{2},p}\boldsymbol{D}_{\boldsymbol{CK}_{gH}}^{\frac{1}{2},p}x(t) = x(t)\ominus_{gH}x(t_0)$$

with

$$x(t)\ominus_{gH}x(t_0) = \boldsymbol{I}_{\boldsymbol{RL}}^{\frac{1}{2},p}f(t) = \frac{p^{1-\alpha}}{\Gamma(\alpha)}\int_{t_0}^{t} s^{p-1}(t^p - s^p)^{\alpha-1}\odot f(s,x(s))ds, \quad t\in(0,1]$$

And $p = 1$ we have,

$$\boldsymbol{I}_{\boldsymbol{RL}}^{\frac{1}{2},1}f(t) = \frac{1}{\Gamma\left(\frac{1}{2}\right)}\odot\int_{0}^{t}(t-s)^{-\frac{1}{2}}\odot\sqrt{\pi}\left(-s^2,0,2s-s^2\right)ds$$
$$= \left(-\frac{16}{15}t^{\frac{5}{2}},0,\frac{8}{3}t^{\frac{3}{2}}-\frac{16}{15}t^{\frac{5}{2}}\right)$$

then

$$x(t)\ominus_{gH}x(t_0) = \left(-\frac{16}{15}t^{\frac{5}{2}},0,\frac{8}{3}t^{\frac{3}{2}}-\frac{16}{15}t^{\frac{5}{2}}\right)$$

Now, two types of the solution we should have, (Fig. 4.4)

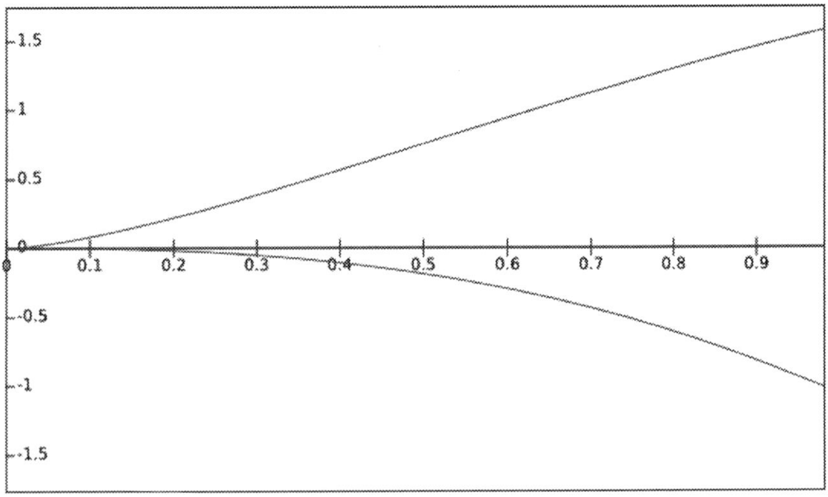

Fig. 4.4 Length of $x(t)\ominus_{gH}x(t_0)$ is increasing

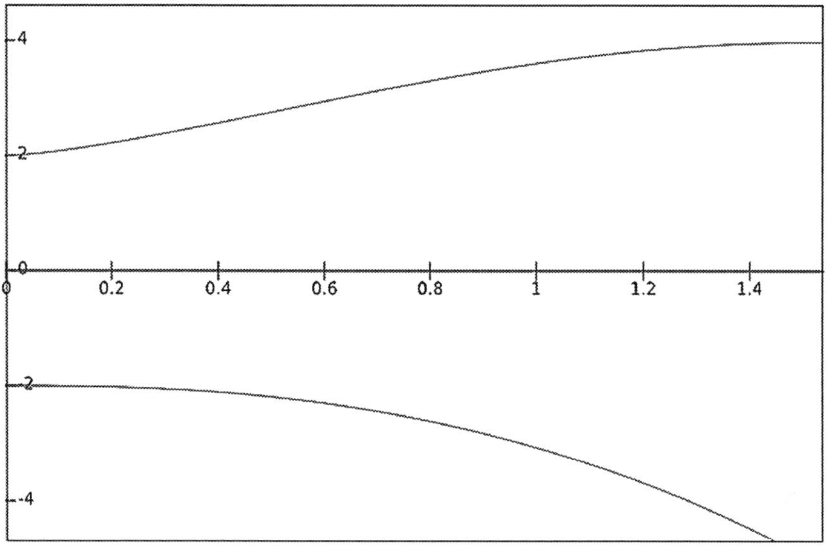

Fig. 4.5 Length of $x_{i-gH}(t)$ is increasing

- $i-gH$ solution

$$x_{i-gH}(t) = (-2,0,2) \oplus \left(-\frac{16}{15}t^{\frac{5}{2}}, 0, \frac{8}{3}t^{\frac{3}{2}} - \frac{16}{15}t^{\frac{5}{2}}\right)$$
$$= \left(-2 - \frac{16}{15}t^{\frac{5}{2}}, 0, \frac{8}{3}t^{\frac{3}{2}} - \frac{16}{15}t^{\frac{5}{2}} + 2\right)$$

- $ii-gH$ solution (Fig. 4.5)

$$x_{ii-gH}(t) = (-2,0,2) \ominus_H (-1)\left(-\frac{16}{15}t^{\frac{5}{2}}, 0, \frac{8}{3}t^{\frac{3}{2}} - \frac{16}{15}t^{\frac{5}{2}}\right)$$
$$= \left(\frac{8}{3}t^{\frac{3}{2}} - \frac{16}{15}t^{\frac{5}{2}} - 2, 0, 2 - \frac{16}{15}t^{\frac{5}{2}}\right)$$

To check the solutions, (Fig. 4.6)

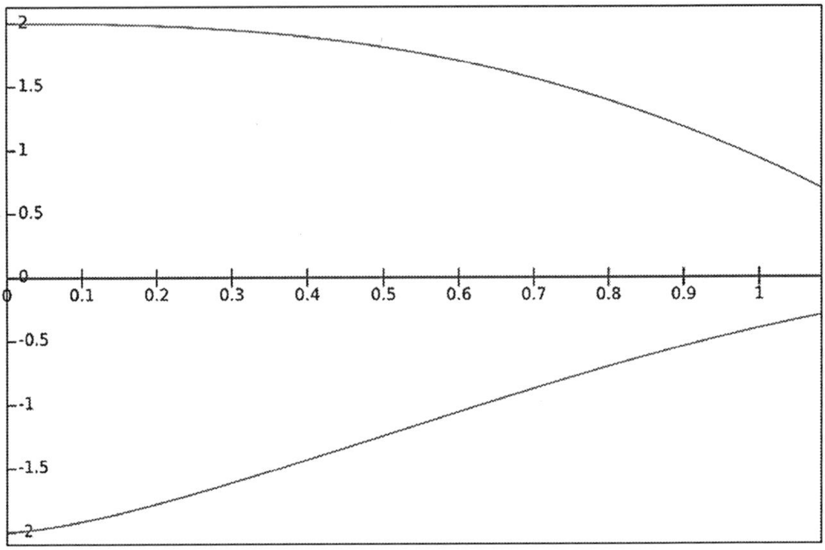

Fig. 4.6 Length of $x_{ii-gH}(t)$ is decreasing

$$\begin{aligned}
\boldsymbol{D}^{\frac{1}{2},1}_{\boldsymbol{CK}_{gH}} x_{i-gH}(t) &= \boldsymbol{D}^{\frac{1}{2},1}_{\boldsymbol{CK}_{gH}}\left(-2-\frac{16}{15}t^{\frac{5}{2}},0,\frac{8}{3}t^{\frac{3}{2}}-\frac{16}{15}t^{\frac{5}{2}}+2\right)\\
&= \left(\boldsymbol{D}^{\frac{1}{2},1}_{\boldsymbol{CK}}\left(-2-\frac{16}{15}t^{\frac{5}{2}}\right),0,\boldsymbol{D}^{\frac{1}{2},1}_{\boldsymbol{CK}}\left(\frac{8}{3}t^{\frac{3}{2}}-\frac{16}{15}t^{\frac{5}{2}}+2\right)\right)\\
&= \left(\frac{-1}{\Gamma\left(\frac{1}{2}\right)}\int_0^t \frac{8s^{\frac{3}{2}}}{3\sqrt{t-s}}ds,0,\frac{1}{\Gamma\left(\frac{1}{2}\right)}\int_0^t \frac{4s^{\frac{1}{2}}-\frac{8}{3}s^{\frac{3}{2}}}{\sqrt{t-s}}ds\right)\\
&= \sqrt{\pi}\left(-t^2,0,2t-t^2\right)
\end{aligned}$$

Then the $i-gH$ solution satisfies the fractional problem. The same process can be done for checking $ii-gH$ solution.

$$\begin{aligned}
\boldsymbol{D}^{\frac{1}{2},1}_{\boldsymbol{CK}_{gH}} x_{ii-gH}(t) &= \boldsymbol{D}^{\frac{1}{2},1}_{\boldsymbol{CK}_{gH}}\left(\frac{8}{3}t^{\frac{3}{2}}-\frac{16}{15}t^{\frac{5}{2}}-2,0,2-\frac{16}{15}t^{\frac{5}{2}}\right)\\
&= \left(\boldsymbol{D}^{\frac{1}{2},1}_{\boldsymbol{CK}}\left(-2-\frac{16}{15}t^{\frac{5}{2}}\right),0,\boldsymbol{D}^{\frac{1}{2},1}_{\boldsymbol{CK}}\left(\frac{8}{3}t^{\frac{3}{2}}-\frac{16}{15}t^{\frac{5}{2}}+2\right)\right)\\
&= \sqrt{\pi}\left(-t^2,0,2t-t^2\right)
\end{aligned}$$

Both are $i-gH$ and $ii-gH$ solutions of the problem because the length of $x(t)\ominus_{gH}x(t_0)$ is increasing.

4.2.1 Existence and Uniqueness of the Solution

In this section the existence and uniqueness results of solution to fuzzy fractional differential equations by using an idea of successive approximations under generalized Lipschitz condition of the right-hand side are investigated. Furthermore, the formula of solution to the linear fuzzy Caputo—Katugampola fractional differential equation is given. Since the real intervals in the level-wise form are used for any arbitrary level, so to reach the aims, we should consider the following theorem in the real case of Caputo fractional derivative.

Theorem—Existence and uniqueness in real fractional differential equation
Consider the initial value problem as follows,

$$\boldsymbol{D}_{CK}^{\alpha,p} u(t) = g\big(t, u(t)\big), \qquad u(t_0) = u_0 = 0\,, \qquad t \in [t_0, T]$$

Let $\eta > 0$ be a given constant and $B(u_0, \eta) = \{u \in R\,; |u - u_0| \leq \eta\}$ is a ball around u_0 with radius η. Also assume a real valued function $g : [t_0, T] \times [0, \eta] \rightarrow R^+$ satisfies the following conditions,

1. $g \in C([t_0, T] \times [0, \eta], R^+)$, $g(t, 0) \equiv 0$, $\exists M_g \geq 0$, $0 \leq g(t, u) \leq M_g$ for all $(t, u) \in [t_0, T] \times [0, \eta]$.
2. $g(t, u)$ is nondecreasing function with respect to x for any $t \in [t_0, T]$.

Then the mentioned fractional problem has at least one solution on $[t_0, T]$ and $u(t) \in B(u_0, \eta)$.

Proof The solution of the mentioned above fractional differential equation is the solution of following fractional integral equation.

$$u(t) = \frac{p^{1-\alpha}}{\Gamma(\alpha)} \int_{t_0}^{t} s^{p-1}(t^p - s^p)^{\alpha-1} g\big(s, u(s)\big) ds$$

The following successive method for approximation of the solution of fractional differential equation is defined,

$$u_{n+1}(t) = \frac{p^{1-\alpha}}{\Gamma(\alpha)} \int_{t_0}^{t} s^{p-1}(t^p - s^p)^{\alpha-1} g\big(s, u_n(s)\big) ds\,, \qquad t \in [t^*, T^*]$$

Such that

$$u_0(t) = \frac{M_g (t^p - a^p)^\alpha}{p^\alpha \Gamma(\alpha + 1)}\,, \quad t_0 < t \leq \left[\left(\frac{\eta \Gamma(\alpha + 1)}{M_g} p^\alpha \right)^{\frac{1}{\alpha}} + a^p \right]^{\frac{1}{p}}, \quad T^* = \min\{t^*, T\}$$

For $n = 0$ and $t \in [t_0, T^*]$ we have,

$$u_1(t) = \frac{p^{1-\alpha}}{\Gamma(\alpha)} \int_{t_0}^{t} s^{p-1}(t^p - s^p)^{\alpha-1} g(s, u_0(s)) ds \leq \frac{M_g(t^p - a^p)^{\alpha}}{p^{\alpha}\Gamma(\alpha+1)} = u_0(t)$$

So, we found,

$$u_1(t) \leq u_0(t)$$

For $n = 1$ and $t \in [t_0, T^*]$,

$$u_2(t) = \frac{p^{1-\alpha}}{\Gamma(\alpha)} \int_{t_0}^{t} s^{p-1}(t^p - s^p)^{\alpha-1} g(s, u_1(s)) ds \leq \frac{M_g(t^p - a^p)^{\alpha}}{p^{\alpha}\Gamma(\alpha+1)} = u_0(t)$$

Since g is a nondecreasing function in u and $u_1(t) \leq u_0(t)$, then $g(s, u_1(s)) \leq g(s, u_0(s))$ thus $u_2(t) \leq u_1(t)$. Now we have

$$u_2(t) \leq u_1(t) \leq u_0(t)$$

Proceeding recursively, we will have

$$u_{n+1}(t) \leq u_n(t) \leq \cdots \leq u_1(t) \leq u_0(t) \leq \eta, \qquad t \in [t_0, T^*]$$

It follows that the sequence $\{u_n(t)\}_n$ is uniformly bounded for $n \geq 0$. On the other hand,

$$\left|\boldsymbol{D}_{CK}^{\alpha,p} u_n(t)\right| = |g(t, u_n(t))| = g(t, u_n(t)) \leq M_g$$

Then in the interval $[t_0, T^*]$, we can use the mean value theorem for $t_1, t_2 \in [t_0, T^*], n \geq 0$,

$$|u_n(t_2) - u_n(t_1)| = \frac{2M_g\left(t_2^p - t_1^p\right)^{\alpha}}{p^{\alpha}\Gamma(\alpha+1)} \leq \frac{2M_g(t_2 - t_1)^{\alpha}}{p^{\alpha}\Gamma(\alpha+1)} \tau^{\alpha(p-1)}, \tau \in [t_1, t_2] \subseteq [t_0, T^*]$$

Therefor the sequence $\{u_n(t)\}_n$ is equi-continuous and then

$$\forall \epsilon > 0, \exists \delta = \left(\frac{\epsilon p^{\alpha}\Gamma(\alpha+1)\tau^{\alpha(p-1)}}{2M_g}\right)^{\frac{1}{\alpha}} > 0, (|t_2 - t_1| < \delta \Rightarrow |u_n(t_2) - u_n(t_1)| < \epsilon)$$

Hence, by the Arzela–Ascoli Theorem and the monotonicity of the sequence $\{u_n(t)\}_n$ we conclude the convergency of the sequence and $\lim_{n \to \infty} u_n(t) = u(t), t \in [t_0, T^*]$.

Theorem—Existence and uniqueness in fuzzy fractional differential equation
Consider the fuzzy initial value problem as follows,

$$D^{\alpha,p}_{CK_{gH}} x(t) = f(t, x(t)), \quad x(t_0) = x_0, \quad t \in [t_0, T], \quad 0 < \alpha < 1$$

Let $\eta > 0$ be a given constant and $B(x_0, \eta) = \{x \in R; |x - x_0| \leq \eta\}$ is a ball around x_0 with radius η. Also assume a fuzzy number valued function $f : [t_0, T] \times [0, \eta] \to \mathbb{F}_R$ satisfies the following conditions,

i. $f \in C([t_0, T] \times [0, \eta], \mathbb{F}_R)$, $\exists M_g \geq 0$, $D_H(g(t, x), 0) \leq M_g$, for all $(t, x) \in [t_0, T] \times B(x_0, \eta)$.
ii. For any $z, w \in B(x_0, \eta), D_H(f(t, z), f(t, w)) \leq g(t, D_H(z, w))$ where $g \in C([t_0, T] \times [0, \eta], R^+)$ in the problem

$$D^{\alpha,p}_{CK} x(t) = g(t, x(t)), \quad x(t_0) = x_0 = 0, \quad t \in [t_0, T]$$

has only solution $x(t) \equiv 0$ on $[t_0, T]$ (the previous theorem of existence and uniqueness in real case).

Then the following successive approximations given by $x_0(t) = x_0$

$$x_n(t) \ominus_{gH} x_0 = \frac{p^{1-\alpha}}{\Gamma(\alpha)} \int_{t_0}^{t} s^{p-1} (t^p - s^p)^{\alpha-1} \odot f(s, x_{n-1}(s)) ds, n = 1, 2, \ldots,$$

Converges to a unique solution of

$$D^{\alpha,p}_{CK_{gH}} x(t) = f(t, x(t)), \quad x(t_0) = x_0, \quad 0 < \alpha < 1$$

on $t \in [t_0, T^*], T^* \in (t_0, T]$, provided that $x_n(t) \ominus_{gH} x_0$ does have increasing length.

Proof Let us consider the point t^*,

$$t_0 < t^* \leq \left[\left(\frac{\eta \Gamma(\alpha + 1)}{M} p^\alpha \right)^{\frac{1}{\alpha}} + a^p \right]^{\frac{1}{p}}, \quad M = \max\{M_f, M_g\}, \quad T^* = \min\{t^*, T\}$$

Now consider the sequence of fuzzy continuous functions $\{x_n(t)\}_n, x_0(t) = x_0, t \in [t_0, T^*]$.

$$x_n(t) \ominus_{gH} x_0 = \frac{p^{1-\alpha}}{\Gamma(\alpha)} \int_{t_0}^{t} s^{p-1} (t^p - s^p)^{\alpha-1} \odot f(s, x_{n-1}(s)) ds, n = 1, 2, \ldots,$$

First, we prove the $x_n(t) \in C([t_0, T^*], B(x_0, \eta))$. To this end, assume $t_1, t_2 \in [t_0, T^*]$ and $t_1 < t_2$.

$$\begin{aligned} &D_H\left(x_n(t_1)\ominus_{gH}x_0, x_n(t_2)\ominus_{gH}x_0\right) \\ &\leq \frac{p^{1-\alpha}}{\Gamma(\alpha)}\int_{t_0}^{t_1} s^{p-1}\left[\left(t_1^p - s^p\right)^{\alpha-1} - \left(t_2^p - s^p\right)^{\alpha-1}\right] D_H(f(s, x_n(s)), 0)ds \\ &+ \frac{p^{1-\alpha}}{\Gamma(\alpha)}\int_{t_1}^{t_2} s^{p-1}\left[\left(t_2^p - s^p\right)^{\alpha-1}\right] D_H(f(s, x_n(s)), 0)ds \end{aligned}$$

And

$$\frac{p^{1-\alpha}}{\Gamma(\alpha)}\int_{t_1}^{t_2} s^{p-1}\left[\left(t_2^p - s^p\right)^{\alpha-1}\right] ds = \frac{p^{-\alpha}}{\Gamma(\alpha+1)}\left(t_2^p - t_1^p\right)^{\alpha}$$

$$\begin{aligned} &\frac{p^{1-\alpha}}{\Gamma(\alpha)}\int_{t_0}^{t_1} s^{p-1}\left[\left(t_1^p - s^p\right)^{\alpha-1} - \left(t_2^p - s^p\right)^{\alpha-1}\right] ds \\ &= \frac{p^{-\alpha}}{\Gamma(\alpha+1)}\left[\left(t_1^p - t_0^p\right)^{\alpha} - \left(t_2^p - t_1^p\right)^{\alpha}\right] \end{aligned}$$

And

$$D_H(f(s, x_n(s)), 0) \leq M_f, D_H\left(x_n(t_1)\ominus_{gH}x_0, x_n(t_2)\ominus_{gH}x_0\right) = D_H(x_n(t_1), x_n(t_2))$$

Hence,

$$D_H(x_n(t_1), x_n(t_2)) \leq \frac{p^{-\alpha}M_f}{\Gamma(\alpha+1)}\left[\left(t_2^p - t_1^p\right)^{\alpha} + \left(t_1^p - t_0^p\right)^{\alpha} + \left(t_2^p - t_1^p\right)^{\alpha}\right]$$

Finally,

$$D_H(x_n(t_1), x_n(t_2)) \leq \frac{2p^{-\alpha}M_f}{\Gamma(\alpha+1)}\left(t_2^p - t_1^p\right)^{\alpha} \leq \eta$$

This means, if $t_2 \to t_1$ then $D_H(x_n(t_1), x_n(t_2)) \to 0$ and follows the function $x_n(t)$ is continuous on $[t_0, T^*]$. In addition, it follows for $n \geq 1, t \in [t_0, T^*]$,

$$x_n(t) \in B(x_0, \eta), x_n(t) \ominus_{gH} x_0 \in B(x_0, \eta)$$

Now, if $x_{n-1}(t) \in B(x_0, \eta)$,

$$D_H\left(x_n(t)\ominus_{gH}x_0, 0\right) \leq \frac{p^{1-\alpha}}{\Gamma(\alpha)}\int_{t_0}^{t} s^{p-1}\left(t^p - s^p\right)^{\alpha-1} D_H(f(s, x_{n-1}(s)), 0)ds$$
$$\leq \frac{p^{-\alpha}M_f}{\Gamma(\alpha+1)}\left(t^p - t_0^p\right)^{\alpha} \leq \eta$$

In conclusion, the fuzzy function $x_n(t) \in B(x_0, \eta)$ for all $n \geq 1$ and all $t \in [t_0, T^*]$. Now our next step is proving the convergence, for $x_n(t), x(t) \in C([t_0, T^*], B(x_0, \eta))$,

$$\lim_{n\to 0} x_n(t) = x(t)$$

To this purpose, we need some relations,

$$\leq \frac{p^{1-\alpha}}{\Gamma(\alpha)}\int_{t_0}^{t} s^{p-1}(t^p - s^p)^{\alpha-1} D_H(f(s, x_n(s)), f(s, x_{n-1}(s))ds$$
$$\leq \frac{p^{1-\alpha}}{\Gamma(\alpha)}\int_{t_0}^{t} s^{p-1}(t^p - s^p)^{\alpha-1} D_H(g(x_n(s), x_{n-1}(s))ds \leq u_n(t) \leq \frac{M_g(t^p - a^p)^{\alpha}}{p^{\alpha}\Gamma(\alpha+1)} \leq \eta$$

Since

$$\boldsymbol{D}^{\alpha,p}_{\boldsymbol{CK}_{gH}} x(t) = x(t)\ominus_{gH}x_0$$

Thus, based on the properties of the distance,

$$D_H\left(\boldsymbol{D}^{\alpha,p}_{\boldsymbol{CK}_{gH}} x_{n+1}(t), \boldsymbol{D}^{\alpha,p}_{\boldsymbol{CK}_{gH}} x_n(t)\right) \leq D_H\left(f(t, x_n(t)), f(t, x_{n-1}(t))\right)$$
$$\leq g\left(D_H(x_n(t), x_{n-1}(t))\right) \leq g(t, u_{n-1}(t))$$

Finally,

$$D_H\left(\boldsymbol{D}^{\alpha,p}_{\boldsymbol{CK}_{gH}} x_{n+1}(t), \boldsymbol{D}^{\alpha,p}_{\boldsymbol{CK}_{gH}} x_n(t)\right) \leq g(t, u_{n-1}(t))$$

Continuing,

$$D_H\left(\boldsymbol{D}^{\alpha,p}_{\boldsymbol{CK}_{gH}} x_n(t), \boldsymbol{D}^{\alpha,p}_{\boldsymbol{CK}_{gH}} x_{n-1}(t)\right) \leq g(t, u_{n-2}(t))$$

Then

$$D_H\left(\boldsymbol{D}^{\alpha,p}_{CK_{gH}}x_{n+1}(t),\boldsymbol{D}^{\alpha,p}_{CK_{gH}}x_{n-1}(t)\right)$$
$$\leq D_H\left(\boldsymbol{D}^{\alpha,p}_{CK_{gH}}x_{n+1}(t),\boldsymbol{D}^{\alpha,p}_{CK_{gH}}x_n(t)\right)+D_H\left(\boldsymbol{D}^{\alpha,p}_{CK_{gH}}x_n(t),\boldsymbol{D}^{\alpha,p}_{CK_{gH}}x_{n-1}(t)\right)$$
$$\leq g(t,u_{n-1}(t))+g(t,u_{n-2}(t))$$

$$D_H\left(\boldsymbol{D}^{\alpha,p}_{CK_{gH}}x_n(t),\boldsymbol{D}^{\alpha,p}_{CK_{gH}}x_1(t)\right)\leq\sum_{i=0}^{n-1}g(t,u_i(t))\to 0$$

Hence,

$$D_H(x_n(t)\ominus_{gH}x_0,x_1(t)\ominus_{gH}x_0)\leq\sum_{i=0}^{n-1}g(t,u_i(t))\to 0$$

And

$$D_H(x_n(t),x_1(t))\leq\sum_{i=0}^{n-1}g(t,u_i(t))\to 0$$

In general, assume for $m\geq n$,

$$D_H(x_m(t),x_n(t))\to 0$$

Then using the definition of the Cauchy sequence the sequence $\{x_n(t)\}_n\to x(t)$.

Uniqueness. To show it, let's suppose that $y(t)$ is another solution of fuzzy fractional differential equation, and assume,

$$D_H(x(t),y(t))=k(t),\boldsymbol{D}^{\alpha,p}_{CK}D_H(x(t),y(t))=\boldsymbol{D}^{\alpha,p}_{CK}k(t)\leq D_H\left(\boldsymbol{D}^{\alpha,p}_{CK}x(t),\boldsymbol{D}^{\alpha,p}_{CK}y(t)\right)$$
$$\leq D_H\left(f(t,x(t)),f(t,y(t))\right)\leq g(t,k(t))$$

Finally, we have,

$$\boldsymbol{D}^{\alpha,p}_{CK}k(t)\leq g(t,k(t))$$

The only solution (because of $\leq$, the maximal solution) is $k(t)\equiv 0$.

Remark In conclusion, if the fuzzy function $f : [t_0, T] \times \mathbb{F}_R \to \mathbb{F}_R$ in the following fuzzy fractional initial value problem,

$$\boldsymbol{D}^{\alpha,p}_{\boldsymbol{CK}_{gH}} x(t) = f(t, x(t)), \quad x(t_0) = x_0, \quad t \in [t_0, T] 0 < \alpha < 1$$

Satisfies in the Lipchitz condition,

$$D_H(f(t, x(t)), f(t, y(t))) \le L D_H(x(t), y(t)), D_H(f(t, x(t)), 0) \le M_f$$

Then the following successive approximations converge uniformly to the unique solution of the problem on $[t_0, T]$, subject to $x_n(t) \ominus_{gH} x_0$ does have increasing length.

$$x_n(t) \ominus_{gH} x_0 = \frac{p^{1-\alpha}}{\Gamma(\alpha)} \int_{t_0}^{t} s^{p-1} (t^p - s^p)^{\alpha-1} \odot f(s, x_{n-1}(s)) ds, \quad n = 1, 2, \ldots,$$

Example Consider

$$\boldsymbol{D}^{\alpha,p}_{\boldsymbol{CK}_{gH}} x(t) = \lambda \odot x(t) \oplus h(t), \quad x(t_0) = x_0, \quad \lambda \in R, \quad t \in (t_0, T]$$

Such that $x(t), h(t) \in C((t_0, T], \mathbb{F}_R)$. Since $\boldsymbol{I}^{\alpha,p}_{\boldsymbol{RL}} \boldsymbol{D}^{\alpha,p}_{\boldsymbol{CK}_{gH}} x(t) = x(t) \ominus_{gH} x(t_0)$ then

$$x(t) \ominus_{gH} x(t_0) = \lambda \boldsymbol{I}^{\alpha,p}_{\boldsymbol{RL}} x(t) \oplus \boldsymbol{I}^{\alpha,p}_{\boldsymbol{RL}} h(t)$$

If consider the $\lambda \ge 0$ then $\lambda \boldsymbol{I}^{\alpha,p}_{\boldsymbol{RL}} x(t) \oplus \boldsymbol{I}^{\alpha,p}_{\boldsymbol{RL}} h(t)$ has increasing length because $x(t)$ and $h(t)$ are two fuzzy number valued functions. Now we can use the successive approximation method,

$$x_n(t) \ominus_{gH} x(t_0) = \lambda \boldsymbol{I}^{\alpha,p}_{\boldsymbol{RL}} x_{n-1}(t) \oplus \boldsymbol{I}^{\alpha,p}_{\boldsymbol{RL}} h(t), \quad n = 1, 2, \ldots,$$

If $n = 1$,
Let us consider $\lambda \ge 0$ and x is $i - gH$ differentiable (increasing length)

$$x_1(t) \ominus_{gH} x(t_0) = x_0 \odot \frac{\lambda \left(t^p - t_0^p\right)^\alpha}{p^\alpha \Gamma(\alpha + 1)} \oplus \boldsymbol{I}^{\alpha,p}_{\boldsymbol{RL}} h(t),$$

Let us consider $\lambda < 0$ and x is $ii - gH$ differentiable (decreasing length)

$$(-1)\left(x(t_0) \ominus_{gH} x_1(t)\right) = x_0 \odot \frac{\lambda \left(t^p - t_0^p\right)^\alpha}{p^\alpha \Gamma(\alpha + 1)} \oplus \boldsymbol{I}^{\alpha,p}_{\boldsymbol{RL}} h(t),$$

If $n = 2$,

Let us consider $\lambda \geq 0$ and x is $i - gH$ differentiable (increasing length)

$$x_2(t) \ominus_{gH} x(t_0) = x_0 \odot \left[\frac{\lambda\left(t^p - t_0^p\right)^\alpha}{p^\alpha \Gamma(\alpha+1)} + \frac{\lambda\left(t^p - t_0^p\right)^{2\alpha}}{p^{2\alpha}\Gamma(2\alpha+1)}\right] \oplus \boldsymbol{I}_{\boldsymbol{RL}}^{\alpha,p} h(t) \oplus \boldsymbol{I}_{\boldsymbol{RL}}^{2\alpha,p} h(t),$$

Let us consider $\lambda < 0$ and x is $ii - gH$ differentiable (decreasing length)

$$(-1)\left(x(t_0) \ominus_{gH} x_2(t)\right) = x_0 \odot \left[\frac{\lambda\left(t^p - t_0^p\right)^\alpha}{p^\alpha \Gamma(\alpha+1)} + \frac{\lambda\left(t^p - t_0^p\right)^{2\alpha}}{p^{2\alpha}\Gamma(2\alpha+1)}\right] \oplus \boldsymbol{I}_{\boldsymbol{RL}}^{\alpha,p} h(t) \oplus \boldsymbol{I}_{\boldsymbol{RL}}^{2\alpha,p} h(t),$$

If it is proceeding to more $n \to \infty$,

$$\begin{aligned} x_n(t) \ominus_{gH} x(t) &= x_0 \odot \sum_{i=1}^{\infty} \frac{\lambda^i\left(t^p - t_0^p\right)^{i\alpha}}{p^{i\alpha}\Gamma(i\alpha+1)} \oplus \int_{t_0}^{t} s^{p-1} \sum_{i=1}^{\infty} \frac{\lambda^{i-1}\left(t^p - s^p\right)^{i\alpha-1}}{p^{i\alpha-1}\Gamma(i\alpha)} \odot h(s)ds \\ &= x_0 \odot \sum_{i=1}^{\infty} \frac{\lambda^i\left(t^p - t_0^p\right)^{i\alpha}}{p^{i\alpha}\Gamma(i\alpha+1)} \oplus \int_{t_0}^{t} s^{p-1} \sum_{i=0}^{\infty} \frac{\lambda^{i}\left(t^p - s^p\right)^{i\alpha+(\alpha-1)}}{p^{i\alpha+(\alpha-1)}\Gamma(i\alpha+\alpha)} \odot h(s)ds \\ &= x_0 \odot \sum_{i=1}^{\infty} \frac{\lambda^i\left(t^p - t_0^p\right)^{i\alpha}}{p^{i\alpha}\Gamma(i\alpha+1)} \oplus \int_{t_0}^{t} s^{p-1} \frac{\left(t^p - s^p\right)^{\alpha-1}}{p^{\alpha-1}} \sum_{i=0}^{\infty} \frac{\lambda^{i}\left(t^p - s^p\right)^{i\alpha}}{p^{i\alpha}\Gamma(i\alpha+\alpha)} \odot h(s)ds \end{aligned}$$

Let us consider $\lambda \geq 0$ and x is $i - gH$ differentiable (increasing length),

$$x(t) = x_0 E_{\alpha,1}\left(\lambda\left(\frac{t^p - t_0^p}{p}\right)^\alpha\right) \oplus \frac{1}{p^{\alpha-1}} \int_{t_0}^{t} s^{p-1}\left(t^p - s^p\right)^{\alpha-1} E_{\alpha,\alpha}\left(\lambda\left(\frac{t^p - s^p}{p}\right)^\alpha\right) \odot h(s)ds$$

Let us consider $\lambda < 0$ and x is $ii - gH$ differentiable (decreasing length),

$$x(t) = x_0 E_{\alpha,1}\left(\lambda\left(\frac{t^p - t_0^p}{p}\right)^\alpha\right) \ominus_H (-1)\frac{1}{p^{\alpha-1}} \int_{t_0}^{t} s^{p-1}\left(t^p - s^p\right)^{\alpha-1} E_{\alpha,\alpha}\left(\lambda\left(\frac{t^p - s^p}{p}\right)^\alpha\right) \odot h(s)ds$$

Such that, the following function is called Mittag-Leffler function,

$$E_{\alpha,\beta}(t) = \sum_{i=0}^{\infty} \frac{t^i}{\Gamma(i\alpha + \beta)}, \quad \alpha > 0, \quad \beta > 0$$

4.2.2 Some Properties of Mittag-Leffler Function

The basic Mittag-Leffler function is defined as,

$$E_{\alpha}(t) = \sum_{i=0}^{\infty} \frac{t^i}{\Gamma(i\alpha + 1)}, \quad \alpha > 0$$

If $\alpha = 1$,

$$E_1(t) = \sum_{i=0}^{\infty} \frac{t^i}{\Gamma(i + 1)} = \sum_{i=1}^{\infty} \frac{t^i}{i!} = e^t$$

Hence $E_{\alpha}(t)$ is a generalization of exponential series. One generalization of $E_{\alpha}(t)$ as a two-parameter generalization of is $E_{\alpha}(t)$ is,

$$E_{\alpha,\beta}(t) = \sum_{i=0}^{\infty} \frac{t^i}{\Gamma(i\alpha + \beta)}, \quad \alpha > 0, \quad \beta > 0$$

A three-parameter generalization of $E_{\alpha}(t)$ is denoted by $E^{\gamma}_{\alpha,\beta}(t)$ and is defined as,

$$E^{\gamma}_{\alpha,\beta}(t) = \sum_{i=0}^{\infty} \frac{(\gamma)_i}{i!\Gamma(i\alpha + \beta)}, \quad \alpha > 0, \quad \beta > 0$$

where $(\gamma)_i$ is the Pochhammer symbol standing for

$$(\gamma)_i = \gamma(\gamma + 1)(\gamma + 2)\ldots(\gamma + i - 1), \quad (\gamma)_0 = 1, \quad \gamma \neq 0$$

Here, no other condition on $(\gamma)_i$, Here γ could be a negative integer also. In that case the series is going to terminate into a polynomial. But, if $(\gamma)_i$ is to be written in terms of a gamma function as

$$(\gamma)_i = \frac{\Gamma(\gamma + i)}{\Gamma(\gamma)}, \quad \gamma > 0$$

If more parameters are to be incorporated, then we can consider

$$E_{\alpha,\beta,\delta_1,\ldots,\delta_q}^{\gamma_1,\gamma_2,\ldots,\gamma_p}(x) = \sum_{i=0}^{\infty} \frac{(\gamma_1)_i \cdots (\gamma_p)_i}{i!(\delta_1)_i \cdots (\delta_q)_i} \frac{x^i}{\Gamma(i\alpha+\beta)}, \quad \alpha > 0, \quad \beta > 0$$

where

$$\delta_j \neq 0, -1, -2, \ldots, \quad j = 1, 2, \ldots, q.$$

No other restrictions on $\gamma_1, \ldots, \gamma_p$ and $\delta_1, \ldots, \delta_q$ are there other than the conditions for the convergence of the series. A δ_j can be a negative integer provided there is a γ_r, a negative integer such that $(\gamma_r)_k = 0$ first before $(\delta_r)_k = 0$, such as $\gamma_2 = -3$ and $\delta_1 = -5$ so that $(\gamma_2)_4 = 0$ and $(\delta_1)_4 \neq 0$.

Example Consider the following fuzzy fractional initial value problem with

$$p = 1, \quad \lambda = 1, \quad \alpha = \frac{1}{2}, \quad h(t) = 0, \quad t \in (0, 1]$$
$$\boldsymbol{D}_{\boldsymbol{CK}_{gH}}^{\frac{1}{2},1} x(t) = x(t), \quad x(t_0, r) = x_0(r) = (2, 2 - r)$$

Such that $x(t) \in C((t_0, T], \mathbb{F}_R)$. Since $\boldsymbol{I}_{\boldsymbol{RL}}^{\frac{1}{2},1} \boldsymbol{D}_{\boldsymbol{CK}_{gH}}^{\frac{1}{2},1} x(t) = x(t) \ominus_{gH} x_0 = \boldsymbol{I}_{\boldsymbol{RL}}^{\alpha,\frac{1}{2}} x(t)$ and it has increasing length then we can use the successive approximation method,

$$x_n(t) \ominus_{gH} x_0 = \boldsymbol{I}_{\boldsymbol{RL}}^{\alpha,\frac{1}{2}} x_{n-1}(t), \quad n = 1, 2, \ldots,$$

The $i - gH$ differentiable solution (with increasing length) is as,

$$x(t) = x_0 \odot E_{\frac{1}{2},1}\left(t^{\frac{1}{2}}\right) = x_0 \odot \left(1 + \frac{2}{\sqrt{\pi}} \int_0^t e^{-s^2} ds\right) e^{t^2}$$

Example Consider the following fuzzy fractional initial value problem with

$$p = 1, \quad \lambda \in \{-1, 1\}, \quad \alpha = \frac{1}{2}, \quad h(t) = c \odot t^2, \quad t \in (0, 1]$$
$$\boldsymbol{D}_{\boldsymbol{CK}_{gH}}^{\frac{1}{2},1} x(t) = \lambda \odot x(t) \oplus h(t), \quad c, x_0 \in \mathbb{F}_R$$

Let us consider $\lambda = 1$ and x is $i - gH$ differentiable,

$$x_{i-gH}(t) = x_0 \odot E_{\alpha,1}\left(t^{\frac{1}{2}}\right) \oplus \int_0^t (t-s)^{-\frac{1}{2}} E_{\frac{1}{2},\frac{1}{2}}\left((t-s)^{\frac{1}{2}}\right) \odot c \odot s^2 ds$$

where

$$E_{\frac{1}{2},1}\left(t^{\frac{1}{2}}\right) = \sum_{i=0}^{\infty} \frac{t^{\frac{i}{2}}}{\Gamma\left(\frac{i}{2}+1\right)},$$

$$E_{\frac{1}{2},\frac{1}{2}}\left((t-s)^{\frac{1}{2}}\right) = \sum_{i=0}^{\infty} \frac{(t-s)^{\frac{i}{2}}}{\Gamma\left(\frac{i}{2}+\frac{1}{2}\right)} = \sqrt{t}\left(1+\frac{2}{\sqrt{\pi}}\int_0^t e^{-s^2}ds\right) + \frac{1}{\Gamma\left(\frac{1}{2}\right)}$$

The solution is obtained as,

$$x_{i-gH}(t) = x_0 \odot \left(1+\frac{2}{\sqrt{\pi}}\int_0^t e^{-s^2}ds\right)e^{t^2} \oplus \int_0^t (t-s)^{-\frac{1}{2}} E_{\frac{1}{2},\frac{1}{2}}\left((t-s)^{\frac{1}{2}}\right) \odot c \odot s^2 ds$$

Now let us consider $\lambda = -1$ and x is $ii-gH$ differentiable,

$$x_{ii-gH}(t) = x_0 \odot \left(1+\frac{2}{\sqrt{\pi}}\int_0^t e^{-s^2}ds\right)e^{t^2} \ominus_H (-1)\int_0^t (t-s)^{-\frac{1}{2}} E_{\frac{1}{2},\frac{1}{2}}\left((t-s)^{\frac{1}{2}}\right) \odot c \odot s^2 ds$$

4.3 Fuzzy Fractional Differential Equations—Laplace Transforms

In this section we suppose that the Laplace operator acts on a fuzzy number valued function and this is the reason we call it fuzzy Laplace transform.

As like as before, let us consider the function x is a fuzzy number valued function and s is a positive real parameter. The fuzzy Laplace transform is defined as follows:

$$X(s) = \mathbf{L}(x(t)) = \int_0^{\infty} e^{-st} \odot x(t)dt, \quad s > 0$$

Or

$$X(s) = \mathbf{L}(x(t)) = \lim_{\tau\to\infty} \int_0^{\tau} e^{-st} \odot x(t)dt, \quad s > 0$$

If we consider the Laplace operator in the level-wise form of $x(t)$ as

$$\boldsymbol{L}(x(t,r)) = X(s,r) = [L(x_l(t,r)), L(x_u(t,r))] = [X_l(s,r), X_u(s,r)]$$

So, the level-wise form of the Laplace operator is as follows,

$$X(s,r) = \lim_{\tau\to\infty} \int_0^\tau e^{-st} \odot x(t,r)dt$$

And

$$[X_l(s,r), X_u(s,r)] = \left[\lim_{\tau\to\infty} \int_0^\tau e^{-st} x_l(t,r)dt, \lim_{\tau\to\infty} \int_0^\tau e^{-st} x_u(t,r)dt\right]$$

Then

$$X_l(s,r) = \lim_{\tau\to\infty} \int_0^\tau e^{-st} x_l(t,r)dt$$

$$X_u(s,r) = \lim_{\tau\to\infty} \int_0^\tau e^{-st} x_u(t,r)dt$$

To define this operator the important condition is, the integral must converge to a real number. It means it should be bounded. However, there are some integrals that are not convergence.

Example Suppose the fuzzy number valued function $x(t) = c \odot e^{t^2}, c \in \mathbb{F}_R$. Then

$$X_l(s,r) = \lim_{\tau\to\infty} \int_0^\tau e^{-st} c_l(r) e^{t^2} dt \to 0$$

$$X_u(s,r) = \lim_{\tau\to\infty} \int_0^\tau e^{-st} c_u(r) e^{t^2} dt \to 0$$

The integral grows without bound for any s as $\tau \to \infty$.

If you remember we defined the absolute value of fuzzy number. Now in the same way we can define the absolute value of a fuzzy number valued function.

The absolute value of the same function in level-wise form is defined as,

$$|x(t,r)| = [\min\{|x_l(t,r)|, |x_u(t,r)|\}, \max\{|x_l(t,r)|, |x_u(t,r)|\}]$$

It can be defined in two cases,

Type I. Type 1 absolute value fuzzy number function

$$|x(t,r)| = [|x_l(t,r)|, |x_u(t,r)|]$$

In another word, if $x_l(t,r) \geq 0$ for all r then x is type 1 absolute value fuzzy number function.

Type II. Type 2 absolute value function

$$|x(t,r)| = [|x_u(t,r)|, |x_l(t,r)|]$$

In another word, if $x_u(t,r) < 0$ for all r then x is type 2 absolute value fuzzy number function. Moreover, the other conditions of a fuzzy number in level-wise form should be satisfied.

Note The absolute value of a fuzzy number function is always a positive fuzzy number valued function.

Example Consider the fuzzy number function $x(t,r) = c[r]e^t$ in the level-wise form where $c[r] = [2+r, 4-r]$. As we know

$$|x_l(t,r)| = |(2+r)e^t| = (2+r)e^t, \quad |x_u(t,r)| = |(4-r)e^t| = (4-r)e^t$$

And

$$|x(t,r)| = [(2+r)e^t, (4-r)e^t]$$

Then then x is type 1 absolute value fuzzy number function.

Example Consider the fuzzy number function $x(t,r) = c[r]e^t$ in the level-wise form where $c[r] = [-4+r, -2-r]$. As we know

$$|x_l(t,r)| = |(-4+r)e^t| = (4-r)e^t, |f_u(t,r)| = |(-2-r)e^t| = (r+2)e^t$$

And

$$|x(t,r)| = [(r+2)e^t, (4-r)e^t]$$

Then then x is type 2 absolute value fuzzy number function.

4.3.1 Definition—Absolutely Convergence

The integral operator in the Laplace transformation,

$$X(s) = \lim_{\tau\to\infty} \int_0^{\tau} e^{-st} \odot x(t)dt$$

Is said to be absolutely convergent if,

$$\lim_{\tau\to\infty} \int_0^{\tau} |e^{-st} \odot x(t)|dt$$

Exists. Considering the level-wise form, both of the following integrals exist.

$$\lim_{\tau\to\infty} \int_0^{\tau} e^{-st}|x_l(t,r)|dt, \lim_{\tau\to\infty} \int_0^{\tau} e^{-st}|x_u(t,r)|dt$$

Theoretically, in order to apply the fuzzy Laplace transform to physical problems, it is necessary to involve the inverse transform. If $X(s) = \boldsymbol{L}(x(t))$ is the Laplace transform the $\boldsymbol{L}^{-1}$ is called as inverse Laplace transform and we have

$$\boldsymbol{L}^{-1}(X(s)) = x(t), \quad t \geq 0$$

where

$$\boldsymbol{L}^{-1}(X(s)) = \frac{1}{2\pi i} \int_{\gamma - i\infty}^{\gamma + i\infty} e^{st} \odot X(s)ds, \quad \gamma \in R$$

As same as the Laplace transform the inverse transform is also a linear transform operator.

4.3.2 Definition—Exponential Order

A fuzzy function $x(t)$ is said to be of exponential order $a > 0$ on $0 \leq t < \infty$ if there exist positive constants K and T such that for all $t > T, D(x(t), 0) \leq Ke^{at}$.

Remark Consider the function $x(t)$ is a fuzzy continuous or peace-wise continuous function on any finite interval, (t_0, T), and of exponential order e^{at},

$$D(X(s),0) = D\left(\int_0^\infty e^{-st} \odot x(t)dt, 0\right) \leq \int_0^\infty e^{-st} \odot D(x(t),0)dt.$$

$$\leq \int_0^\infty e^{-st} \odot Ke^{at}dt = K\int_0^\infty e^{(a-s)t}dt = \frac{K}{s-a}$$

then the fuzzy Laplace transform exists for all $s > a$.

4.3.3 Some Properties of Laplace

1. For two fuzzy functions x, y subject to

$$\begin{aligned} \mathbf{L}(a \odot x(t) \oplus b \odot y(t)) &= \int_0^\infty e^{-st} \odot (a \odot x(t) \oplus b \odot y(t))dt \\ &= a \odot \int_0^\infty e^{-st} \odot x(t)dt \oplus b \odot \int_0^\infty e^{-st} \odot y(t)dt \\ &= a \odot \mathbf{L}(x(t)) \oplus b \odot \mathbf{L}(y(t)) \end{aligned}$$

In level-wise form,

- If $a, b \geq 0$
  Consider $a \odot x(t) \oplus b \odot y(t) = h(t)$

$$h(t,r) = [a \cdot L(x_l(t,r)) + b \cdot L(y_l(t,r)), a \cdot L(x_u(t,r)) + b \cdot L(y_u(t,r))]$$

- $$a, b \leq 0$$

$$h(t,r) = [a \cdot L(x_u(t,r)) + b \cdot L(y_u(t,r)), a \cdot L(x_l(t,r)) + b \cdot L(y_l(t,r))]$$

- $a \geq 0, b \leq 0$

$$h(t,r) = [a \cdot L(x_l(t,r)) + b \cdot L(y_u(t,r)), a \cdot L(x_u(t,r)) + b \cdot L(y_l(t,r))]$$

And also,

$$\mathbf{L}(x(t)) = X(s), \quad \mathbf{L}(y(t)) = Y(s)$$

Then

$$\begin{aligned}\mathbf{L}^{-1}(a \odot X(s) \oplus b \odot Y(s)) &= a \odot \mathbf{L}^{-1}(X(s)) \oplus b \odot \mathbf{L}^{-1}(Y(s)) \\ &= a \odot x(t) \oplus b \odot y(t)\end{aligned}$$

2. For any $\lambda \in R$

$$\mathbf{L}(\lambda \odot x(t)) = \lambda \odot \int_0^\infty e^{-st} \odot x(t)dt = \lambda \odot \mathbf{L}(x(t))$$

In level-wise form,

- $\lambda \geq 0$
  Suppose, $\lambda \odot x(t) = h(t)$,

$$L(h(t,r)) = [\lambda L(x_l(t,r)), \lambda L(x_u(t,r))]$$

- $\lambda < 0$

$$h(t,r) = [\lambda L(x_u(t,r)), \lambda L(x_l(t,r))]$$

A generalized version of this property can be explained for non-negative function instead of λ. Suppose, $y(t) \odot x(t) = h(t)$,

$$\mathbf{L}(y(t) \odot x(t)) = y(t) \odot \int_0^\infty e^{-st} \odot x(t)dt = y(t) \odot \mathbf{L}(x(t))$$

- $y(t) \geq 0$

$$L(h(t,r)) = [\lambda y(t) L(x_l(t,r)), \lambda y(t) L(x_u(t,r))]$$

- $y(t) < 0$

$$L(h(t,r)) = [y(t)L(x_u(t,r)), y(t)L(x_l(t,r))]$$

Derivative property of Laplace

Assume that $x(t)$ is a fuzzy continuous function for $t \geq 0$ and of exponential e^{at} and $x'_{gH}(t)$ is a peace-wise continuous in every finite closed interval. If the type of gH-differentiability does not change then for any $s > a$ we have,

- $i - gH$ differentiability

$$\mathbf{L}\left(x'_{i-gH}(t)\right) = s \odot X(s) \ominus_H x(0)$$

- $ii - gH$ differentiability

$$\mathbf{L}\left(x'_{ii-gH}(t)\right) = (-1)x(0) \ominus_H (-1)s \odot X(s)$$

To show the first item, according to the definition of Laplace,

$$\mathbf{L}\left(x'_{i-gH}(t)\right) = \lim_{\tau \to \infty} \int_0^{\tau} e^{-st} \odot x'_{i-gH}(t)dt$$

Now we are going to investigate the result of integral. Since, the function $e^{-st} > 0$ and its derivative is negative, $-se^{-st} < 0$, then

$$\begin{aligned}(e^{-st} \odot x(t))'_{i-gH} &= e^{-st} \odot x'_{i-gH}(t) \ominus_H (-1)(e^{-st})'' \odot x(t) \\ &= e^{-st} \odot x'_{i-gH}(t) \ominus_H se^{-st} \odot x(t)\end{aligned}$$

Because, the level-wise form of the right-side is,

$$\begin{aligned}&\left[e^{-st}x'_l(t,r) - se^{-st} \odot x_l(t,r), e^{-st}x'_u(t,r) - se^{-st} \odot x_u(t,r)\right] \\ &= \left[(e^{-st}x_l(t,r))', (e^{-st}x_u(t,r))'\right]\end{aligned}$$

And $\left[(e^{-st}x_l(t,r))', (e^{-st}x_u(t,r))'\right]$ is the level-wise form of $(e^{-st} \odot x(t))'_{i-gH}$.

Now,

$$\int_0^\tau \left(e^{-st} \odot x(t)\right)'_{i-gH} dt = \int_0^\tau \left(e^{-st} \odot x'_{i-gH}(t) \ominus_H s e^{-st} \odot x(t)\right) dt$$

$$= \int_0^\tau e^{-st} \odot x'_{i-gH}(t) dt \ominus_H s \odot \int_0^\tau e^{-st} \odot x(t) dt$$

On the other hand,

$$\int_0^\tau \left(e^{-st} \odot x(t)\right)'_{i-gH} dt = e^{-s\tau} \odot x(\tau) \ominus_H x(0)$$

By substituting we have,

$$e^{-s\tau} \odot x(\tau) \ominus_H x(0) = \int_0^\tau e^{-st} \odot x'_{i-gH}(t) dt \ominus_H s \odot \int_0^\tau e^{-st} \odot x(t) dt$$

Based on the definition of H-difference,

$$\int_0^\tau e^{-st} \odot x'_{i-gH}(t) dt = e^{-s\tau} \odot x(\tau) \ominus_H x(0) \oplus s \odot \int_0^\tau e^{-st} \odot x(t) dt$$

$$\begin{aligned} \boldsymbol{L}\left(x'_{i-gH}(t)\right) &= \lim_{\tau \to \infty} \int_0^\tau e^{-st} \odot x'_{i-gH}(t) dt \\ &= \lim_{\tau \to \infty} \left(e^{-s\tau} \odot x(\tau) \ominus_H x(0) \oplus s \odot \int_0^\tau e^{-st} \odot x(t) dt\right) \\ &= \lim_{\tau \to \infty} \left(e^{-s\tau} \odot x(\tau) \ominus_H x(0)\right) \oplus s \odot \lim_{\tau \to \infty} \int_0^\tau e^{-st} \odot x(t) dt \\ &= s \odot X(s) \ominus_H x(0) \end{aligned}$$

Finally, the proof is completed.

$$\boldsymbol{L}\left(x'_{i-gH}(t)\right) = s \odot X(s) \ominus_H x(0)$$

For proving the second item, again according to the definition of Laplace,

$$\mathbf{L}\left(x'_{ii-gH}(t)\right) = \lim_{\tau\to\infty} \int_0^\tau e^{-st} \odot x'_{ii-gH}(t)dt$$

It is enough to find integral. Since, the function $e^{-st} > 0$ and its derivative is negative, $-se^{-st} < 0$, then

$$\begin{aligned}(e^{-st} \odot x(t))'_{ii-gH} &= e^{-st} \odot x'_{ii-gH}(t) \oplus (e^{-st})' \odot x(t) \\ &= e^{-st} \odot x'_{ii-gH}(t) \oplus (-1)se^{-st} \odot x(t)\end{aligned}$$

Because, the level-wise form of the right-side is,

$$\begin{aligned}&\left[e^{-st}x'_u(t,r) - se^{-st} \odot x_u(t,r), e^{-st}x'_l(t,r) - se^{-st} \odot x_l(t,r)\right] \\ &= \left[(e^{-st}x_u(t,r))', (e^{-st}x_l(t,r))'\right]\end{aligned}$$

And $\left[(e^{-st}x_u(t,r))', (e^{-st}x_l(t,r))'\right]$ is the level-wise form of $(e^{-st} \odot x(t))'_{ii-gH}$. Now,

$$\begin{aligned}\int_0^\tau (e^{-st} \odot x(t))'_{ii-gH}dt &= \int_0^\tau \left(e^{-st} \odot x'_{ii-gH}(t) \oplus (-1)se^{-st} \odot x(t)\right)dt \\ &= \int_0^\tau e^{-st} \odot x'_{ii-gH}(t)dt \oplus (-1)s \odot \int_0^\tau e^{-st} \odot x(t)dt\end{aligned}$$

On the other hand,

$$\int_0^\tau (e^{-st} \odot x(t))'_{ii-gH}dt = (-1)x(0) \ominus_H (-1)e^{-s\tau} \odot x(\tau)$$

By substituting we have,

$$\begin{aligned}&(-1)x(0) \ominus_H (-1)e^{-s\tau} \odot x(\tau) \\ &= \int_0^\tau e^{-st} \odot x'_{ii-gH}(t)dt \oplus (-1)s \odot \int_0^\tau e^{-st} \odot x(t)dt\end{aligned}$$

Based on the definition of H-difference in type 2,

$$\int_0^{\tau} e^{-st} \odot x'_{ii-gH}(t)dt$$
$$= (-1)x(0)\ominus_H(-1)e^{-s\tau} \odot x(\tau)\ominus(-1)s \odot \int_0^{\tau} e^{-st} \odot x(t)dt$$

$$\begin{aligned}
\mathbf{L}\left(x'_{ii-gH}(t)\right) &= \lim_{\tau\to\infty} \int_0^{\tau} e^{-st} \odot x'_{ii-gH}(t)dt \\
&= \lim_{\tau\to\infty}\left((-1)x(0)\ominus_H(-1)e^{-s\tau} \odot x(\tau)\ominus(-1)s \odot \int_0^{\tau} e^{-st} \odot x(t)dt\right) \\
&= (-1)x(0)\ominus(-1)s \odot X(s)
\end{aligned}$$

Finally, the proof is completed.

$$\mathbf{L}\left(x'_{ii-gH}(t)\right) = (-1)x(0)\ominus(-1)s \odot X(s)$$

4.3.4 Convolution Theorem

If $x(t)$ is a fuzzy peace-wise continuous function on $[0, \infty]$ and of exponential order a, then

$$\mathbf{L}((x \cdot y)(t)) = \mathbf{L}(x(t)) \odot \mathbf{L}(y(t))$$

where $y(t)$ is a peace-wise continuous real function on $[0, \infty)$.

To prove,

$$\begin{aligned}
\mathbf{L}(x(t)) \odot \mathbf{L}(y(t)) &= \left(\int_0^{\infty} e^{-s\tau} \odot x(\tau)d\tau\right) \odot \left(\int_0^{\infty} e^{-s\sigma} \odot y(\sigma)d\sigma\right) \\
&= \int_0^{\infty}\left(\int_0^{\infty} e^{-s(\tau+\sigma)} \odot x(\tau)d\tau\right) \odot y(\sigma)d\sigma
\end{aligned}$$

Let us to hold τ fixed in the interior integral, substituting $t = \tau + \sigma$ and $d\sigma = dt$, we obtain

$$\begin{aligned}
\mathbf{L}(x(t)) \odot \mathbf{L}(y(t)) &= \int_0^\infty \left(\int_\sigma^\infty e^{-st} \odot x(\tau) \odot y(t-\tau)dt \right) d\tau \\
&= \int_0^\infty \int_\sigma^\infty e^{-st} \odot x(\tau) \odot y(t-\tau)dtd\tau \\
&= \int_0^\infty e^{-st} \odot \left(\int_0^t x(t-\sigma) \odot y(\sigma)d\tau \right) d\sigma \\
&= \mathbf{L}((x \cdot y)(t))
\end{aligned}$$

One of the important functions occurring in some electrical systems is the delay and it can be displayed as a unit step function like,

$$u_a(t) := u(t-a) = \begin{cases} 1, & t \geq a \\ 0, & t < a \end{cases}$$

For instance, in an electric circuit for a voltage at a particular time $t = a$ we write such a situation using unit step functions as Fig. 4.7.

$$V(t) = u(t) - u(t-a)$$

It is a shifted unit step. It is clear that $u(t) = u(t-a) = 1$ and $V(t) = 0$ for $t \geq a \geq 0$ and $u(t) = 1, u(t-a) = 0$ and $V(t) = 1$ for $a > t \geq 0$ (Fig. 4.8).

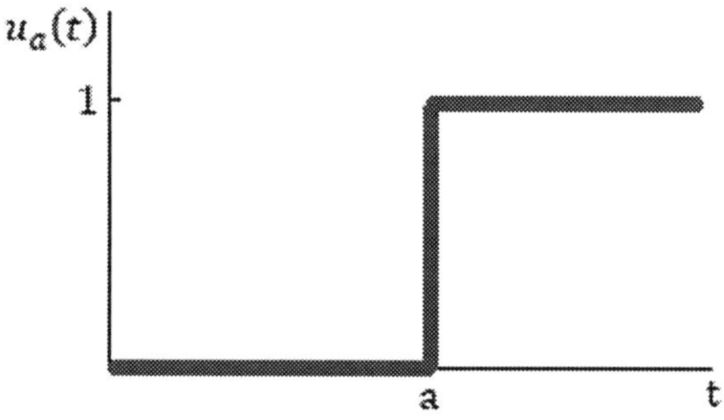

Fig. 4.7 Unit step function $u_a(t)$

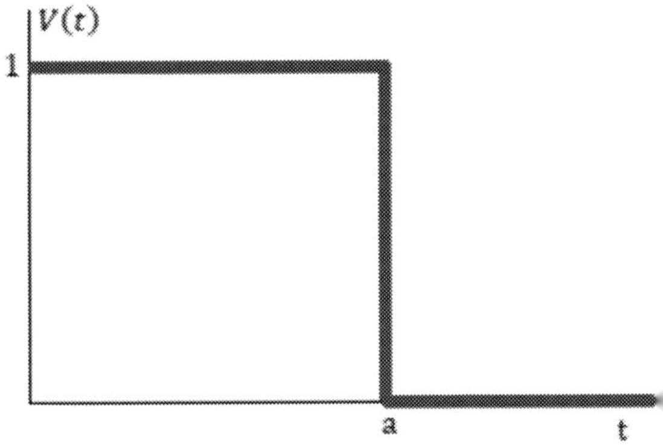

Fig. 4.8 Shifted unit step function $V(t)$

4.3.5 First Translation Theorem

If $X(s) = \mathbf{L}(x(t))$ for $s > a$ then $X(s-a) = \mathbf{L}(e^{at} \odot x(t))$ such that a is a real number.

The proof is clear from the definition of Laplace transform,

$$X(s-a) = \int_0^\infty e^{-(s-a)t} \odot x(t)dt = \int_0^\infty e^{-st}e^{at} \odot x(t)dt = \mathbf{L}(e^{at} \odot x(t))$$

4.3.6 Second Translation Theorem

If $X(s) = \mathbf{L}(x(t))$ for $s > a \geq 0$ then

$$e^{as} \odot X(s) = \mathbf{L}(u_a(t) \odot x(t-a))$$

According to the definition,

$$\mathbf{L}(u_a(t) \odot x(t-a)) = \int_0^\infty e^{-st}u_a(t) \odot x(t-a)dt$$

Since $u_a(t) = 0$ for $0 < t < a$ and $u_a(t) = 1$ for $t \geq a$ then

$$\mathbf{L}(u_a(t) \odot x(t-a)) = \int_a^\infty e^{-st} \odot x(t-a)dt$$

Let us suppose that $t - a = \tau$

$$\int_a^\infty e^{-st} \odot x(t-a)dt = e^{-sa} \odot \int_a^\infty e^{-s\tau} \odot x(\tau)d\tau = e^{-sa} \odot X(s)$$

Finally

$$\boldsymbol{L}(u_a(t) \odot x(t-a)) = e^{-sa} \odot X(s)$$

4.3.7 Remark—Laplace Forms of Fractional Derivatives

If the fuzzy function $x(t)$ is continuous and integrable on $[t_0, T]$ then its RL-gH and C-gH derivatives are explained as follows,

RL-gH derivative,

$$\boldsymbol{L}\left(\boldsymbol{D}^\alpha_{\boldsymbol{RL}_{gH}} x(t)\right) = s^\alpha \odot \boldsymbol{L}(x(t)) \ominus_{gH} \boldsymbol{L}\left(\boldsymbol{D}^{\alpha-1}_{\boldsymbol{RL}_{gH}} x(t_0)\right)$$

where

$$\boldsymbol{D}^\alpha_{\boldsymbol{RL}_{gH}} x(t) = \frac{1}{\Gamma(1-\alpha)} \odot \frac{d}{dt} \int_{t_0}^t \frac{x(\tau)}{(t-\tau)^\alpha} d\tau, \quad 0<\alpha<1$$

$$\boldsymbol{L}\left(\boldsymbol{D}^\alpha_{\boldsymbol{RL}_{gH}} x(t)\right) = \int_0^\infty e^{-st} \odot \boldsymbol{D}^\alpha_{\boldsymbol{RL}_{gH}} x(t)dt$$

And

$$\begin{aligned}\boldsymbol{L}\left(\boldsymbol{D}^\alpha_{\boldsymbol{RL}_{gH}} x(t,r)\right) &= \left[L\left(\boldsymbol{D}^\alpha_{\boldsymbol{RL}} x_l(t,r)\right), L\left(\boldsymbol{D}^\alpha_{\boldsymbol{RL}} x_u(t,r)\right)\right] \\ &= \int_0^\infty e^{-st}\left[\boldsymbol{D}^\alpha_{\boldsymbol{RL}} x_l(t,r), \boldsymbol{D}^\alpha_{\boldsymbol{RL}} x_u(t,r)\right] dt \\ &= \left[\int_0^\infty e^{-st} \boldsymbol{D}^\alpha_{\boldsymbol{RL}} x_l(t,r)dt, \int_0^\infty e^{-st} \boldsymbol{D}^\alpha_{\boldsymbol{RL}} x_u(t,r)dt\right]\end{aligned}$$

Finally,

$$L\big(\boldsymbol{D}_{\boldsymbol{RL}}^{\alpha} x_l(t,r)\big) = \int_0^\infty e^{-st} \boldsymbol{D}_{\boldsymbol{RL}}^{\alpha} x_l(t,r)dt$$

$$L\big(\boldsymbol{D}_{\boldsymbol{RL}}^{\alpha} x_u(t,r)\big) = \int_0^\infty e^{-st} \boldsymbol{D}_{\boldsymbol{RL}}^{\alpha} x_u(t,r)dt$$

- If $x(t)$ is $i - RL_{gH}$ differentiable,

$$\boldsymbol{L}\left(\boldsymbol{D}_{\boldsymbol{RL}_{i-gH}}^{\alpha} x(t)\right) = s^\alpha \odot \boldsymbol{L}(x(t)) \ominus_H \boldsymbol{L}\left(\boldsymbol{D}_{\boldsymbol{RL}_{i-gH}}^{\alpha-1} x(t_0)\right)$$

- If $x(t)$ is $ii - RL_{gH}$ differentiable,

$$\boldsymbol{L}\left(\boldsymbol{D}_{\boldsymbol{RL}_{ii-gH}}^{\alpha} x(t)\right) = (-1)\boldsymbol{L}\left(\boldsymbol{D}_{\boldsymbol{RL}_{ii-gH}}^{\alpha-1} x(t_0)\right) \ominus_H (-1) s^\alpha \odot \boldsymbol{L}(x(t))$$

It can be investigated in level-wise form very easily.

- If $x(t)$ is $i - RL_{gH}$ differentiable,

The level-wise form of right side is,

$$\begin{aligned}
&s^\alpha \odot \boldsymbol{L}(x(t,r)) \ominus_H \boldsymbol{L}\left(\boldsymbol{D}_{\boldsymbol{RL}_{i-gH}}^{\alpha-1} x(t_0,r)\right) \\
&= \left[s^\alpha L(x_l(t,r)) - L\big(\boldsymbol{D}_{\boldsymbol{RL}}^{\alpha-1} x_l(t_0,r)\big), s^\alpha L(x_u(t,r)) - L\big(\boldsymbol{D}_{\boldsymbol{RL}}^{\alpha-1} x_u(t_0,r)\big)\right] \\
&= \left[L\big(\boldsymbol{D}_{\boldsymbol{RL}}^{\alpha} x_l(t,r)\big), L\big(\boldsymbol{D}_{\boldsymbol{RL}}^{\alpha} x_u(t,r)\big)\right] = \boldsymbol{L}\left(\boldsymbol{D}_{\boldsymbol{RL}_{i-gH}}^{\alpha} x(t,r)\right)
\end{aligned}$$

Because

$$\boldsymbol{L}\left(\boldsymbol{D}_{\boldsymbol{RL}_{i-gH}}^{\alpha} x(t,r)\right) = \left[L\big(\boldsymbol{D}_{\boldsymbol{RL}}^{\alpha} x_l(t,r)\big), L\big(\boldsymbol{D}_{\boldsymbol{RL}}^{\alpha} x_u(t,r)\big)\right]$$

- If $x(t)$ is $ii - RL_{gH}$ differentiable,

The level-wise form of right side is,

$$\begin{aligned}
&(-1)\boldsymbol{L}\left(\boldsymbol{D}_{\boldsymbol{RL}_{ii-gH}}^{\alpha-1} x(t_0,r)\right) \ominus_H (-1) s^\alpha \odot \boldsymbol{L}(x(t,r)) \\
&= \left[-L\big(\boldsymbol{D}_{\boldsymbol{RL}}^{\alpha-1} x_l(t_0,r)\big) + s^\alpha L(x_l(t,r)), -L\big(\boldsymbol{D}_{\boldsymbol{RL}}^{\alpha-1} x_u(t_0,r)\big) + s^\alpha L(x_u(t,r))\right] \\
&= \left[L\big(\boldsymbol{D}_{\boldsymbol{RL}}^{\alpha} x_l(t,r)\big), L\big(\boldsymbol{D}_{\boldsymbol{RL}}^{\alpha} x_u(t,r)\big)\right] = \boldsymbol{L}\left(\boldsymbol{D}_{\boldsymbol{RL}_{i-gH}}^{\alpha} x(t,r)\right)
\end{aligned}$$

Because,

$$L\left(D^{\alpha}_{RL_{i-gH}}x(t,r)\right) = \left[L\left(D^{\alpha}_{RL}x_l(t,r)\right), L\left(D^{\alpha}_{RL}x_u(t,r)\right)\right]$$

Caputo-gH derivative

Also some similar relation about the Laplace transform of Caputo-gH derivative can be obtained as follow,

$$D^{\alpha}_{C_{gH}}x(t) = \frac{1}{\Gamma(1-\alpha)} \odot \int_{t_0}^{t} \frac{x'_{gH}(\tau)}{(s-\tau)^{\alpha}} d\tau$$

$$L\left(D^{\alpha}_{C_{gH}}x(t)\right) = \int_{0}^{\infty} e^{-st} \odot D^{\alpha}_{C_{gH}}x(t)dt$$

And

$$\begin{aligned} L\left(D^{\alpha}_{C_{gH}}x(t,r)\right) &= \left[L\left(D^{\alpha}_{C}x_l(t,r)\right), L\left(D^{\alpha}_{C}x_u(t,r)\right)\right] \\ &= \int_{0}^{\infty} e^{-st}\left[D^{\alpha}_{C}x_l(t,r), D^{\alpha}_{C}x_u(t,r)\right]dt \\ &= \left[\int_{0}^{\infty} e^{-st}D^{\alpha}_{C}x_l(t,r)dt, \int_{0}^{\infty} e^{-st}D^{\alpha}_{C}x_u(t,r)dt\right] \end{aligned}$$

Finally,

$$L\left(D^{\alpha}_{C}x_l(t,r)\right) = \int_{0}^{\infty} e^{-st}D^{\alpha}_{C}x_l(t,r)dt$$

$$L\left(D^{\alpha}_{C}x_u(t,r)\right) = \int_{0}^{\infty} e^{-st}D^{\alpha}_{C}x_u(t,r)dt$$

- $i-gH$ differentiable

$$D^{\alpha}_{C_{i-gH}}x(s,r) = \left[D^{\alpha}_{C}x_l(s,r), D^{\alpha}_{C}x_u(s,r)\right]$$

$$L\left(D^{\alpha}_{C_{i-gH}}x(t)\right) = s^{\alpha} \odot L(x(t)) \ominus_H s^{\alpha-1}x(t_0)$$

- $ii - gH$ differentiable

$$\begin{aligned}
\boldsymbol{D}^{\alpha}_{\boldsymbol{C}_{ii\text{-}gH}} x(s,r) &= \left[\boldsymbol{D}^{\alpha}_{\boldsymbol{C}} x_u(s,r), \boldsymbol{D}^{\alpha}_{\boldsymbol{C}} x_l(s,r)\right] \\
\boldsymbol{L}\left(\boldsymbol{D}^{\alpha}_{\boldsymbol{C}_{i\text{-}gH}} x(t)\right) &= (-1)s^{\alpha-1}x(t_0) \ominus_H (-1)s^{\alpha} \odot \boldsymbol{L}(x(t))
\end{aligned}$$

The process to show these two cases is similar to the RL-derivative. It can be shown by level-wise form.

Laplace—Fuzzy fractional differential equation—RL derivative

Consider the following fractional differential equation with fuzzy initial value and RL-derivative,

$$\begin{cases} \boldsymbol{D}^{\alpha}_{\boldsymbol{RL}_{gH}} x(t) = f(t, x(t)) \\ \boldsymbol{D}^{\alpha-1}_{\boldsymbol{RL}_{gH}} x(t_0) = x_0 \end{cases}$$

The approach is, using the Laplace transform. To this end, we take the Laplace operator form both sides of the equation. So we have,

$$\begin{cases} L\left(\boldsymbol{D}^{\alpha}_{\boldsymbol{RL}_{gH}} x(t,r)\right) = L(f(t, x(t))) \\ L\left(\boldsymbol{D}^{\alpha-1}_{\boldsymbol{RL}_{gH}} x(t_0)\right) = L(x_0) \end{cases}$$

On the other hand,

$$\boldsymbol{L}\left(\boldsymbol{D}^{\alpha}_{\boldsymbol{RL}_{gH}} x(t)\right) = s^{\alpha} \odot \boldsymbol{L}(x(t)) \ominus_{gH} \boldsymbol{L}\left(\boldsymbol{D}^{\alpha-1}_{\boldsymbol{RL}_{gH}} x(t_0)\right)$$

So,

$$\begin{cases} s^{\alpha} \odot L(x(t)) \ominus_{gH} L\left(\boldsymbol{D}^{\alpha-1}_{\boldsymbol{RL}_{gH}} x(t_0)\right) = L(f(t, x(t))) \\ L\left(\boldsymbol{D}^{\alpha-1}_{\boldsymbol{RL}_{gH}} x(t_0)\right) = L(x_0) \end{cases}$$

Finally,

$$s^{\alpha} \odot \boldsymbol{L}(x(t)) \ominus_{gH} \boldsymbol{L}(x_0) = \boldsymbol{L}(f(t, x(t)))$$

In two types of gH-difference,

- $i - gH$ difference

$$s^{\alpha} \odot \boldsymbol{L}(x(t)) \ominus_{H} \boldsymbol{L}(x_0) = \boldsymbol{L}(f(t, x(t)))$$

In the level-wise form,

$$s^{\alpha}L(x_l(t,r)) - L\big(x_{0,l}(r)\big) = L(f_l(t,x(t,r)))$$
$$s^{\alpha}L(x_u(t,r)) - L\big(x_{0,u}(r)\big) = L(f_u(t,x(t,r)))$$

$$s^{\alpha}L(x_l(t,r)) = L\big(x_{0,l}(r)\big) + L(f_l(t,x(t,r)))$$
$$s^{\alpha}L(x_u(t,r)) = L\big(x_{0,u}(r)\big) + L(f_u(t,x(t,r)))$$

- $ii-gH$ difference

$$\boldsymbol{L}(x_0) = s^{\alpha} \odot \boldsymbol{L}(x(t)) \oplus (-1)\boldsymbol{L}(f(t,x(t)))$$

In the level-wise form,

$$L\big(x_{0,l}(r)\big) = s^{\alpha}L(x_l(t,r)) - L(f_u(t,x_l(t,r),x_u(t,r)))$$
$$L\big(x_{0,u}(r)\big) = s^{\alpha}L(x_u(t,r)) - L(f_l(t,x_l(t,r),x_u(t,r)))$$

$$s^{\alpha}L(x_l(t,r)) = L\big(x_{0,l}(r)\big) + L(f_u(t,x_l(t,r),x_u(t,r)))$$
$$s^{\alpha}L(x_u(t,r)) = L\big(x_{0,u}(r)\big) + L(f_l(t,x_l(t,r),x_u(t,r)))$$

where

$$f_l(t,x_l(t,r),x_u(t,r)) = \min\{f(t,u)|u \in x(t,r)\}$$
$$f_u(t,x_l(t,r),x_u(t,r)) = \max\{f(t,u)|u \in x(t,r)\}$$

Example Consider the following fractional differential equation with fuzzy initial value,

$$\begin{cases} \boldsymbol{D}^{\alpha}_{\boldsymbol{RL}_{gH}}x(t) = \lambda \odot x(t), \quad t \in [0,1], \quad 0<\alpha<1 \\ \boldsymbol{D}^{\alpha-1}_{\boldsymbol{RL}_{gH}}x(t_0) = x_0 \in \mathbb{F}_R \end{cases}$$

In general, we found that

$$s^{\alpha} \odot \boldsymbol{L}(x(t)) \ominus_{gH} \boldsymbol{L}(x_0) = \boldsymbol{L}(f(t,x(t)))$$
$$\boldsymbol{L}(\lambda \odot x(t)) = s^{\alpha} \odot \boldsymbol{L}(x(t)) \ominus_{gH} \boldsymbol{L}(x_0)$$

$$\boldsymbol{L}(\lambda \odot x(t)) = \lambda \odot \int_0^{\infty} e^{-st} \odot x(t)dt = \lambda \odot \boldsymbol{L}(x(t))$$

Then

$$\lambda \odot \boldsymbol{L}(x(t)) = s^\alpha \odot \boldsymbol{L}(x(t)) \ominus_{gH} \boldsymbol{L}(x_0)$$

Case 1. $\lambda \geq 0$

$$s^\alpha L(x_l(t,r)) = L\big(x_{0,l}(r)\big) + \lambda L(x_l(t,r))$$
$$s^\alpha L(x_u(t,r)) = L\big(x_{0,u}(r)\big) + \lambda L(x_u(t,r))$$

This is exactly the case $i - gH$ difference. The solution can be obtained as,

$$(s^\alpha - \lambda)L(x_l(t,r)) = L\big(x_{0,l}(r)\big), (s^\alpha - \lambda)L(x_u(t,r)) = L\big(x_{0,u}(r)\big)$$

Taking inverse Laplace,

$$x_l(t,r) = L^{-1}\left(\frac{1}{s^\alpha - \lambda}\right)x_{0,l}(r), \quad x_u(t,r) = L^{-1}\left(\frac{1}{s^\alpha - \lambda}\right)x_{0,u}(r)$$
$$x_l(t,r) = t^{\alpha-1}E_{\alpha,\alpha}(\lambda t^\alpha)x_{0,l}(r), \quad x_u(t,r) = t^{\alpha-1}E_{\alpha,\alpha}(\lambda t^\alpha)x_{0,u}(r)$$

Case 2. $\lambda < 0$

$$s^\alpha L(x_l(t,r)) = L\big(x_{0,l}(r)\big) + L(\lambda x_u(t,r))$$
$$s^\alpha L(x_u(t,r)) = L\big(x_{0,u}(r)\big) + L(\lambda x_l(t,r))$$

Since $\lambda < 0$ then

$$s^\alpha L(x_l(t,r)) = L\big(x_{0,l}(r)\big) + \lambda L(x_l(t,r))$$
$$s^\alpha L(x_u(t,r)) = L\big(x_{0,u}(r)\big) + \lambda L(x_u(t,r))$$

For $0 \leq r \leq 1$ the same solution is found.

Example Consider the following fractional differential equation with fuzzy initial value,

$$\begin{cases} \boldsymbol{D}^\alpha_{\boldsymbol{RL}_{gH}} x(t) = (-1) \odot x(t) + t + 1, \quad t \in [0,1], \quad 0 < \alpha < 1 \\ \boldsymbol{D}^{\alpha-1}_{\boldsymbol{RL}_{gH}} x(t_0) = x_0 \in \mathbb{F}_R \end{cases}$$

In general, we found that

$$\boldsymbol{L}((-1) \odot x(t) \oplus t \oplus 1) = s^\alpha \odot \boldsymbol{L}(x(t)) \ominus_{gH} \boldsymbol{L}(x_0)$$

On the other hand, based on the linearity property of the Laplace transform,

$$\boldsymbol{L}((-1)\odot x(t)\oplus t\oplus 1)=(-1)\odot \boldsymbol{L}(x(t))\oplus \boldsymbol{L}(t)\oplus \boldsymbol{L}(1)$$

By substituting,

$$(-1)\odot \boldsymbol{L}(x(t))\oplus \boldsymbol{L}(t)\oplus \boldsymbol{L}(1)=s^{\alpha}\odot \boldsymbol{L}(x(t))\ominus_{gH}\boldsymbol{L}(x_0)$$

- $i-gH$ difference

$$(-1)\odot \boldsymbol{L}(x(t))\oplus \boldsymbol{L}(t)\oplus \boldsymbol{L}(1)=s^{\alpha}\odot \boldsymbol{L}(x(t))\ominus_{H}\boldsymbol{L}(x_0)$$

In the level-wise form,

$$\begin{aligned}
-L(x_l(t,r))+L(t)+L(1)&=s^{\alpha}L(x_l(t,r))-L(x_{0,l}(r))\\
-L(x_u(t,r))+L(t)+L(1)&=s^{\alpha}L(x_u(t,r))-L(x_{0,u}(r))
\end{aligned}$$

Then

$$\begin{aligned}
(1+s^{\alpha})L(x_l(t,r))&=L(t)+L(1)+L(x_{0,l}(r))\\
(1+s^{\alpha})L(x_u(t,r))&=L(t)+L(1)+L(x_{0,u}(r))
\end{aligned}$$

$$\begin{aligned}
L(x_l(t,r))&=\frac{L(t)}{(1+s^{\alpha})}+\frac{L(1)}{(1+s^{\alpha})}+\frac{1}{(1+s^{\alpha})}L(x_{0,l}(r))\\
L(x_u(t,r))&=\frac{L(t)}{(1+s^{\alpha})}+\frac{L(1)}{(1+s^{\alpha})}+\frac{1}{(1+s^{\alpha})}L(x_{0,u}(r))
\end{aligned}$$

By taking inverse Laplace,

$$\begin{aligned}
x_l(t,r)&=L^{-1}\left(\frac{1}{s^2(1+s^{\alpha})}\right)+L^{-1}\left(\frac{1}{s(1+s^{\alpha})}\right)+L^{-1}\left(\frac{x_{0,l}(r)}{1+s^{\alpha}}\right)\\
x_u(t,r)&=L^{-1}\left(\frac{1}{s^2(1+s^{\alpha})}\right)+L^{-1}\left(\frac{1}{s(1+s^{\alpha})}\right)+L^{-1}\left(\frac{x_{0,u}(r)}{1+s^{\alpha}}\right)
\end{aligned}$$

$$x_l(t,r)=t^{\alpha-1}E_{\alpha,\alpha}(\lambda t^{\alpha})x_{0,l}(r)+\int_0^t (t-u)^{\alpha-1}E_{\alpha,\alpha}(\lambda(t-u)^{\alpha})(u+1)du$$

$$x_u(t,r)=t^{\alpha-1}E_{\alpha,\alpha}(\lambda t^{\alpha})x_{0,u}(r)+\int_0^t (t-u)^{\alpha-1}E_{\alpha,\alpha}(\lambda(t-u)^{\alpha})(u+1)du$$

For $0\leq r\leq 1$.

4.3.8 Fuzzy Fourier Transform Operator

Consider the function $x : R \to \mathbb{F}_R$ is the fuzzy valued function. The fuzzy Fourier transform of $x(t)$ denoted by $(\mathcal{F}\{x(t)\} : R \to \mathbb{F}_C)$ is given by the following integral,

$$\mathcal{F}\{x(t)\} = \frac{1}{\sqrt{2\pi}} \int_{-\infty}^{\infty} x(t) \odot e^{-iwt} dt = F(w)$$

Here $\mathbb{F}_C$ is the set of all fuzzy nu90mbers on complex numbers.

In the level-wise form,

$$\mathcal{F}\{x(t,r)\} = \left[\frac{1}{\sqrt{2\pi}} \int_{-\infty}^{\infty} x_l(t,r) e^{-iwt} dt, \frac{1}{\sqrt{2\pi}} \int_{-\infty}^{\infty} x_u(t,r) e^{-iwt} dt \right] = F(w,r)$$

$$\mathcal{F}\{x_l(t,r)\} = \frac{1}{\sqrt{2\pi}} \int_{-\infty}^{\infty} x_l(t,r) e^{-iwt} dt \ , \mathcal{F}\{x_u(t,r)\} = \frac{1}{\sqrt{2\pi}} \int_{-\infty}^{\infty} x_l(t,r) e^{-iwt} dt$$

where,

$$\mathcal{F}\{x(t,r)\} = [\mathcal{F}(x_l(t,r)), \mathcal{F}(x_u(t,r))]$$

4.3.9 Existence of Fourier Transform

Suppose that the fuzzy valued function $x(t)$ is fuzzy absolutely integrable on $(-\infty, \infty)$ the fuzzy Fourier transform $\mathcal{F}\{x(t)\}$ exists.

Using the distance it will be proved and

$$D_H(\mathcal{F}\{x(t)\}, 0) = D_H\left(\frac{1}{\sqrt{2\pi}} \int_{-\infty}^{\infty} x(w) \odot e^{-iwt} dt, 0 \right)$$

$$\leq \frac{1}{\sqrt{2\pi}} \int_{-\infty}^{\infty} D_H\left(x(w) \odot e^{-iwt}, 0\right) dt \leq \frac{1}{\sqrt{2\pi}} \int_{-\infty}^{\infty} \left|e^{-iwt}\right| D_H(x(t), 0) dt$$

Since $\left|e^{-iwt}\right| = 1$,

$$D_H(\mathcal{F}\{x(t)\}, 0) \leq \frac{1}{\sqrt{2\pi}} \int_{-\infty}^{\infty} D_H(x(t), 0) dt < \infty$$

The proof is completed.

Fuzzy inverse Fourier transform

If $F(w)$ is the fuzzy Fourier transform of $x(t)$, then the fuzzy inverse Fourier transform of $F(w)$ is defined as,

$$\mathcal{F}^{-1}\{F(w)\} = \frac{1}{\sqrt{2\pi}} \int_{-\infty}^{\infty} x(w) \odot e^{iwt} dw = x(t)$$

In the same way we can show that if $F(w)$ fuzzy absolutely integrable then the fuzzy inverse Fourier transform $\mathcal{F}^{-1}\{F(w)\}$ exists.

Similar to the fuzzy Laplace transforms, the fuzzy Fourier transformations are linear and this came from the linearity property of the fuzzy Riemann integral.

Remark—Fuzzy Fourier transform of first derivative

Let $x(t)$ be fuzzy continuous, fuzzy absolutely integrable and converge to zero as $|t| \to 0$. Furthermore, let $x'_{gH}(t)$ is fuzzy absolutely integrable on $(-\infty, \infty)$. Then

- $i - gH$ differentiability

$$\mathcal{F}\left\{x'_{i-gH}(t)\right\} = (-iw) \odot \mathcal{F}\{x(t)\}$$

- $ii - gH$ differentiability

$$\mathcal{F}\left\{x'_{ii-gH}(t)\right\} = (-1)\ominus_H(iw) \odot \mathcal{F}\{x(t)\}$$

Because, in case $i - gH$ differentiability,

$$\mathcal{F}\left\{x'_{i-gH}(t)\right\} = \frac{1}{\sqrt{2\pi}} \int_{-\infty}^{\infty} x'_{i-gH}(t) \odot e^{-iwt} dt$$

Since $e^{-iwt} > 0$ and $(e^{-iwt})' < 0$ then

$$\int_{-\tau}^{\tau} x'_{i-gH}(t) \odot e^{-iwt} dt = \left[x(t) \odot e^{-iwt}\right]_{\tau} \ominus_H \left[x(t) \odot e^{-iwt}\right]_{-\tau}$$

$$\oplus \int_{-\tau}^{\tau} (e^{-iwt})' \odot x(t) dt$$

So,

$$\lim_{\tau\to\infty}\frac{1}{\sqrt{2\pi}}\int_{-\tau}^{\tau} x'_{i-gH}(t)\odot e^{-iwt}dt = (-iw)\lim_{\tau\to\infty}\frac{1}{\sqrt{2\pi}}\int_{-\tau}^{\tau} x(t)\odot e^{-iwt}dt$$
$$= (-iw)\odot \mathcal{F}\{x(t)\}$$

Because, $\lim_{\tau\to\infty} x(t) = 0$.
In case $ii-gH$ differentiability,

$$\mathcal{F}\left\{x'_{ii-gH}(t)\right\} = \frac{1}{\sqrt{2\pi}}\int_{-\infty}^{\infty} x'_{ii-gH}(t)\odot e^{-iwt}dt$$

Since $e^{-iwt} > 0$ and $(e^{-iwt})' < 0$ then

$$\int_{-\tau}^{\tau} x'_{i-gH}(t)\odot e^{-iwt}dt =$$

$$= \left[x(t)\odot(-1)e^{-iwt}\right]_{-\tau}\ominus_H\left[x(t)\odot(-1)e^{-iwt}\right]_{\tau}\ominus_H\int_{-\tau}^{\tau}(e^{-iwt})'\odot x(t)dt$$

So,

$$\lim_{\tau\to\infty}\frac{1}{\sqrt{2\pi}}\int_{-\tau}^{\tau} x'_{i-gH}(t)\odot e^{-iwt}dt = (-1)\ominus_H(iw)\lim_{\tau\to\infty}\frac{1}{\sqrt{2\pi}}\int_{-\tau}^{\tau} x(t)\odot e^{-iwt}dt$$
$$= (-1)\ominus_H(iw)\odot\mathcal{F}\{x(t)\}$$

Example If $x(t) = c\odot\delta(t)$ where δ is a real Dirac function. Therefore

$$\mathcal{F}\{c\odot\delta(t)\} = \frac{1}{\sqrt{2\pi}}\int_{-\infty}^{\infty} c\odot\delta(t)\odot e^{-iwt}dt = c\odot\frac{1}{\sqrt{2\pi}}\int_{-\infty}^{\infty}\delta(t)\odot e^{-iwt}dt = c$$

Because

$$\frac{1}{\sqrt{2\pi}}\int_{-\infty}^{\infty}\delta(t)\odot e^{-iwt}dt = 1$$

4.4 Fuzzy Solutions of Time-Fractional Problems

The main purpose of this section is to obtain an analytical solution for the time-fractional fuzzy equation. To do this, the time-fractional equation is transformed into an algebraic equation using the fuzzy Laplace and Fourier transforms. In this study, the fractional derivatives are described in the Caputo gH-differentiability. This section examines the explicit and fundamental solutions of the following fuzzy time fractional problem,

$$\boldsymbol{D}^{\alpha}_{C_{gH}} u(t,x) = \lambda \odot \frac{\partial u(t,x)}{\partial x}, \quad t > 0, \quad -\infty < x < \infty$$

where $0 \neq \lambda \in R, 0 < \alpha < 1$ and

$$\boldsymbol{D}^{\alpha}_{C_{gH}} u(t,x) = \frac{1}{\Gamma(1-\alpha)} \odot \int_{t_0}^{t} (t-\tau)^{\alpha} \odot \frac{\partial u(\tau,x)}{\partial \tau} d\tau$$

where p_{gH} differentiability of the solution is defined as,

$$\frac{\partial u(\tau,x)}{\partial \tau} = \lim_{h \to 0} \frac{u(\tau+h,x) \ominus_{gH} u(\tau,x)}{h}$$

Partial gH-differentiability

The fuzzy number valued function of two variables $u(t,x) \in \mathbb{F}_R$ is called gH-partial differentiable $p - gH$ at the point $(t,x) \in \mathbb{D}$ with respect to t and x and denoted by $\partial_{t_{gH}} u(t,x)$ and $\partial_{x_{gH}} u(t,x)$ if

$$\partial_{t_{gH}} u(t,x) = \lim_{h \to 0} \frac{f(t+h,x) \ominus_{gH} f(t,x)}{h}$$

$$\partial_{x_{gH}} u(t,x) = \lim_{k \to 0} \frac{f(t,x+k) \ominus_{gH} f(t,x)}{k}$$

Provided to both derivatives $\partial_{t_{gH}} u(t,x)$ and $\partial_{x_{gH}} u(t,x)$ are fuzzy number valued functions not fuzzy sets.

Level-wise form of Partial gH-differentiability

Suppose that the fuzzy number valued function $u(t,x) \in \mathbb{F}_R$ is $p - gH$ differentiable at the point (t,x) with respect to t and $u_l(t,x,r), u_u(t,x,r)$ are real valued functions and partial differentiable with respect to t. We say,

- $f(t,x)$ is $(i-p-gH)$ differentiable with respect to t at (t,x) if

$$\partial_{t,i-gH}u(t,x,r) = [\partial_t u_l(t,x,r), \partial_t u_u(t,x,r)]$$

- $f(t,x)$ is $(ii-p-gH)$ differentiable with respect to t at (t,x) if

$$\partial_{t,ii-gH}u(t,x,r) = [\partial_t u_u(t,x,r), \partial_t u_l(t,x,r)]$$

Please note that in each cases the conditions of the definition in level-wise form should be satisfied.

4.4.1 Fuzzy Explicit Solution of the Time-Fractional Problem

Now, we investigate the fuzzy explicit solution of the fuzzy linear partial fractional differential Equation with the following fuzzy boundary conditions,

$$\lim_{x\to\pm\infty} u(t,x) = 0, \quad u(0^+,x) = g(x)$$

Suppose that $u(t,x)$ is the fuzzy explicit solution of fuzzy time fractional problem provided that the types of p_{gH} differentiability with respect to t and x are the same.

Consider the fuzzy Laplace transform of the solution with respect to t, for fixed x,

$$\mathbf{L}_t(u(t,x)) = \int_0^{\infty} e^{-st} \odot u(t,x)dt = U_t(s,x)$$

The Laplace inverse of $U_t(s,x)$ with respect to first component is defined as,

$$\mathbf{L}_s^{-1}(U_t(s,x)) = \frac{1}{2\pi i}\int_{\gamma-i\infty}^{\gamma+i\infty} e^{st} \odot U_t(s,x)ds, \quad \gamma \in R$$

Such that

$$\mathbf{L}_s^{-1}\mathbf{L}_t(u(t,x)) = u(t,x)$$

Also consider fuzzy Fourier transform with respect to x for fixed $t > 0$,

$$\mathcal{F}_x\{u(t,x)\} = \frac{1}{\sqrt{2\pi}} \int_{-\infty}^{\infty} u(t,x) \odot e^{-iwx} dx = \mathcal{U}_x(t,w)$$

It inverse with respect to the second component is also as,

$$\mathcal{F}_w^{-1}\{\mathcal{U}_w(t,w)\} = \frac{1}{\sqrt{2\pi}} \int_{-\infty}^{\infty} \mathcal{U}_w(t,w) \odot e^{iwt} dw$$

Such that

$$\mathcal{F}_w^{-1}\mathcal{F}_x(u(t,x)) = u(t,x)$$

Let $\mathcal{LF}$ denotes the space of all fuzzy number valued functions $u(t,x)$ such that the fuzzy Laplace transform and the fuzzy Fourier transform exist with the following notation,

$$\begin{aligned}\mathcal{F}_x\mathbf{L}_t(u(t,x)) = \mathcal{F}_x(\mathbf{L}_t(u(t,x))) &= \frac{1}{\sqrt{2\pi}} \int_{-\infty}^{\infty} \mathbf{L}_t(u(t,x)) \odot e^{-iwx} dx \\ &= \frac{1}{\sqrt{2\pi}} \int_{-\infty}^{\infty} \left(\int_{0}^{\infty} e^{-st} \odot u(t,x) dt \right) \odot e^{-iwx} dx \\ &= \frac{1}{\sqrt{2\pi}} \int_{-\infty}^{\infty} \int_{0}^{\infty} e^{-(s+w)t} \odot u(t,x) dt dx := v(s,w)\end{aligned}$$

Note Let $g(x)$ be a fuzzy number valued function such that, $\mathcal{F}(g(x)) = G(w)$. If the fuzzy solution $u(t,x) \in \mathcal{LF}$ of

$$\mathbf{D}^{\alpha}_{C_{gH}} u(t,x) = \lambda \odot \frac{\partial u(t,x)}{\partial x}$$

is $i - gH$ Caputo differentiability with respect to t and $i - gH$ partial differentiability with respect to x, then it satisfies the following relation,

$$\mathbf{L}_t^{-1}\mathcal{F}_x^{-1}(s^{\alpha}) \odot u(x,t) \ominus_H (-1)\lambda(iw) \odot u(x,t) = \mathbf{L}_t^{-1}\mathcal{F}_x^{-1}\left(s^{\alpha-1}G(w)\right)$$

Subject to the integrals and difference exist. In this case if the fuzzy Laplace transform is applied for both sides,

$$\mathbf{L}_t\left(\mathbf{D}^{\alpha}_{\mathbf{C}_{i-gH}}u(t,x)\right) = s^{\alpha} \odot \mathbf{L}_t(u(t,x))\ominus_H s^{\alpha-1}g(x)$$

Because $u(0^+,x) = g(x)$

$$\mathbf{L}_t\left(\lambda\partial_{x_{i-gH}}u(t,x)\right) = \lambda\partial_{x_{i-gH}}\mathbf{L}_t(u(t,x))$$

Since

$$\mathbf{L}_t(u(t,x)) = \int\limits_0^{\infty} e^{-st} \odot u(t,x)dt = U_t(s,x)$$

Then

$$\mathbf{L}_t\left(\lambda\partial_{x_{i-gH}}u(t,x)\right) = \lambda\partial_{x_{i-gH}}U_t(s,x)$$

Finally,

$$s^{\alpha} \odot \mathbf{L}_t(u(t,x))\ominus_H s^{\alpha-1}g(x) = \lambda\partial_{x_{i-gH}}U_t(s,x)$$

Applying fuzzy Fourier to both sides,

$$s^{\alpha} \odot \mathcal{F}_x\mathbf{L}_t(u(t,x))\ominus_H s^{\alpha-1}\mathcal{F}_x(g(x)) = \lambda\mathcal{F}_x\left(\partial_{x_{i-gH}}U_t(s,x)\right)$$

Since,

$$\mathcal{F}_x\left\{\partial_{x_{i-gH}}U_t(s,x)\right\} = (-iw) \odot \mathcal{F}_x\{U_t(s,x)\}$$

By substituting,

$$s^{\alpha} \odot \mathcal{F}_x\mathbf{L}_t(u(t,x))\ominus_H s^{\alpha-1}G(w) = \lambda(-iw) \odot \mathcal{F}\{U_t(s,x)\}$$

$$s^{\alpha} \odot \mathcal{F}_x\mathbf{L}_t(u(t,x))\ominus_H s^{\alpha-1}G(w) = \lambda(-iw) \odot \mathcal{F}_x\{\mathbf{L}_t(u(t,x))\}$$

Based on the definition of H-difference,

$$s^{\alpha} \odot \mathcal{F}_x\mathbf{L}_t(u(t,x))\ominus_H(-1)\lambda(iw) \odot \mathcal{F}_x\mathbf{L}_t(u(t,x)) = s^{\alpha-1}G(w)$$

Now first, the inverse Fourier the inverse Laplace are applied, then

$$\mathbf{L}_t^{-1}\mathcal{F}_x^{-1}(s^{\alpha}) \odot u(x,t)\ominus_H(-1)\lambda(iw) \odot u(x,t) = \mathbf{L}_t^{-1}\mathcal{F}_x^{-1}\left(s^{\alpha-1}G(w)\right)$$

Now, assume the solution is $ii-gH$ Caputo differentiability with respect to t and $ii-gH$ partial differentiability with respect to x, then it satisfies the following relation,

$$s^{\alpha-1} \odot \mathbf{L}_t^{-1}\mathcal{F}_x^{-1}(G(w)) = s^{\alpha} \odot u(t,x)\ominus_H(iw) \odot u(t,x)$$

Subject to the integrals and difference exist. In this case if the fuzzy Laplace transform is applied for both sides,

$$\mathbf{L}_t\left(\mathbf{D}^{\alpha}_{C_{ii-gH}}u(t,x)\right) = (-1)s^{\alpha-1} \odot g(x)\ominus_H(-1)s^{\alpha} \odot \mathbf{L}_t(u(t,x))$$

Because $u(0^+,x) = g(x)$

$$\mathbf{L}_t\left(\lambda\partial_{x_{ii-gH}}u(t,x)\right) = \lambda\partial_{x_{ii-gH}}U_t(s,x)$$

Finally,

$$(-1)s^{\alpha-1} \odot g(x)\ominus_H(-1)s^{\alpha} \odot \mathbf{L}_t(u(t,x)) = \lambda \odot \partial_{x_{ii-gH}}U_t(s,x)$$

Applying fuzzy Fourier to both sides,

$$(-1)s^{\alpha-1} \odot \mathcal{F}(g(x))\ominus_H(-1)s^{\alpha} \odot \mathcal{F}\mathbf{L}_t(u(t,x)) = \lambda \odot \mathcal{F}\left(\partial_{x_{ii-gH}}U_t(s,x)\right)$$

Since,

$$\mathcal{F}\left\{\partial_{x_{ii-gH}}U_t(s,x)\right\} = \ominus_H(-1)(iw) \odot \mathcal{F}\{U_t(s,x)\}$$

By substituting,

$$\begin{aligned}
&(-1)s^{\alpha-1} \odot G(w)\ominus_H(-1)s^{\alpha} \odot \mathcal{F}_x\mathbf{L}_t(u(t,x)) = \ominus_H(-1)(iw) \odot \mathcal{F}_x\{U_t(s,x)\}\\
&s^{\alpha-1} \odot G(w)\ominus_H s^{\alpha} \odot \mathcal{F}_x\mathbf{L}_t(u(t,x)) = \ominus_H(iw) \odot \mathcal{F}_x\mathbf{L}_t(u(t,x))\\
&s^{\alpha-1} \odot G(w)\ominus_H s^{\alpha} \odot v(s,w) = \ominus_H(iw) \odot v(s,w)
\end{aligned}$$

Based on the definition of H-difference,

$$s^{\alpha-1} \odot G(w) = s^{\alpha} \odot \mathcal{F}_x\mathbf{L}_t(u(t,x))\ominus_H(iw) \odot \mathcal{F}_x\mathbf{L}_t(u(t,x))$$

Now first, the inverse Fourier the inverse Laplace are applied,

$$s^{\alpha-1} \odot \mathbf{L}_t^{-1}\mathcal{F}_x^{-1}(G(w)) = s^{\alpha} \odot u(t,x)\ominus_H(iw) \odot u(t,x)$$

4.5 Fuzzy Impulsive Fractional Differential Equations

In this section, the concept of fuzzy fractional impulsive differential equations is considered, and its solution is going to be determined under some conditions. In the next chapter several numerical methods will be introduced. To this end, consider the following fuzzy fractional differential equations with not instantaneous impulsive, or impulsive fractional differential equations with fuzzy initial value.

$$\begin{cases} \boldsymbol{D}^{\alpha,p}_{\boldsymbol{CK}_{gH}}x(t) = f(t,x(t)), & t \in (t_k, s_k], \quad k = 0,1,2,\ldots,m \\ x(t) = I_k(t,x(t)), & t \in (s_{k-1}, t_k], \quad k = 1,2,\ldots,m \\ x(t_0) = x_0 \end{cases}$$

where $p > 0$, $0 < \alpha < 1$ and the functions $f : [t_0, T] \times \mathbb{F}_R \to \mathbb{F}_R$ and $I_k : (t_k, s_k] \times \mathbb{F}_R \to \mathbb{F}_R$ are jointly continuous at the points $t_0 < s_0 < t_1 < s_1 \cdots < t_m < s_m = T$. The fuzzy solution of the fuzzy impulsive problem $x(t)$ is a peace-wise continuous function on $(t_k, s_k]$ and $[t_0, T]$ and Caputo-Katugampola generalized fractional differentiable. By this assumption, the values $x(t_k^-)$ and $x(t_k^+)$ exist and they are equal, $x(t_k^-) = x(t_k^+)$. Indeed, in general, $x(t_k^+) \ominus_H x(t_k^-) = I_k(t, x(t))$. In the first discussion of Sect. 4.2, we discussed about the following remark about the equivalency of the solution of two differential and integral equations. In the other word, the continuous fuzzy function $x(t)$ is the solution of

$$\boldsymbol{D}^{\alpha,p}_{\boldsymbol{CK}_{gH}}x(t) = f(t,x(t)), \quad x(t_0) = x_0, \quad t \in [t_0, T], \quad 0 < \alpha < 1$$

If and only if, $x(t)$ satisfies the following integral equation,

$$x(t) \ominus_{gH} x(t_0) = \frac{p^{1-\alpha}}{\Gamma(\alpha)} \int_{t_0}^{t} s^{p-1}(t^p - s^p)^{\alpha-1} f(s, x(s))ds, \quad t \in [t_0, T]$$

Note Based on the definition of the gH-difference in the integral equation,

$$x(t) \ominus_{gH} x(t_0) = \frac{p^{1-\alpha}}{\Gamma(\alpha)} \int_{t_0}^{t} s^{p-1}(t^p - s^p)^{\alpha-1} f(s, x(s))ds, \quad t \in [t_0, T]$$

- In case the gH-difference is $i - gH$ difference,

$$x(t) = x(t_0) \oplus \frac{p^{1-\alpha}}{\Gamma(\alpha)} \int_{t_0}^{t} s^{p-1}(t^p - s^p)^{\alpha-1} f(s, x(s))ds, \quad t \in [t_0, T]$$

- In case the gH-difference is $ii - gH$ difference,

$$x(t) = x(t_0) \ominus_H (-1) \frac{p^{1-\alpha}}{\Gamma(\alpha)} \int_{t_0}^{t} s^{p-1} (t^p - s^p)^{\alpha-1} f(s, x(s)) ds, \quad t \in [t_0, T]$$

In a similar way and assuming the jointly continuity of $x(t)$, we can show the same corresponding relation for the following fractional equations.

Note The jointly continuous and gH-differentiable fuzzy function $x(t)$ is the solution of the following CK-fractional differential equations with fuzzy initial values

$$\begin{cases} \boldsymbol{D}^{\alpha,p}_{\boldsymbol{CK_{gH}}} x(t) = f(t, x(t)), & t \in [t_0, T] \\ x(t_1) = x_1 \in \mathbb{F}_R, & t_1 > t_0 \\ x(t_0) = x_0 \in \mathbb{F}_R \end{cases}$$

If and only if, the same function $x(t)$ is the solution of the following fuzzy fractional integral equation,

$$x(t) \ominus_{gH} x_*(t_1) = \frac{p^{1-\alpha}}{\Gamma(\alpha)} \int_{t_0}^{t} s^{p-1} (t^p - s^p)^{\alpha-1} f(s, x(s)) ds, \quad t \in (t_0, T]$$

where $x(t) \ominus_{gH} x_*(t_1)$ exists and

$$x_*(t_1) \ominus_{gH} x_1 = \frac{p^{1-\alpha}}{\Gamma(\alpha)} \int_{t_0}^{t_1} s^{p-1} (t_1^p - s^p)^{\alpha-1} f(s, x(s)) ds, \quad t \in (t_0, T]$$

To show the assertion, it is enough to take the RL-derivative both sides of

$$\boldsymbol{D}^{\alpha,p}_{\boldsymbol{CK_{gH}}} x(t) = f(t, x(t))$$

Since we have this relation in the previous Chap. 3,

$$\boldsymbol{I}^{\alpha,p}_{\boldsymbol{RL}} \boldsymbol{D}^{\alpha,p}_{\boldsymbol{CK_{gH}}} x(t) = x(t) \ominus_{gH} x(t_0)$$

then

$$x(t) \ominus_{gH} x(t_0) = \boldsymbol{I}^{\alpha,p}_{\boldsymbol{RL}} f(t, x(t))$$

Also, we have,

$$\boldsymbol{I}_{\boldsymbol{RL}}^{\alpha,p} f(t,x(t)) = \frac{p^{1-\alpha}}{\Gamma(\alpha)} \odot \int_{t_0}^{t} s^{p-1}(t^p - s^p)^{\alpha-1} \odot f(s,x(s))ds$$

Therefor,

$$x(t) \ominus_{gH} x(t_0) = \frac{p^{1-\alpha}}{\Gamma(\alpha)} \odot \int_{t_0}^{t} s^{p-1}(t^p - s^p)^{\alpha-1} \odot f(s,x(s))ds$$

The proof is completed.

Remark Based on the definition of gH-difference the following items can be concluded,

- Case 1, $x(t) \ominus_{i-gH} x_*(t_1)$ and $x_*(t_1) \ominus_{i-gH} x_1$.

$$x(t) = x_*(t_1) \oplus \frac{p^{1-\alpha}}{\Gamma(\alpha)} \int_{t_0}^{t} s^{p-1}(t^p - s^p)^{\alpha-1} \odot f(s,x(s))ds$$

$$x_*(t_1) = x_1 \oplus \frac{p^{1-\alpha}}{\Gamma(\alpha)} \int_{t_0}^{t_1} s^{p-1}(t_1^p - s^p)^{\alpha-1} \odot f(s,x(s))ds$$

Then

$$x(t) = x_1 \oplus \frac{p^{1-\alpha}}{\Gamma(\alpha)} \int_{t_0}^{t_1} s^{p-1}(t_1^p - s^p)^{\alpha-1} \odot f(s,x(s))ds$$

$$\oplus \frac{p^{1-\alpha}}{\Gamma(\alpha)} \int_{t_0}^{t} s^{p-1}(t^p - s^p)^{\alpha-1} \odot f(s,x(s))ds$$

- Case 2, $x(t) \ominus_{i-gH} x_*(t_1)$ and $x_*(t_1) \ominus_{ii-gH} x_1$

$$x(t) = x_*(t_1) \oplus \frac{p^{1-\alpha}}{\Gamma(\alpha)} \int_{t_0}^{t} s^{p-1}(t^p - s^p)^{\alpha-1} \odot f(s,x(s))ds$$

$$x_*(t_1) = x_1 \ominus_H (-1) \frac{p^{1-\alpha}}{\Gamma(\alpha)} \int_{t_0}^{t_1} s^{p-1}(t_1^p - s^p)^{\alpha-1} \odot f(s,x(s))ds$$

Then

$$x(t) = x_1 \ominus_H (-1)\frac{p^{1-\alpha}}{\Gamma(\alpha)}\int_{t_0}^{t_1} s^{p-1}\left(t_1^p - s^p\right)^{\alpha-1} \odot f(s, x(s))ds$$

$$\oplus \frac{p^{1-\alpha}}{\Gamma(\alpha)}\int_{t_0}^{t} s^{p-1}(t^p - s^p)^{\alpha-1} \odot f(s, x(s))ds$$

- Case 3, $x(t) \ominus_{ii-gH} x_*(t_1)$ and $x_*(t_1) \ominus_{i-gH} x_1$

$$x(t) = x_*(t_1) \ominus_H (-1)\frac{p^{1-\alpha}}{\Gamma(\alpha)}\int_{t_0}^{t} s^{p-1}(t^p - s^p)^{\alpha-1} \odot f(s, x(s))ds$$

$$x_*(t_1) = x_1 \oplus \frac{p^{1-\alpha}}{\Gamma(\alpha)}\int_{t_0}^{t_1} s^{p-1}\left(t_1^p - s^p\right)^{\alpha-1} \odot f(s, x(s))ds$$

Then

$$x(t) = x_1 \oplus \frac{p^{1-\alpha}}{\Gamma(\alpha)}\int_{t_0}^{t_1} s^{p-1}\left(t_1^p - s^p\right)^{\alpha-1} \odot f(s, x(s))ds$$

$$\ominus_H (-1)\frac{p^{1-\alpha}}{\Gamma(\alpha)}\int_{t_0}^{t} s^{p-1}(t^p - s^p)^{\alpha-1} \odot f(s, x(s))ds$$

- Case 4, $x(t) \ominus_{ii-gH} x_*(t_1)$ and $x_*(t_1) \ominus_{ii-gH} x_1$

$$x(t) = x_*(t_1) \ominus_H (-1)\frac{p^{1-\alpha}}{\Gamma(\alpha)}\int_{t_0}^{t} s^{p-1}(t^p - s^p)^{\alpha-1} \odot f(s, x(s))ds$$

$$x_*(t_1) = x_1 \ominus_H (-1)\frac{p^{1-\alpha}}{\Gamma(\alpha)}\int_{t_0}^{t_1} s^{p-1}\left(t_1^p - s^p\right)^{\alpha-1} \odot f(s, x(s))ds$$

Then

$$x(t) = x_1 \ominus_H (-1)\frac{p^{1-\alpha}}{\Gamma(\alpha)}\int_{t_0}^{t_1} s^{p-1}\left(t_1^p - s^p\right)^{\alpha-1} \odot f(s, x(s))ds$$

$$\ominus_H (-1)\frac{p^{1-\alpha}}{\Gamma(\alpha)}\int_{t_0}^{t} s^{p-1}(t^p - s^p)^{\alpha-1} \odot f(s, x(s))ds$$

Now we are going to discuss the solution of extended conditions on our fuzzy impulsive fractional differential equation.

Where $p > 0$, $0 < \alpha < 1$ and the functions $f : [t_0, T] \times \mathbb{F}_R \to \mathbb{F}_R$ and $I_k : (t_k, s_k] \times \mathbb{F}_R \to \mathbb{F}_R$ are jointly continuous at the points $t_0 < s_0 < t_1 < s_1 \cdots < t_m < s_m = T$.

Remark With the same conditions on the jointly continuous fuzzy functions and for > 0, $0 < \alpha < 1, f : [t_0, T] \times \mathbb{F}_R \to \mathbb{F}_R$, and $I_k : (t_k, s_k] \times \mathbb{F}_R \to \mathbb{F}_R$, the peace-wise function $x(t)$ is the solution of the fuzzy fractional impulsive equation,

$$\begin{cases} \boldsymbol{D}_{\boldsymbol{CK}_{gH}}^{\alpha,p} x(t) = f(t, x(t)), & t \in (t_k, s_k], \quad k = 0, 1, 2, \ldots, m \\ x(t) = I_k(t, x(t)), & t \in (s_{k-1}, t_k], \quad k = 1, 2, \ldots, m \\ x(t_0) = x_0 \end{cases}$$

If and only if, $x(t)$ satisfies the following equations and conditions.

$$\begin{cases} x(t) \ominus_{gH} x_0 = \frac{p^{1-\alpha}}{\Gamma(\alpha)} \int_{t_0}^{t} s^{p-1}(t^p - s^p)^{\alpha-1} \odot f(s, x(s))ds, & t \in (t_0, s_0] \\ x(t) = I_k(t, x(t)), & t \in (s_{k-1}, t_k] \\ x(t) \ominus_{gH} x_*(t_k) = \frac{p^{1-\alpha}}{\Gamma(\alpha)} \int_{t_0}^{t} s^{p-1}(t^p - s^p)^{\alpha-1} \odot f(s, x(s))ds, t \in (t_k, s_k] \end{cases}$$

where $x(t_k^+) \ominus_H x(t_k^-) = I_k(t, x(t))$ and

$$x_*(t_k) = I_k(t_k, x(t_k)) \ominus_H \frac{p^{1-\alpha}}{\Gamma(\alpha)} \int_{t_0}^{t_k} s^{p-1}(t^p - s^p)^{\alpha-1} f(s, x(s))ds$$

If $length\left(x(t) \ominus_{gH} x_*(t_k)\right)$ is increasing function or $x(t) \ominus_{i-gH} x_*(t_k)$ exists on $(t_k, s_k]$. Also,

$$x_*(t_k) = I_k(t_k, x(t_k)) \oplus (-1)\frac{p^{1-\alpha}}{\Gamma(\alpha)} \int_{t_0}^{t_k} s^{p-1}(t^p - s^p)^{\alpha-1} \odot f(s, x(s))ds$$

If $length\big(x(t)\ominus_{gH}x_*(t_k)\big)$ is decreasing function or $x(t)\ominus_{ii-gH}x_*(t_k)$ exists on $(t_k, s_k]$.

Proof We know, the continuous fuzzy function $x(t)$ is the solution of

$$\boldsymbol{D}^{\alpha,p}_{\boldsymbol{CK}_{gH}}x(t) = f(t, x(t)), \quad x(t_0) = x_0, \quad t \in [t_0, T], \quad 0<\alpha<1$$

If and only if, $x(t)$ satisfies the following integral equation,

$$x(t)\ominus_{gH}x(t_0) = \frac{p^{1-\alpha}}{\Gamma(\alpha)}\int_{t_0}^{t} s^{p-1}(t^p - s^p)^{\alpha-1}\odot f(s, x(s))ds, \quad t \in [t_0, T]$$

And on the interval $(s_0, t_1]$ we define the function $x(t_1) = I_1(t_1, x(t_1))$, so for $t \in (t_1, s_1]$ we have the following impulsive equation,

$$\begin{cases} \boldsymbol{D}^{\alpha,p}_{\boldsymbol{CK}_{gH}}x(t) = f(t, x(t)), & t \in (t_1, s_1] \\ x(t_1) = I_1(t_1, x(t_1)) \end{cases}$$

And it is equivalent to

$$x(t)\ominus_{gH}x_*(t_1) = \frac{p^{1-\alpha}}{\Gamma(\alpha)}\int_{t_0}^{t} s^{p-1}(t^p - s^p)^{\alpha-1}\odot f(s, x(s))ds, \quad t \in (t_1, s_1]$$

Where

$$x_*(t_1) = I_1(t_1, x(t_1))\ominus_H \frac{p^{1-\alpha}}{\Gamma(\alpha)}\int_{t_0}^{t_1} s^{p-1}(t^p - s^p)^{\alpha-1}\odot f(s, x(s))ds$$

If $length\big(x(t)\ominus_{gH}x_0\big)$ is increasing function or $x(t)\ominus_{i-gH}x_0$ exists on $(t_1, s_1]$. Also,

$$x_*(t_1) = I_1(t_1, x(t_1)) \oplus (-1)\frac{p^{1-\alpha}}{\Gamma(\alpha)}\int_{t_0}^{t_1} s^{p-1}(t^p - s^p)^{\alpha-1}\odot f(s, x(s))ds$$

If $length\big(x(t)\ominus_{gH}x_0\big)$ is decreasing function or $x(t)\ominus_{ii-gH}x_0$ exists on $(t_1, s_1]$.

Also, on the interval $(s_1, t_2]$ we define the function $x(t_2) = I_2(t_2, x(t_2))$, so for $t \in (t_2, s_2]$ we have the following impulsive equation,

$$\begin{cases} \boldsymbol{D}^{\alpha,p}_{\boldsymbol{CK}_{gH}}x(t) = f(t, x(t)), & t \in (t_2, s_2] \\ x(t_2) = I_2(t_2, x(t_2)) \end{cases}$$

And it is equivalent to

$$x(t)\ominus_{gH}x_*(t_2) = \frac{p^{1-\alpha}}{\Gamma(\alpha)}\int_{t_0}^{t_2} s^{p-1}(t^p - s^p)^{\alpha-1}\odot f(s,x(s))ds, \quad t\in(t_2,s_2]$$

Where

$$x_*(t_2) = I_2(t_2,x(t_2))\ominus_H \frac{p^{1-\alpha}}{\Gamma(\alpha)}\int_{t_0}^{t_2} s^{p-1}(t^p - s^p)^{\alpha-1}\odot f(s,x(s))ds$$

If $length\big(x(t)\ominus_{gH}x_0\big)$ is increasing function or $x(t)\ominus_{i-gH}x_0$ exists on $(t_2,s_2]$. Also,

$$x_*(t_2) = I_2(t_2,x(t_2)) \oplus (-1)\frac{p^{1-\alpha}}{\Gamma(\alpha)}\int_{t_0}^{t_2} s^{p-1}(t^p - s^p)^{\alpha-1}\odot f(s,x(s))ds$$

If $length\big(x(t)\ominus_{gH}x_0\big)$ is decreasing function or $x(t)\ominus_{ii-gH}x_0$ exists on $(t_2,s_2]$. By proceeding the process for $t\in(s_{k-1},t_k], k=3,4,\ldots,m$, $x(t_k)=I_k(t_k,x(t_k))$, so for $t\in(t_k,s_k]$ we have the following impulsive equation,

$$\begin{cases} \boldsymbol{D}^{\alpha,p}_{\boldsymbol{CK}_{gH}}x(t) = f(t,x(t)), & t\in(t_k,s_k] \\ x(t_k) = I_k(t_k,x(t_k)) \end{cases}$$

And it is equivalent to

$$x(t)\ominus_{gH}x_*(t_k) = \frac{p^{1-\alpha}}{\Gamma(\alpha)}\int_{t_0}^{t_k} s^{p-1}(t^p - s^p)^{\alpha-1}\odot f(s,x(s))ds, \quad t\in(t_k,s_k]$$

where

$$x_*(t_k) = I_k(t_k,x(t_k))\ominus_H \frac{p^{1-\alpha}}{\Gamma(\alpha)}\int_{t_0}^{t_k} s^{p-1}(t^p - s^p)^{\alpha-1}\odot f(s,x(s))ds$$

If $length\big(x(t)\ominus_{gH}x_0\big)$ is increasing function or $x(t)\ominus_{i-gH}x_0$ exists on $(t_k,s_k]$. Also,

$$x_*(t_k) = I_k(t_k,x(t_k)) \oplus (-1)\frac{p^{1-\alpha}}{\Gamma(\alpha)}\int_{t_0}^{t_k} s^{p-1}(t^p - s^p)^{\alpha-1}\odot f(s,x(s))ds$$

If $length\big(x(t)\ominus_{gH}x_0\big)$ is decreasing function or $x(t)\ominus_{ii-gH}x_0$ exists on $(t_k,s_k]$. Thus, the proof is completed.

4.6 Concrete Solution of Fractional Differential Equations

In this section, the fuzzy linear fractional differential equations under Riemann–Liouville gH-differentiability as the following fuzzy initial value problems are studied.

$$\begin{cases} \boldsymbol{D}^{\alpha}_{\boldsymbol{RL}_{gH}} x(t) = \lambda \odot x(t) \oplus y(t), & t \in [t_o, T], \quad 0<\alpha<1 \\ \boldsymbol{D}^{\alpha-1}_{\boldsymbol{RL}_{gH}} x(t_0) = x_0 \in \mathbb{F}_R \end{cases}$$

where $x(t)$, is continuous fuzzy number valued function that belongs to the space of all Lebesque integrable fuzzy number valued functions on $[t_0, T]$. Also $y(t) \in \mathbb{F}_R$.

To this end, some of the previous results on solutions of these equations are concreted. The new solutions by using the fractional hyperbolic functions and their properties are obtained, in details.

4.6.1 Fractional Hyperbolic Functions

Here, the fractional hyperbolic functions and their properties that will be used in the next sections are pointed out. As it is mentioned in the chapter three, the Mittag-Leffler function frequently used in the solutions of fractional order systems and it is defined as,

$$E_{\alpha,\beta}(t) = \sum_{k=0}^{\infty} \frac{t^k}{\Gamma(\alpha k + \beta)}, \quad \alpha > 0, \quad \beta > 0$$

The fractional hyperbolic functions are defined as,

$$\cos h_{\alpha,\beta}(t) = \sum_{k=0}^{\infty} \frac{t^{2k}}{\Gamma(2\alpha k + \beta)} = E_{2\alpha,\beta}(t^2), \quad \alpha > 0, \quad \beta > 0$$

$$\sin h_{\alpha,\beta}(t) = \sum_{k=0}^{\infty} \frac{t^{2k+1}}{\Gamma(2\alpha k + \alpha + \beta)} = tE_{2\alpha,\alpha+\beta}(t^2), \quad \alpha > 0, \quad \beta > 0$$

where $\cos h_{\alpha,\beta}(t) \geq 0$ and even function for all $t \in R$ and $\sin h_{\alpha,\beta}(t)$ is an odd function for all $t \in R$ and,

$$\begin{cases} \sin h_{\alpha,\beta}(t) \geq 0, t \geq 0 \\ \cos h_{\alpha,\beta}(t) < 0, t < 0 \end{cases}$$

Note Please note that in case, $\alpha = \beta$, $\cos h_{\alpha,\alpha}(t) := Ch_{\alpha}(t)$ and $\sin h_{\alpha,\alpha}(t) := Sh_{\alpha}(t)$, and for all $t \in R$ and $0<\alpha \leq 1$, the function $E_{\alpha,\alpha}(t) > 0$.

Property It is easy to see that,

$$Ch_\alpha(t) + Sh_\alpha(t) = E_{\alpha,\alpha}(t), \quad Ch_\alpha(t) - Sh_\alpha(t) = E_{\alpha,\alpha}(-t)$$
$$\cos h(t) = E_{2,1}\left(t^2\right) = Ch_1(t), \quad \sin h(t) = tE_{2,2}\left(t^2\right) = Sh_1(t)$$

4.6.2 Some Derivation Rules for the Fractional Hyperbolic Functions

Suppose that $x(t) \in [t_0, T]$, $\lambda \in R$ and $\gamma \leq \alpha$ we have,

- $\mathbf{D}^{\gamma}_{\mathbf{RL}_{gH}}(t^{\alpha-1}Ch_\alpha(\lambda t^\alpha)) = \begin{cases} t^{\alpha-\gamma-1}\cos h_{\alpha,\alpha-\gamma}(\lambda t^\alpha), & \alpha > \gamma \\ \lambda t^{\alpha-1}Sh_\alpha(\lambda t^\alpha), & \alpha = \gamma \end{cases}$
- $\mathbf{D}^{\gamma}_{\mathbf{RL}_{gH}}(t^{\alpha-1}Sh_\alpha(\lambda t^\alpha)) = \begin{cases} t^{\alpha-\gamma-1}\sin h_{\alpha,\alpha-\gamma}(\lambda t^\alpha), & \alpha > \gamma \\ \lambda t^{\alpha-1}Ch_\alpha(\lambda t^\alpha), & \alpha = \gamma \end{cases}$

The proofs are straightforward, based on the definition of RL-derivative,

$$\mathbf{D}^{\alpha}_{\mathbf{RL}_{gH}}x(t) = \frac{1}{\Gamma(1-\alpha)} \odot \frac{d}{dt}\int_{t_0}^{t} \frac{x(\tau)}{(t-\tau)^\alpha}d\tau$$

Now,

$$\begin{aligned}\mathbf{D}^{\alpha}_{\mathbf{RL}_{gH}}(t^{\alpha-1}Ch_\alpha(\lambda t^\alpha)) &= \frac{1}{\Gamma(1-\gamma)} \odot \frac{d}{dt}\int_{t_0}^{t} \frac{\tau^{\alpha-1}Ch_\alpha(\lambda\tau^\alpha)}{(t-\tau)^\gamma}d\tau \\ &= \frac{d}{dt}\left(\sum_{k=0}^{\infty}\frac{\lambda^{2k}t^{2k\alpha+\alpha-\gamma}}{\Gamma(2k\alpha+\alpha-\gamma+1)}\right) \\ &= \begin{cases} \sum_{k=0}^{\infty}\frac{\lambda^{2k}t^{2k\alpha-\alpha-\gamma-1}}{\Gamma(2k\alpha+\alpha-\gamma)}, \alpha > \gamma \\ \sum_{k=0}^{\infty}\frac{\lambda^{2k-2}t^{2k\alpha-2\alpha-1}}{\Gamma(2k\alpha+2\alpha)}, \alpha = \gamma \end{cases} = \begin{cases} t^{\alpha-\gamma-1}\cos h_{\alpha,\alpha-\gamma}(\lambda t^\alpha), & \alpha > \gamma \\ \lambda t^{\alpha-1}Sh_\alpha(\lambda t^\alpha), & \alpha = \gamma \end{cases}\end{aligned}$$

The proof of the second case is similar to the first case.

Note As an immediate results,

$$\mathbf{D}^{\alpha}_{\mathbf{RL}_{gH}}(t^{\alpha-1}Ch_\alpha(\lambda t^\alpha)) = \begin{cases} t^{\alpha-\gamma-1}E_{\alpha,\alpha-\gamma}(\lambda t^\alpha), & \alpha > \gamma \\ \lambda t^{\alpha-1}E_{\alpha,\alpha}(\lambda t^\alpha), & \alpha = \gamma \end{cases}$$

Now, we are going to find the concrete solution of our mentioned fractional differential equation. Consider,

$$\begin{cases} \boldsymbol{D}^{\alpha}_{\boldsymbol{RL}_{gH}} x(t) = \lambda \odot x(t) \oplus y(t), \quad t \in [t_o, T], \quad 0 < \alpha < 1 \\ \boldsymbol{D}^{\alpha-1}_{\boldsymbol{RL}_{gH}} x(t_0) = x_0 \in \mathbb{F}_R \end{cases}$$

Regarding to the definition of gH-differentiability, we have two types of differentiability and also two types of solutions, $i - RL_{gH}$ solution and $ii - RL_{gH}$ solution. Here, our discussion and strategy is using length function. First, we are going to cover the concept of types of differentiability in accordance with the definition of length function.

Remark

- $length\left(\boldsymbol{D}^{\alpha}_{\boldsymbol{RL}_{gH}} x(t)\right) = \boldsymbol{D}^{\alpha}_{\boldsymbol{RL}_{gH}} length(x(t))$ iff $x(t)$ is $i - RL_{gH}$ differentiable
- $length\left(\boldsymbol{D}^{\alpha}_{\boldsymbol{RL}_{gH}} x(t)\right) = -\boldsymbol{D}^{\alpha}_{\boldsymbol{RL}_{gH}} length(x(t))$ iff $x(t)$ is $ii - RL_{gH}$ differentiable

To show the propositions, first we suppose that $x(t)$ is $i - RL_{gH}$ differentiable, then

$$\begin{aligned} \boldsymbol{D}^{\alpha}_{\boldsymbol{RL}} length(x(t)) &= \boldsymbol{D}^{\alpha}_{\boldsymbol{RL}}(x_u(t,r) - x_l(t,r)) = \boldsymbol{D}^{\alpha}_{\boldsymbol{RL}} x_u(t,r) - \boldsymbol{D}^{\alpha}_{\boldsymbol{RL}} x_l(t,r) \\ &= \boldsymbol{D}^{\alpha}_{\boldsymbol{RL}_u} x(t,r) - \boldsymbol{D}^{\alpha}_{\boldsymbol{RL}_l} x(t,r) = length\left(\boldsymbol{D}^{\alpha}_{\boldsymbol{RL}_{gH}} x(t)\right) \end{aligned}$$

And if $x(t)$ is $ii - RL_{gH}$ differentiable, then

$$\begin{aligned} \boldsymbol{D}^{\alpha}_{\boldsymbol{RL}} length(x(t)) &= \boldsymbol{D}^{\alpha}_{\boldsymbol{RL}}(x_u(t,r) - x_l(t,r)) = \boldsymbol{D}^{\alpha}_{\boldsymbol{RL}} x_u(t,r) - \boldsymbol{D}^{\alpha}_{\boldsymbol{RL}} x_l(t,r) \\ &= \boldsymbol{D}^{\alpha}_{\boldsymbol{RL}_l} x(t,r) - \boldsymbol{D}^{\alpha}_{\boldsymbol{RL}_u} x(t,r) = -length\left(\boldsymbol{D}^{\alpha}_{\boldsymbol{RL}_{gH}} x(t)\right) \end{aligned}$$

Now converse,

$$\begin{aligned} length\left(\boldsymbol{D}^{\alpha}_{\boldsymbol{RL}} x(t)\right) = \max&\left\{\boldsymbol{D}^{\alpha}_{\boldsymbol{RL}} x_l(t,r), \boldsymbol{D}^{\alpha}_{\boldsymbol{RL}} x_u(t,r)\right\} \\ - \min&\left\{\boldsymbol{D}^{\alpha}_{\boldsymbol{RL}} x_l(t,r), \boldsymbol{D}^{\alpha}_{\boldsymbol{RL}} x_u(t,r)\right\} \geq 0 \end{aligned}$$

If you suppose that

$$\begin{aligned} \max\left\{\boldsymbol{D}^{\alpha}_{\boldsymbol{RL}} x_l(t,r), \boldsymbol{D}^{\alpha}_{\boldsymbol{RL}} x_u(t,r)\right\} = \boldsymbol{D}^{\alpha}_{\boldsymbol{RL}} x_u(t,r) \\ \min\left\{\boldsymbol{D}^{\alpha}_{\boldsymbol{RL}} x_l(t,r), \boldsymbol{D}^{\alpha}_{\boldsymbol{RL}} x_u(t,r)\right\} = \boldsymbol{D}^{\alpha}_{\boldsymbol{RL}} x_l(t,r) \end{aligned}$$

Thus, $\boldsymbol{D}^{\alpha}_{\boldsymbol{RL}} x_u(t,r) - \boldsymbol{D}^{\alpha}_{\boldsymbol{RL}} x_l(t,r) \geq 0$ and it means $length\left(\boldsymbol{D}^{\alpha}_{\boldsymbol{RL}_{gH}} x(t)\right) \geq 0$ and it points out the differential is defined as $i - RL_{gH}$ differentiability.

If you suppose that

$$\max\left\{\boldsymbol{D}^{\alpha}_{\boldsymbol{RL}}x_l(t,r), \boldsymbol{D}^{\alpha}_{\boldsymbol{RL}}x_u(t,r)\right\} = \boldsymbol{D}^{\alpha}_{\boldsymbol{RL}}x_l(t,r)$$
$$\min\left\{\boldsymbol{D}^{\alpha}_{\boldsymbol{RL}}x_l(t,r), \boldsymbol{D}^{\alpha}_{\boldsymbol{RL}}x_u(t,r)\right\} = \boldsymbol{D}^{\alpha}_{\boldsymbol{RL}}x_u(t,r)$$

Thus, $\boldsymbol{D}^{\alpha}_{\boldsymbol{RL}}x_u(t,r) - \boldsymbol{D}^{\alpha}_{\boldsymbol{RL}}x_l(t,r) \leq 0$ and it means $length\left(\boldsymbol{D}^{\alpha}_{\boldsymbol{RL_{gH}}}x(t)\right) \leq 0$ and it points out the differential is defined as $ii - RL_{gH}$ differentiability.

Concerning, $length\left(\boldsymbol{D}^{\alpha}_{\boldsymbol{RL_{gH}}}x(t)\right) = 0$ at some point t, it can be concluded that the derivative must be a scalar, $\boldsymbol{D}^{\alpha}_{\boldsymbol{RL_{gH}}}x(t) = \{c\}$. Therefor,

$$\boldsymbol{D}^{\alpha}_{\boldsymbol{RL}}length(x(t)) = 0 = length\left(\boldsymbol{D}^{\alpha}_{\boldsymbol{RL_{gH}}}x(t)\right)$$

To discuss the solution of

$$\begin{cases} \boldsymbol{D}^{\alpha}_{\boldsymbol{RL}}x(t) = \lambda \odot x(t) \oplus y(t), \quad t \in [t_o, T], \quad 0<\alpha<1 \\ \boldsymbol{D}^{\alpha-1}_{\boldsymbol{RL_{gH}}}x(t_0) = x_0 \in \mathbb{F}_R \end{cases}$$

The sign of λ does have important role. If it is positive $\lambda > 0$ we have $i - RL_{gH}$ solution and in case $\lambda<0$ we have $ii - RL_{gH}$ solution.

- $i - RL_{gH}$ solution for $\lambda > 0$

$$x(t) = t^{\alpha-1}E_{\alpha,\alpha}(\lambda t^{\alpha}) \odot x_0 \oplus \int_{t_0}^{t} (t-\tau)^{\alpha-1}E_{\alpha,\alpha}(\lambda(t-\tau)^{\alpha}) \odot y(\tau)d\tau$$

- $ii - RL_{gH}$ solution for $\lambda<0$

$$x(t) = t^{\alpha-1}E_{\alpha,\alpha}(\lambda t^{\alpha}) \odot x_0 \ominus_H (-1) \int_{t_0}^{t} (t-\tau)^{\alpha-1}E_{\alpha,\alpha}(\lambda(t-\tau)^{\alpha}) \odot y(\tau)d\tau$$

To show that $x(t)$ is the solution, we use the length function in each case. Since the Mittag-Leffler function is positive for $0<\alpha<1$ then

- $i - RL_{gH}$ solution for $\lambda > 0$

$$\begin{aligned} length(x(t)) &= t^{\alpha-1}E_{\alpha,\alpha}(\lambda t^{\alpha})length(x_0) \\ &+ \int_{t_0}^{t} (t-\tau)^{\alpha-1}E_{\alpha,\alpha}(\lambda(t-\tau)^{\alpha})length(y(\tau))d\tau \end{aligned}$$

- $ii - RL_{gH}$ solution for $\lambda < 0$

$$length(x(t)) = t^{\alpha-1} E_{\alpha,\alpha}(\lambda t^{\alpha}) length(x_0) \ominus_H (-1) \int_{t_0}^{t} (t-\tau)^{\alpha-1} E_{\alpha,\alpha}(\lambda (t-\tau)^{\alpha}) length(y(\tau)) d\tau$$

Using the RL-derivative we have,

- $i - RL_{gH}$ solution for $\lambda > 0$

$$\begin{aligned} &\boldsymbol{D}^{\alpha}_{\boldsymbol{RL_{gH}}} length(x(t)) \\ &\quad + \lambda \int_{t_0}^{t} (t-\tau)^{\alpha-1} E_{\alpha,\alpha}(\lambda (t-\tau)^{\alpha}) length(y(\tau)) d\tau \\ &+ length(y(t)) = \lambda length(y(\tau)) + length(y(t)) \\ &= length(\lambda \odot x(t) \oplus y(t)) = length\left(\boldsymbol{D}^{\alpha}_{\boldsymbol{RL_{gH}}} x(t)\right) \end{aligned}$$

- $ii - RL_{gH}$ solution for $\lambda < 0$

$$\begin{aligned} &= \lambda t^{\alpha-1} E_{\alpha,\alpha}(\lambda t^{\alpha}) length(x_0) + \lambda \int_{t_0}^{t} (t-\tau)^{\alpha-1} E_{\alpha,\alpha}(\lambda (t-\tau)^{\alpha}) length(y(\tau)) d\tau \\ &\quad - length(y(t)) = \lambda length(y(\tau)) - length(y(t)) \\ &= -(-\lambda length(y(\tau)) + length(y(t))) = -length(\lambda \odot x(t) \oplus y(t)) \\ &= -length\left(\boldsymbol{D}^{\alpha}_{\boldsymbol{RL_{gH}}} x(t)\right) \end{aligned}$$

The proof is completed.

Note Please notice that, if the H-difference in $ii - RL_{gH}$ solution does not exist then, based on the above mentioned results, we cannot say anything about the existence of solution for the problem in case $\lambda < 0$. Furthermore, the behavior of the solution function should reflect the real behavior of a system. Therefore, if the solution of the equation is not unique, then we may sometimes choose a better solution between two solutions, for example, we can study the real system and choose the solution which has better reflects from behavior of the system. So it would be better to find the $i - RL_{gH}$ solution for $\lambda < 0$ and $ii - RL_{gH}$ solution for $\lambda > 0$. To this end, we first, search the solution functions as the following result.

Remark If in the problem

$$\begin{cases} D^{\alpha}_{RL_{gH}} x(t) = \lambda \odot x(t) \oplus y(t), & t \in [t_o, T], \quad 0<\alpha<1 \\ D^{\alpha-1}_{RL_{gH}} x(t_0) = x_0 \in \mathbb{F}_R \end{cases}$$

The initial condition is a symmetric fuzzy number, $x_{0,l}(r) = -x_{0,u}(r)$ and also suppose $y(t) = g(x) \odot x_0$ where $g(x)$, is a real continuous function on $[t_0, T]$. Then

- $i - RL_{gH}$ solution for $\lambda<0$

$$x_1(t) = t^{\alpha-1} E_{\alpha,\alpha}(-\lambda t^{\alpha}) \odot x_0 \oplus \int_{t_0}^{t} (t-\tau)^{\alpha-1} E_{\alpha,\alpha}(-\lambda (t-\tau)^{\alpha}) \odot y(\tau) d\tau$$

- $ii - RL_{gH}$ solution for $\lambda > 0$

$$x_2(t) = t^{\alpha-1} E_{\alpha,\alpha}(-\lambda t^{\alpha}) \odot x_0 \ominus_H (-1) \int_{t_0}^{t} (t-\tau)^{\alpha-1} E_{\alpha,\alpha}(-\lambda (t-\tau)^{\alpha}) \odot y(\tau) d\tau$$

Provided that the H-difference exists. In fact, it seems that $x_1(t)$ is $i - RL_{gH}$ differentiable for $-\lambda > 0$ or $\lambda<0$ and $x_2(t)$ is $ii - RL_{gH}$ differentiable for $-\lambda<0$ or $\lambda > 0$. It is so easy to verify that $x_i(t, r) = [x_l(t, r), x_u(t, r)], i = 1, 2$ is symmetric because x_0 is symmetric. The level-wise form of $i - RL_{gH}$ solution,

$$x_{1,l}(t, r) = t^{\alpha-1} E_{\alpha,\alpha}(-\lambda t^{\alpha}) \odot x_{0,l}(r) \oplus \int_{t_0}^{t} (t-\tau)^{\alpha-1} E_{\alpha,\alpha}(-\lambda (t-\tau)^{\alpha}) \odot y_l(\tau, r) d\tau$$

$$x_{1,u}(t, r) = t^{\alpha-1} E_{\alpha,\alpha}(-\lambda t^{\alpha}) \odot x_{0,u}(r) \oplus \int_{t_0}^{t} (t-\tau)^{\alpha-1} E_{\alpha,\alpha}(-\lambda (t-\tau)^{\alpha}) \odot y_u(\tau, r) d\tau$$

The corresponding differential equations on $t \in [t_o, T]$ are as,

$$\begin{cases} D^{\alpha}_{RL} x_{1,l}(t, r) = -\lambda \odot x_{1,l}(t, r) + y_l(t, r) \\ D^{\alpha-1}_{RL} x_{1,l}(t_0, r) = x_{0,l}(r) \end{cases}$$
$$\begin{cases} D^{\alpha}_{RL} x_{1,u}(t, r) = -\lambda \odot x_{1,u}(t, r) + y_u(t, r) \\ D^{\alpha-1}_{RL} x_{1,u}(t_0, r) = x_{0,u}(r) \end{cases}$$

Since $x_l(t,r) = -x_u(t,r)$ then

$$\begin{cases} \mathbf{D}_{RL}^{\alpha} x_{1,l}(t,r) = \lambda \odot x_{1,u}(t,r) + y_l(t,r) \\ \mathbf{D}_{RL}^{\alpha} x_{1,u}(t,r) = \lambda \odot x_{1,l}(t,r) + y_u(t,r) \\ \mathbf{D}_{RL}^{\alpha-1} x_{1,l}(t_0,r) = x_{0,l}(r) \\ \mathbf{D}_{RL}^{\alpha-1} x_{1,u}(t_0,r) = x_{0,u}(r) \end{cases}$$

We know that this a system of fractional differential equations and the solutions are the solution of the main problem when $\lambda < 0$. The same process can be done for the $ii - RL_{gH}$ solution, and we have

$$\begin{cases} \mathbf{D}_{RL}^{\alpha} x_{2,l}(t,r) = \lambda \odot x_{2,u}(t,r) + y_l(t,r) \\ \mathbf{D}_{RL}^{\alpha} x_{2,u}(t,r) = \lambda \odot x_{2,l}(t,r) + y_u(t,r) \\ \mathbf{D}_{RL}^{\alpha-1} x_{2,l}(t_0,r) = x_{0,l}(r) \\ \mathbf{D}_{RL}^{\alpha-1} x_{2,u}(t_0,r) = x_{0,u}(r) \end{cases}$$

And the solution of this system is the solution of our problem in case $\lambda > 0$. The proof is completed.

Note Here we mentioned that the initial condition is a symmetric fuzzy number, $x_{0,l}(r) = -x_{0,u}(r)$ and also suppose $y(t) = g(x) \odot x_0$ with a real continuous function $g(x)$.

Now assume that these two conditions are not met. In this situation, with considering three cases for λ we should change the forms of $x_1(t), x_2(t)$ as follows,

$$x_1(t) = t^{\alpha-1}(Ch_\alpha(\lambda t^\alpha) \odot x_0 \ominus_H Sh_\alpha(\lambda t^\alpha) \odot x_0)$$
$$\oplus \int_{t_0}^{t} (t-\tau)^{\alpha-1}(Ch_\alpha(\lambda(t-\tau)^\alpha) \odot y(\tau) \ominus_H Sh_\alpha(\lambda(t-\tau)^\alpha) \odot y(\tau))d\tau$$

We connect the solution $x_1(t)$ to vector $c = (c_1, c_2, c_3, c_4), |c_i| = 1, i = 1,2,3,4$.

$$x_1(t;c) = t^{\alpha-1}(c_1 Ch_\alpha(\lambda t^\alpha) \odot x_0 \ominus_H c_2 Sh_\alpha(\lambda t^\alpha) \odot x_0)$$
$$\oplus \int_{t_0}^{t} (t-\tau)^{\alpha-1}(c_3 Ch_\alpha(\lambda(t-\tau)^\alpha) \odot y(\tau) \ominus_H c_4 Sh_\alpha(\lambda(t-\tau)^\alpha) \odot y(\tau))d\tau$$

This is true because, $length(x_1(t)) = length(x_1(t;c)), x > 0$. It means that $x_1(t;c)$ is $i-gH$ solution as well. Since $\lambda<0$,

$$x_{1,l}(t,r;c) = t^{\alpha-1}\left(Ch_\alpha(\lambda t^\alpha)(c_1 \odot x_0)_l(r) - Sh_\alpha(\lambda t^\alpha)(c_2 \odot x_0)_l(r)\right) + \int_{t_0}^{t} (t-\tau)^{\alpha-1}\left(Ch_\alpha(\lambda(t-\tau)^\alpha)(c_3 \odot y(\tau))_l(r) - Sh_\alpha(\lambda(t-\tau)^\alpha)(c_4 \odot y(\tau))_l(r)\right)d\tau$$

Taking the derivative,

$$\begin{aligned} &\boldsymbol{D}_{\boldsymbol{RL}}^{\alpha} x_{1,l}(t,r;c) \\ &= \lambda\Big\{t^{\alpha-1}\left(Sh_\alpha(\lambda t^\alpha)(c_1 \odot x_0)_l(r) - Ch_\alpha(\lambda t^\alpha)(c_2 \odot x_0)_l(r)\right) \\ &+ \int_{t_0}^{t} (t-\tau)^{\alpha-1}\left(Sh_\alpha(\lambda(t-\tau)^\alpha)(c_3 \odot y(\tau))_l(r) - Ch_\alpha(\lambda(t-\tau)^\alpha)(c_4 \odot y(\tau))_l(r)\right)d\tau\Big\} \\ &+ c_3 y_l(t,r) \end{aligned}$$

The same process can be done for $x_{1,u}(t)$.

$$\begin{aligned} &\boldsymbol{D}_{\boldsymbol{RL}}^{\alpha} x_{1,u}(t,r;c) \\ &= \lambda\Big\{t^{\alpha-1}\left(Sh_\alpha(\lambda t^\alpha)(c_1 \odot x_0)_u(r) - Ch_\alpha(\lambda t^\alpha)(c_2 \odot x_0)_u(r)\right) \\ &+ \int_{t_0}^{t} (t-\tau)^{\alpha-1}\left(Sh_\alpha(\lambda(t-\tau)^\alpha)(c_3 \odot y(\tau))_u(r) - Ch_\alpha(\lambda(t-\tau)^\alpha)(c_4 \odot y(\tau))_u(r)\right)d\tau\Big\} \\ &+ c_3 y_u(t,r) \end{aligned}$$

Please note that, if $c_1 = c_3 = 1, c_2 = c_4 = -1$ then this is exactly the following system,

$$\begin{cases} \boldsymbol{D}_{\boldsymbol{RL}}^{\alpha} x_{1,l}(t,r) = \lambda x_{1,u}(t,r) + y_l(t,r) \\ \boldsymbol{D}_{\boldsymbol{RL}}^{\alpha} x_{1,u}(t,r) = \lambda x_{1,l}(t,r) + y_u(t,r) \\ \boldsymbol{D}_{\boldsymbol{RL}}^{\alpha-1} x_{1,l}(t_0,r) = x_{0,l}(r) \\ \boldsymbol{D}_{\boldsymbol{RL}}^{\alpha-1} x_{1,u}(t_0,r) = x_{0,u}(r) \end{cases}$$

Indeed we proved the following theorem or here we call it remark.

Remark If $x(t)$ is $i-RL_{gH}$ solution for $\lambda<0$ the it is as the following form,

$$x_1(t) = t^{\alpha-1}\left(Ch_\alpha(\lambda t^\alpha) \odot x_0 \ominus_H Sh_\alpha(\lambda t^\alpha) \odot x_0\right) \oplus \int_{t_0}^{t} (t-\tau)^{\alpha-1}\left(Ch_\alpha(\lambda(t-\tau)^\alpha) \odot y(\tau) \oplus Sh_\alpha(\lambda(t-\tau)^\alpha) \odot y(\tau)\right)d\tau$$

In case $\lambda > 0$ and $ii - RL_{gH}$ differentiability, similar to case 1 ($\lambda<0$) we have

$$x_2(t) = t^{\alpha-1}E_{\alpha,\alpha}(-\lambda t^{\alpha}) \odot x_0 \ominus_H (-1) \int_{t_0}^{t} (t-\tau)^{\alpha-1} E_{\alpha,\alpha}(-\lambda(t-\tau)^{\alpha}) \odot y(\tau)d\tau$$

And connecting the vector $c = (c_1, c_2, c_3, c_4), |c_i| = 1, i = 1, 2, 3, 4$ we have,

$$\begin{aligned} x_2(t;c) &= t^{\alpha-1}(Ch_{\alpha}(\lambda t^{\alpha})c_1 \odot x_0 \ominus_H Sh_{\alpha}(\lambda t^{\alpha})c_2 \odot x_0) \\ &\quad \ominus_H \int_{t_0}^{t} (t-\tau)^{\alpha-1}(Ch_{\alpha}(\lambda(t-\tau)^{\alpha})c_3 \odot y(\tau) \ominus_H Sh_{\alpha}(\lambda(t-\tau)^{\alpha})c_4 \odot y(\tau))d\tau \end{aligned}$$

Subject to the H-difference exists. Now by differentiating,

$$\begin{aligned} &\boldsymbol{D}_{\boldsymbol{RL}}^{\alpha} x_{2,l}(t,r;c) \\ &= \lambda\{t^{\alpha-1}(Sh_{\alpha}(\lambda t^{\alpha})(c_1 \odot x_0)_l(r) - Ch_{\alpha}(\lambda t^{\alpha})(c_2 \odot x_0)_l(r)) \\ &+ \int_{t_0}^{t} (t-\tau)^{\alpha-1}(-Sh_{\alpha}(\lambda(t-\tau)^{\alpha})(c_3 \odot y(\tau))_l(r) + Ch_{\alpha}(\lambda(t-\tau)^{\alpha})(c_4 \odot y(\tau))_l(r))d\tau \Big\} \\ &- c_3 y_l(t,r) \end{aligned}$$

And for the upper function,

$$\begin{aligned} &\boldsymbol{D}_{\boldsymbol{RL}}^{\alpha} x_{2,u}(t,r;c) \\ &= \lambda\{t^{\alpha-1}(Sh_{\alpha}(\lambda t^{\alpha})(c_1 \odot x_0)_u(r) - Ch_{\alpha}(\lambda t^{\alpha})(c_2 \odot x_0)_u(r)) \\ &+ \int_{t_0}^{t} (t-\tau)^{\alpha-1}(-Sh_{\alpha}(\lambda(t-\tau)^{\alpha})(c_3 \odot y(\tau))_u(r) + Ch_{\alpha}(\lambda(t-\tau)^{\alpha})(c_4 \odot y(\tau))_u(r))d\tau \Big\} \\ &- c_3 y_u(t,r) \end{aligned}$$

Please note that, if $c_1 = c_4 = 1, c_2 = c_3 = -1$ then this is exactly the following system,

$$\begin{cases} \boldsymbol{D}_{\boldsymbol{RL}}^{\alpha} x_{2,l}(t,r) = \lambda x_{2,u}(t,r) + y_l(t,r) \\ \boldsymbol{D}_{\boldsymbol{RL}}^{\alpha} x_{2,u}(t,r) = \lambda x_{2,l}(t,r) + y_u(t,r) \\ \boldsymbol{D}_{\boldsymbol{RL}}^{\alpha-1} x_{2,l}(t_0,r) = x_{0,l}(r) \\ \boldsymbol{D}_{\boldsymbol{RL}}^{\alpha-1} x_{2,u}(t_0,r) = x_{0,u}(r) \end{cases}$$

Indeed we proved the following theorem or here we call it remark.

Remark If $x(t)$ is $ii - RL_{gH}$ solution for $\lambda > 0$ the it is as the following form,

$$x_2(t) = t^{\alpha-1}(Ch_\alpha(\lambda t^\alpha) \odot x_0 \ominus_H (-1) Sh_\alpha(\lambda t^\alpha) \odot x_0)$$
$$\ominus_H (-1) \int_{t_0}^{t} (t-\tau)^{\alpha-1} (Ch_\alpha(\lambda(t-\tau)^\alpha) \odot y(\tau) \ominus_H Sh_\alpha(\lambda(t-\tau)^\alpha) \odot y(\tau)) d\tau$$

Note Please note that $Ch_\alpha(\lambda t^\alpha) \odot x_0 \ominus_H (-1) Sh_\alpha(\lambda t^\alpha) \odot x_0$ is always exists for $\lambda > 0$ because

$$length(Ch_\alpha(\lambda t^\alpha) \odot x_0) = Ch_\alpha(\lambda t^\alpha) \odot length(x_0)$$

Is greater than

$$length((-1) Sh_\alpha(\lambda t^\alpha) \odot x_0) = Sh_\alpha(\lambda t^\alpha) \odot length(x_0)$$

The reason is,

$$Ch_\alpha(\lambda t^\alpha) - Sh_\alpha(\lambda t^\alpha) = E_{\alpha,\alpha}(-\lambda t^\alpha) > 0$$

The case $\lambda = 0$ is very easy and similar to the previous discussion in chapter three. The problem is as,

$$\begin{cases} \boldsymbol{D}^{\alpha}_{\boldsymbol{RL_{gH}}} x(t) = y(t), \quad t \in [t_o, T], \quad 0<\alpha<1 \\ \boldsymbol{D}^{\alpha-1}_{\boldsymbol{RL_{gH}}} x(t_0) = x_0 \in \mathbb{F}_R \end{cases}$$

And the solutions,

- $i - RL_{gH}$ solution

$$x_1(t) = t^{\alpha-1} \odot x_0 \oplus \int_{t_0}^{t} (t-\tau)^{\alpha-1} \odot y(\tau) d\tau$$

- $ii - RL_{gH}$ solution

$$x_2(t) = t^{\alpha-1} \odot x_0 \ominus_H (-1) \int_{t_0}^{t} (t-\tau)^{\alpha-1} \odot y(\tau) d\tau$$

Example Consider the following fractional differential equation with fuzzy initial value,

$$\begin{cases} \boldsymbol{D}^{\alpha}_{\boldsymbol{RL_{gH}}} x(t) = \lambda \odot x(t), \quad t \in [0,1], \quad 0<\alpha<1 \\ \boldsymbol{D}^{\alpha-1}_{\boldsymbol{RL_{gH}}} x(t_0) = x_0 \in \mathbb{F}_R \end{cases}$$

The solution is as

$$x(t) = t^{\alpha-1} E_{\alpha,\alpha}(\lambda t^{\alpha}) \odot x_0$$

And

$$\boldsymbol{D}^{\alpha}_{\boldsymbol{RL_{gH}}} x(t) = \lambda t^{\alpha-1} E_{\alpha,\alpha}(\lambda t^{\alpha}) \odot x_0$$
$$length\left(\boldsymbol{D}^{\alpha}_{\boldsymbol{RL_{gH}}} x(t)\right) = \lambda t^{\alpha-1} E_{\alpha,\alpha}(\lambda t^{\alpha}) \odot length(x_0) = \lambda length(x(t))$$

So the solution $i - LR_{gH}$ differentiable if $\lambda \geq 0$ and $ii - LR_{gH}$ differentiable if $\lambda < 0$. Also to find $i - LR_{gH}$ for $\lambda < 0$ and $ii - LR_{gH}$ for $\lambda \geq 0$,

Case 1. $\lambda < 0$ and $i - RL_{gH}$ differentiability

$$x_1(t) = t^{\alpha-1}(Ch_{\alpha}(\lambda t^{\alpha}) \odot x_0 \oplus Sh_{\alpha}(\lambda t^{\alpha}) \odot x_0)$$

Case 2. $\lambda > 0$ and $ii - RL_{gH}$ differentiability

$$x_1(t) = t^{\alpha-1}(Ch_{\alpha}(\lambda t^{\alpha}) \odot x_0 \ominus_H (-1) Sh_{\alpha}(\lambda t^{\alpha}) \odot x_0)$$

In special case, $\alpha = 0.5, \lambda = -1$ and $\boldsymbol{D}^{-0.5}_{\boldsymbol{RL_{gH}}} x(0) = (1+r, 3-r)$ then

$$t^{\alpha-1} Ch_{\alpha}(\lambda t^{\alpha}) = \frac{1}{\sqrt{\pi t}} + \frac{1}{2}\left(e^t \operatorname{erf}\left(-\sqrt{t}\right) - e^t \operatorname{erf}\left(\sqrt{t}\right)\right)$$
$$t^{\alpha-1} Sh_{\alpha}(\lambda t^{\alpha}) = \frac{1}{2}\left(e^t \operatorname{erf}\left(-\sqrt{t}\right) - e^t \operatorname{erf}\left(\sqrt{t}\right)\right)$$

Example Consider the following fractional differential equation with fuzzy initial value,

$$\begin{cases} \boldsymbol{D}^{\alpha}_{\boldsymbol{RL_{gH}}} x(t) = (-1) \odot x(t) + t + 1, \quad t \in [0,1], \quad 0<\alpha<1 \\ \boldsymbol{D}^{\alpha-1}_{\boldsymbol{RL_{gH}}} x(t_0) = x_0 \in \mathbb{F}_R \end{cases}$$

The $i - RL_{gH}$ solution is,

$$x_1(t) = t^{\alpha-1}(Ch_\alpha(t^\alpha) \odot x_0 \ominus_H Sh_\alpha(t^\alpha) \odot x_0)$$
$$\oplus \int_{t_0}^{t} (t-\tau)^{\alpha-1}(Ch_\alpha((t-\tau)^\alpha) \odot (\tau+1) \oplus Sh_\alpha((t-\tau)^\alpha) \odot (\tau+1))d\tau$$

Chapter 5
Numerical Solution of Fuzzy Fractional Differential Equations

5.1 Introduction

In this chapter, first, some numerical methods such as generalized fuzzy fractional Taylor's expansion and its application entitled fuzzy fractional Euler's method are presented for fuzzy-valued function in the sense of Caputo differentiability. Then the fuzzy impulsive fractional differential equations are going to be considered for numerically solving by some semi analytically methods. Also the fuzzy fractional differential equation is solved by applying some other derivatives such as ABC derivative by using numerical methods.

5.2 Preliminaries

To discuss the subject, we need some definitions and other concepts.

Definition—Partial ordering
For two fuzzy numbers $A, B \in \mathbb{F}_R$,

$$A \preccurlyeq B \Leftrightarrow A_l(r) \leq B_l(r) \ \& \ A_u(r) \leq B_u(r)$$

And

$$A \prec B \Leftrightarrow A_l(r) < B_l(r) \ \& \ A_u(r) < B_u(r)$$

For any $0 \leq r \leq 1$.

Properties—Partial ordering

- $A \preccurlyeq B \Rightarrow -B \preccurlyeq -A$
- $A \preccurlyeq B \ \& \ B \preccurlyeq A \Rightarrow A = B$

T. Allahviranloo, *Fuzzy Fractional Differential Operators and Equations*, Studies in Fuzziness and Soft Computing 397, https://doi.org/10.1007/978-3-030-51272-9_5

- If $x, y : [a, b] \to \mathbb{F}_R$ & $x(t) \preccurlyeq y(t)$ then $\int_a^b x(t)dt \preccurlyeq \int_a^b y(t)dt$

All the properties can be proved by using the level-wise form of fuzzy numbers and fuzzy number valued functions.

To prove fuzzy mean value theorems for Riemann-Liouville integral, the fuzzy intermediate value theorem and fuzzy mean value theorem for integrals are needed, so we first prove these theorems.

Remark Suppose that $x(t)$ is a continuous fuzzy number valued function on $[a, b]$ and there exists a fuzzy number γ such that $x(a) \preccurlyeq \gamma \preccurlyeq x(b)$ then there exists at least $c \in [a, b]$ such that $x(c) = \gamma$.

To prove, define $S = \{t | t \in [a, b], x(t) \preccurlyeq \gamma\}$, this set is bounded above by b and non-empty. Then it does have a supremum and suppose $c = \sup S$. First, assume that $x(c) \succ \gamma$ then by definition,

$$x_l(c, r) > \gamma_l(r) \ \& \ x_u(c, r) > \gamma_u(r)$$

Since x is continuous,

$$\forall \epsilon > 0, \exists \delta > 0, \forall x(|t - c| < \delta \Rightarrow D_H(x(t), x(c)) < \epsilon)$$

Now, for a fixed $r \in [0, 1]$ let $\epsilon = \min\{x_l(c, r) - \gamma_l(r), x_u(c, r) - \gamma_u(r)\}$ then from the definition of Husdorff distance,

$$|x_l(t, r) - x_l(c, r)| \langle \epsilon \Rightarrow x_l(t, r) \rangle x_l(c, r) - \epsilon \geq \gamma_l(r)$$

$$|x_u(t, r) - x_u(c, r)| \langle \epsilon \Rightarrow x_u(t, r) \rangle x_u(c, r) - \epsilon \geq \gamma_u(r)$$

Thus for any $t \epsilon (c - \delta, c + \delta)$ it is deduced that $x_l(t, r) > \gamma_l(r)$ and $x_u(t, r) > \gamma_u(r)$. On the other hand we know that there is no point in $(c - \delta, \delta]$ such that $x_l(t, r) > \gamma_l(r), x_u(t, r) > \gamma_u(r)$ such that $c = \sup S$ and this is contradiction. Therefore $x(c) \preccurlyeq \gamma$. Now consider $x(c) \prec \gamma$, again by continuity and definition of distance we have,

$$|x_l(t, r) - x_l(c, r)| < \epsilon \Rightarrow x_l(t, r) < x_l(c, r) + \epsilon \leq \gamma_l(r)$$

$$|x_u(t, r) - x_u(c, r)| < \epsilon \Rightarrow x_u(t, r) < x_u(c, r) + \epsilon \leq \gamma_u(r)$$

Thus for any $t \in (c - \delta, c + \delta)$ it is deduced that $x_l(t, r) > \gamma_l(r)$ and $x_u(t, r) > \gamma_u(r)$. Thus $c + \frac{\delta}{2} \in S$ which is contradiction. Hence for any $r \in [0, 1]$,

$$x_l(t, r) > \gamma_l(r), \quad x_u(t, r) > \gamma_u(r) \Rightarrow x(c) \succcurlyeq \gamma$$

Finally,

$$x(c) \preccurlyeq \gamma \;\&\; x(c) \succcurlyeq \gamma \Rightarrow x(c) = \gamma$$

Remark Suppose $x : [a, b] \to \mathbb{F}_R$ is a continuous fuzzy number valued function and for all $t \in [a, b]$, $x(a) \preccurlyeq x(t) \preccurlyeq x(b)$. Moreover assume $y : [a, b] \to R$ is an integrable real function on (a, b), then there exists at least one $c \in (a, b)$ such that

$$\int_a^b x(t) \odot y(t) dt = x(c) \odot \int_a^b y(t) dt$$

To show the assertion, first consider $y(t) \geq 0$,

$$x(a) \odot y(t) \preccurlyeq x(t) \odot y(t) \preccurlyeq x(b) \odot y(t)$$

Then

$$\int_a^b x(a) \odot y(t) dt \preccurlyeq \int_a^b x(t) \odot y(t) dt \preccurlyeq \int_a^b x(b) \odot y(t) dt$$

Therefore

$$x(a) \odot \int_a^b y(t) dt \preccurlyeq \int_a^b x(t) \odot y(t) dt \preccurlyeq x(b) \odot \int_a^b y(t) dt$$

$$x(a) \preccurlyeq \frac{\int_a^b x(t) \odot y(t) dt}{\int_a^b y(t) dt} \preccurlyeq x(b)$$

In accordance with the previous Remark, there is at least $c \in [a, b]$,

$$x(c) = \frac{\int_a^b x(t) \odot y(t) dt}{\int_a^b y(t) dt}$$

In case $y(t) < 0$,

$$x(a) \odot y(t) \succcurlyeq x(t) \odot y(t) \succcurlyeq x(b) \odot y(t)$$

So the process is similar to the previous one.

5.2.1 *Fuzzy Mean Value Theorem for Riemann-Liouville Integral*

Suppose $x : [a,b] \to \mathbb{F}_R$ is a continuous fuzzy number valued function and for all $t \in [a,b]$, $x(a) \preccurlyeq x(t) \preccurlyeq x(b)$. Moreover assume $y : [a,b] \to R$ is an integrable real function on (a,b), such that does not change the sign in the open interval then there exists at least one $c \in (a,b)$ such that

$$\boldsymbol{I}_{RL}^{\alpha} x(t) \odot y(t) = x(c) \odot \boldsymbol{I}_{RL}^{\alpha} y(t)$$

To prove, by the definition we have,

$$\boldsymbol{I}_{RL}^{\alpha} x(t) \odot y(t) = \frac{1}{\Gamma(\alpha)} \odot \int_a^t (t-\tau)^{\alpha-1} \odot x(\tau) \odot y(\tau) d\tau$$

If we consider

$$\frac{1}{\Gamma(\alpha)(t-\tau)^{1-\alpha}} \odot y(\tau) = \widetilde{y}(t)$$

So the $\widetilde{y}(t)$ is continuous and integrable on open interval. Also it does not change its sign on the interval,

$$\begin{aligned}
\boldsymbol{I}_{RL}^{\alpha} x(t) \odot y(t) &= \int_a^t x(\tau) \odot \widetilde{y}(t) d\tau = x(c) \odot \int_a^t \widetilde{y}(t) d\tau \\
&= x(c) \odot \frac{1}{\Gamma(\alpha)} \odot \int_a^t (t-\tau)^{\alpha-1} \odot y d\tau = \boldsymbol{I}_{RL}^{\alpha} y(t)
\end{aligned}$$

The proof is completed.

5.3 Fuzzy Fractional Taylor's Expansion with Caputo gH-Derivative

Here we are going to consider the Taylor expansion of a fuzzy number valued function concerning the series result is a fuzzy number too. To this end, we need some discussion on the relations of Riemann integral and gH-derivative.

Extended Integral Relation

Let $x_{gH}^{(n)}(t)$ is a continuous fuzzy number valued function for any $s \in (a,b)$, the following integral equations are valid.

Note Please note that,

$$\begin{aligned} x_{i-gH}(t,r) &= [x_l(t,r), x_u(t,r)] \\ &= \ominus_H(-1)[x_u(t,r), x_l(t,r)] := \ominus_H(-1)x_{ii-gH}(t,r) \end{aligned}$$

Item 1 Consider $x_{gH}^{(n)}(t), n = 1, 2, \ldots, m$ is $(i - gH)-$ differentiable and the type of differentiability does not change on (a, b).

$$\begin{aligned} \left(FR\int_a^s x_{gH}^{(n)}(t)dt\right)[r] &= FR\int_a^s \left(x_{gH}^{(n)}(t)\right)[r]dt \\ &= \left[R\int_a^s x_l^{(n)}(t,r)dt, R\int_a^s x_u^{(n)}(t,r)dt\right] \\ &= \left[x_l^{(n-1)}(s,r) - x_l^{(n-1)}(a,r), x_u^{(n-1)}(s,r) - x_u^{(n-1)}(a,r)\right] \\ &= \left[x_l^{(n-1)}(s,r), x_u^{(n-1)}(s,r)\right] - \left[x_l^{(n-1)}(a,r), x_u^{(n-1)}(a,r)\right] \\ &= \left(x_{gH}^{(n-1)}(s)\right)[r] - \left(x_{gH}^{(n-1)}(a)\right)[r] \end{aligned}$$

Then it is concluded,

$$FR\int_a^s x_{gH}^{(n)}(t)dt = x_{gH}^{(n-1)}(s) \ominus_H x_{gH}^{(n-1)}(a)$$

$$x_{gH}^{(n-1)}(s) = x_{gH}^{(n-1)}(a) \oplus FR\int_a^s x_{gH}^{(n)}(t)dt$$

Item 2 Consider $x_{gH}^{(n)}(t), n = 1, 2, \ldots, m$ is $(ii - gH)$—differentiable and the type of differentiability do not change on (a, b).

$$\left(FR\int_a^s x_{gH}^{(n)}(t)dt\right)[r] = FR\int_a^s x_{gH}^{(n)}(t,r)dt$$
$$= \left[R\int_a^s x_u^{(n)}(t,r)dt, R\int_a^s x_l^{(n)}(t,r)dt\right]$$
$$= \left[x_u^{(n-1)}(s,r) - x_u^{(n-1)}(a,r), x_l^{(n-1)}(s,r) - x_l^{(n-1)}(a,r)\right]$$
$$= \left[x_u^{(n-1)}(s,r), x_l^{(n-1)}(s,r)\right] - \left[x_u^{(n-1)}(a,r), x_l^{(n-1)}(a,r)\right]$$
$$= \ominus_H(-1)x_{gH}^{(n-1)}(s,r) \ominus_H \left(\ominus_H(-1)x_{gH}^{(n-1)}(a,r)\right)$$
$$= \ominus_H(-1)x_{gH}^{(n-1)}(a,r) \oplus (-1)x_{gH}^{(n-1)}(s,r)$$

Then it is concluded,

$$FR\int_a^s x_{gH}^{(n)}(t)dt = (-1)x_{gH}^{(n-1)}(s,r) \ominus_H (-1)x_{gH}^{(n-1)}(a,r)$$

Item 3 Consider $x_{gH}^{(n)}(t)$ is $(i-gH)-$ differentiable and $x_{gH}^{(n-1)}(t)$ is $(ii-gH)-$ differentiable. We have,

$$x_{gH}^{(n-1)}(t,r) = \left[x_u^{(n-1)}(t,r), x_l^{(n-1)}(t,r)\right]$$

$$\ominus_H(-1)FR\int_a^s x_{gH}^{(n)}(t,r)dt = \left[R\int_a^s x_u^{(n)}(t,r), R\int_a^s x_l^{(n)}(t,r)\right]$$

$$\ominus_H(-1)FR\int_a^s x_{gH}^{(n)}(t,r)dt = \left[x_u^{(n-1)}(s,r) - x_u^{(n-1)}(a,r), x_l^{(n-1)}(s,r) - x_l^{(n-1)}(a,r)\right]$$
$$= \left[x_u^{(n-1)}(s,r), x_l^{(n-1)}(s,r)\right] - \left[x_u^{(n-1)}(a,r), x_l^{(n-1)}(a,r)\right]$$

Then

$$\ominus_H(-1)FR\int_a^s x_{gH}^{(n)}(t)dt = x_{gH}^{(n-1)}(s) \ominus_H x_{gH}^{(n-1)}(a)$$

Then it is concluded,

$$FR\int_a^s x_{gH}^{(n)}(t)dt = \ominus_H(-1)x_{gH}^{(n-1)}(s) \oplus (-1)x_{gH}^{(n-1)}(a)$$

Item 4 Consider $x_{gH}^{(n)}(t)$ is $(ii - gH)$—differentiable and $x_{gH}^{(n-1)}(t)$ is $(i - gH)$—differentiable then

$$x_{gH}^{(n-1)}(t,r) = \left[x_l^{(n-1)}(t,r), x_u^{(n-1)}(t,r)\right]$$

$$FR\int_a^s x_{gH}^{(n)}(t,r)dt = \ominus_H(-1)\left[R\int_a^s x_l^{(n)}(t,r), R\int_a^s x_u^{(n)}(t,r)\right]$$

$$FR\int_a^s x_{gH}^{(n)}(t,r)dt = \left[R\int_a^s x_u^{(n)}(t,r), R\int_a^s x_l^{(n)}(t,r)\right]$$

$$FR\int_a^s x_{gH}^{(n)}(t,r)dt = \left[x_u^{(n-1)}(s,r) - x_u^{(n-1)}(a,r), x_l^{(n-1)}(s,r) - x_l^{(n-1)}(a,r)\right]$$

$$= \left[x_u^{(n-1)}(s,r), x_l^{(n-1)}(s,r)\right] - \left[x_u^{(n-1)}(a,r), x_l^{(n-1)}(a,r)\right]$$

Then

$$FR\int_a^s x_{gH}^{(n)}(t)dt = x_{gH}^{(n-1)}(s) \ominus_H x_{gH}^{(n-1)}(a)$$

Then the same result is concluded,

$$FR\int_a^s x_{gH}^{(n)}(t)dt = \ominus_H(-1)x_{gH}^{(n-1)}(s) \oplus (-1)x_{gH}^{(n-1)}(a)$$

Concerning to the concept and generalization of fuzzy fractional Taylor expansion we should discuss the relation of fuzzy fractional RL-integration and Caputo derivative.

Remark If the function $x(t)$ is $i - gH$ or $ii - gH$ differentiable on $(a, b]$, for $0 < \alpha < 1$ we have,

- $\boldsymbol{I}_{\boldsymbol{RL}}^{\alpha}\boldsymbol{D}_{\boldsymbol{C_{gH}}}^{\alpha}x(t) = x(t) \ominus_{gH} x(a)$
- $\boldsymbol{D}_{\boldsymbol{C_{gH}}}^{\alpha}\boldsymbol{I}_{\boldsymbol{RL}}^{\alpha}x(t) = x(t)$

The first item,

$$\begin{aligned} \boldsymbol{I}_{\boldsymbol{RL}}^{\alpha}\boldsymbol{D}_{\boldsymbol{C}_{gH}}^{\alpha}x(t) &= \boldsymbol{I}_{\boldsymbol{RL}}^{\alpha}\left(\boldsymbol{I}_{\boldsymbol{RL}}^{(1-\alpha)}\frac{d}{ds}x\right)(t) \\ &= \left(\boldsymbol{I}_{\boldsymbol{RL}}^{1}\frac{d}{ds}x\right)(t) = \int_a^t x'(s)ds = x(t)\ominus_{gH} x(a) \end{aligned}$$

Because in the definition,

$$\boldsymbol{I}_{\boldsymbol{RL}}^{\alpha}x(t) = \frac{1}{\Gamma(\alpha)}\odot\int_a^t (t-s)^{\alpha-1}\odot x(s)ds$$

If $\alpha = 1$ then,

$$\boldsymbol{I}_{\boldsymbol{RL}}^{1}x(t) = \int_a^t x(s)ds$$

Now if $x(t) = x'(s)$ then

$$\boldsymbol{I}_{\boldsymbol{RL}}^{1}x'_{gH}(t) = \int_a^t x'_{gH}(s)ds = x(t)\ominus_{gH} x(a)$$

To prove the second item, we have the following relations,
For $x(t_0)$ as a constant fuzzy, we have

$$\boldsymbol{I}_{\boldsymbol{RL}}^{(1-\alpha)}x(t_0) = \frac{1}{\Gamma(1-\alpha)}\odot\int_a^t (t-s)^{-\alpha}\odot x(a)ds = \frac{(t-a)^{1-\alpha}}{\Gamma(2-\alpha)}\odot x(a)$$

And if $\alpha :\rightarrow 1+\alpha$,

$$\begin{aligned} \boldsymbol{D}_{\boldsymbol{RL}_{gH}}^{\alpha}x(t_0) &= \boldsymbol{I}_{\boldsymbol{RL}}^{-\alpha}y(t) = \frac{(t-a)^{-\alpha}}{\Gamma(1-\alpha)}\odot x(a) \\ \boldsymbol{D}_{\boldsymbol{C}_{gH}}^{\alpha}x(t) &= \boldsymbol{D}_{\boldsymbol{RL}_{gH}}^{\alpha}\left(x(t)\ominus_{gH} x(a)\right) = \boldsymbol{D}_{\boldsymbol{RL}_{gH}}^{\alpha}x(t)\ominus_{gH}\frac{(t-a)^{-\alpha}}{\Gamma(1-\alpha)}\odot x(a) \end{aligned}$$

Now by substituting $\boldsymbol{I}_{\boldsymbol{RL}}^{\alpha}x(t)\rightarrow x(t)$ we get,

$$\boldsymbol{D}_{\boldsymbol{C}_{gH}}^{\alpha}\boldsymbol{I}_{\boldsymbol{RL}}^{\alpha}x(t) = \boldsymbol{D}_{\boldsymbol{RL}_{gH}}^{\alpha}\boldsymbol{I}_{\boldsymbol{RL}}^{\alpha}x(t)\ominus_{gH}\frac{(t-a)^{-\alpha}}{\Gamma(1-\alpha)}\odot\left(\boldsymbol{I}_{\boldsymbol{RL}}^{\alpha}x\right)(a)$$

If we show $\left(\boldsymbol{I}_{\boldsymbol{RL}}^{\alpha}x\right)(a) = 0$ the proof is completed. To do this, we use the distance.

$$D_H\left(\left(\boldsymbol{I}_{\boldsymbol{RL}}^{\alpha}x\right)(a), 0\right) \leq \frac{k}{\Gamma(\alpha)} \int_{t_0}^{t} (t-s)^{\alpha-1} ds = k \frac{(t-a)^{\alpha}}{\Gamma(1+\alpha)}$$

where $\sup_{t\in[t_0,T]} D_H(x(t), 0) = k$. Then

$$D_H\left(\left(\boldsymbol{I}_{\boldsymbol{RL}}^{\alpha}x\right)(a), 0\right) \leq k \frac{(t-a)^{\alpha}}{\Gamma(1+\alpha)} \to 0$$

Then

$$\boldsymbol{D}_{\boldsymbol{C}_{gH}}^{\alpha} \boldsymbol{I}_{\boldsymbol{RL}}^{\alpha} x(t) = \boldsymbol{D}_{\boldsymbol{RL}_{gH}}^{\alpha} \boldsymbol{I}_{\boldsymbol{RL}}^{\alpha} x(t) = x(t)$$

5.3.1 *Fuzzy Fractional Taylor Expansion*

In this section, the fuzzy fractional Taylor expansion and then its generalized version is going to be explained. Assume $\alpha > 0, \alpha \notin \mathbb{N}, m = [\alpha] + 1$ and $x_{gH}^{(m)}(t)$ is a continuous fuzzy number valued function. Then we have the following cases,

Case 1 If $x(t)$ and its higher order derivatives are $i-gH$ differentiable on $[a, b]$ then

$$x(t) = x(a) \oplus \sum_{k=1}^{m-1} \frac{(t-a)^k}{k!} \odot x_{gH}^{(k)}(a) \oplus \boldsymbol{I}_{\boldsymbol{RL}}^{\alpha} \boldsymbol{D}_{\boldsymbol{C}_{gH}}^{\alpha} x(t)$$

Case 2 If $x(t)$ and its higher order derivatives are $ii-gH$ differentiable on $[a, b]$ then

$$x(t) = x(a) \ominus_H (-1) \sum_{k=1}^{m-1} \frac{(t-a)^k}{k!} \odot x_{gH}^{(k)}(a) \ominus_H (-1) \boldsymbol{I}_{\boldsymbol{RL}}^{\alpha} \boldsymbol{D}_{\boldsymbol{C}_{gH}}^{\alpha} x(t)$$

Case 3 If $x(t)$ is $i-gH$ differentiable and its higher-order derivatives are changing in every other type periodically then

$$x(t) = x(a) \ominus_H (-1) \sum_{k=1,even}^{m-1} \frac{(t-a)^k}{k!} \odot x_{gH}^{(k)}(a) \oplus \sum_{k=1,odd}^{m-1} \frac{(t-a)^k}{k!} \odot x_{gH}^{(k)}(a)$$
$$\oplus \boldsymbol{I}_{\boldsymbol{RL}}^{\alpha} \boldsymbol{D}_{\boldsymbol{C}_{gH}}^{\alpha} x(t)$$

Case 4 If $x(t)$ is $ii - gH$ differentiable before $\xi \in [a,b]$ and $i - gH$ differentiable after ξ, and if $x'(t)$ is $i - gH$ differentiable before $\sigma \in [\xi, b]$ and $ii - gH$ differentiable after σ, moreover $x_{gH}^{(k)}(t), k = 2, 3, \ldots, m$ are $i - gH$ differentiable on $[a,b]$, then

$$x(t) = \begin{cases} x(a) \oplus \sum_{k=1}^{m-1} \frac{(t-a)^k}{k!} \odot x_{gH}^{(k)}(a) \oplus \boldsymbol{I}_{\boldsymbol{RL}}^{\alpha} \boldsymbol{D}_{\boldsymbol{C}_{gH}}^{\alpha} x(t), a \le x \le \xi \\ x(a) \ominus_H (-1) \sum_{k=1}^{m-1} \frac{(t-a)^k}{k!} \odot x_{gH}^{(k)}(a) \ominus_H (-1) \boldsymbol{I}_{\boldsymbol{RL}}^{\alpha} \boldsymbol{D}_{\boldsymbol{C}_{gH}}^{\alpha} x(t), \xi \le x \le \sigma \\ x(a) \oplus (x-a) \odot x'_{gH}(a) \ominus_H (-1) \sum_{k=2}^{m-1} \frac{(t-a)^k}{k!} \odot x_{gH}^{(k)}(a) \ominus_H (-1) \boldsymbol{I}_{\boldsymbol{RL}}^{\alpha} \boldsymbol{D}_{\boldsymbol{C}_{gH}}^{\alpha} x(t), \\ \qquad \sigma \le x \le b \end{cases}$$

Case 5 If $x(t)$ is $i - gH$ differentiable before $\xi \in [a,b]$ and $ii - gH$ differentiable after ξ, and if $x'(t)$ is $ii - gH$ differentiable before $\sigma \in [\xi, b]$ and $i - gH$ differentiable after σ, moreover $x_{gH}^{(k)}(t), k = 2, 3, \ldots, m$ are $ii - gH$ differentiable on $[a,b]$, then

$$x(t) = \begin{cases} x(a) \ominus_H (-1) \sum_{k=1}^{m-1} \frac{(t-a)^k}{k!} \odot x_{gH}^{(k)}(a) \ominus_H (-1) \boldsymbol{I}_{\boldsymbol{RL}}^{\alpha} \boldsymbol{D}_{\boldsymbol{C}_{gH}}^{\alpha} x(t), a \le x \le \xi \\ x(a) \oplus \sum_{k=1}^{m-1} \frac{(t-a)^k}{k!} \odot x_{gH}^{(k)}(a) \oplus \boldsymbol{I}_{\boldsymbol{RL}}^{\alpha} \boldsymbol{D}_{\boldsymbol{C}_{gH}}^{\alpha} x(t), \xi \le x \le \sigma \\ x(a) \ominus_H (-1)(x-a) \odot x'_{gH}(a) \oplus \sum_{k=2}^{m-1} \frac{(t-a)^k}{k!} \odot x_{gH}^{(k)}(a) \ominus_H (-1) \boldsymbol{I}_{\boldsymbol{RL}}^{\alpha} \boldsymbol{D}_{\boldsymbol{C}_{gH}}^{\alpha} x(t), \\ \qquad \sigma \le x \le b \end{cases}$$

To prove all cases, we need to stablish the following relations, since $x_{gH}^{(m)}(t)$ is continuous fuzzy number functions and x is gH-Caputo differentiable, then

$$\boldsymbol{I}_{\boldsymbol{RL}}^{\alpha} \boldsymbol{D}_{\boldsymbol{C}_{gH}}^{\alpha} x(t) = \boldsymbol{I}_{\boldsymbol{RL}}^{\alpha} \boldsymbol{I}_{\boldsymbol{RL}}^{-\alpha} x(t) = \boldsymbol{I}_{\boldsymbol{RL}}^{\alpha} \left(\boldsymbol{I}_{\boldsymbol{RL}}^{m-\alpha} x_{gH}^{(m)}(t) \right) = \boldsymbol{I}_{\boldsymbol{RL}}^{\alpha} \boldsymbol{I}_{\boldsymbol{RL}}^{m-\alpha} x_{gH}^{(m)}(t) = \boldsymbol{I}_{\boldsymbol{RL}}^{m} x_{gH}^{(m)}(t)$$

Case 1 ***Proof*** All derivatives $x_{gH}^{(k)}, k = 1, 2, \ldots, m$ are $i - gH$ differentiable.

$$\boldsymbol{I}_{\boldsymbol{RL}}^{\alpha}\boldsymbol{D}_{C_{gH}}^{\alpha}x(t) = \boldsymbol{I}_{\boldsymbol{RL}}^{m}x_{gH}^{(m)}(t) = \boldsymbol{I}_{\boldsymbol{RL}}^{m-1}x_{gH}^{(m-1)}(t) \ominus_H \boldsymbol{I}_{\boldsymbol{RL}}^{m-1}x_{gH}^{(m-1)}(a)$$

Because

$$x_{gH}^{(m-1)}(t) = x_{gH}^{(m-1)}(a) \oplus FR\int_a^t x_{gH}^{(m)}(\tau)d\tau$$

In the operator form it is,

$$FR\int_a^s x_{gH}^{(m)}(t)dt = \boldsymbol{I}_{\boldsymbol{RL}}^{1}x_{gH}^{(m)}(t) = x_{gH}^{(m-1)}(t) \ominus_H x_{gH}^{(m-1)}(a)$$

And

$$\boldsymbol{I}_{\boldsymbol{RL}}^{m-1}x_{gH}^{(m-1)}(t) \ominus_H \boldsymbol{I}_{\boldsymbol{RL}}^{m-1}x_{gH}^{(m-1)}(a) = \boldsymbol{I}_{\boldsymbol{RL}}^{m-2}x_{gH}^{(m-2)}(t) \ominus_H \boldsymbol{I}_{\boldsymbol{RL}}^{m-2}x_{gH}^{(m-2)}(a) \ominus_H \boldsymbol{I}_{\boldsymbol{RL}}^{m-1}x_{gH}^{(m-1)}(a)$$
$$\boldsymbol{I}_{\boldsymbol{RL}}^{m-2}x_{gH}^{(m-2)}(t) \ominus_H \boldsymbol{I}_{\boldsymbol{RL}}^{m-2}x_{gH}^{(m-2)}(a) \ominus_H \boldsymbol{I}_{\boldsymbol{RL}}^{m-1}x_{gH}^{(m-1)}(a)$$
$$= \boldsymbol{I}_{\boldsymbol{RL}}^{m-3}x_{gH}^{(m-3)}(t) \ominus_H \boldsymbol{I}_{\boldsymbol{RL}}^{m-3}x_{gH}^{(m-3)}(a) \ominus_H \boldsymbol{I}_{\boldsymbol{RL}}^{m-2}x_{gH}^{(m-2)}(a) \ominus_H \boldsymbol{I}_{\boldsymbol{RL}}^{m-1}x_{gH}^{(m-1)}(a)$$

By substituting then,

$$\begin{aligned}\boldsymbol{I}_{\boldsymbol{RL}}^{\alpha}\boldsymbol{D}_{C_{gH}}^{\alpha}x(t) = \boldsymbol{I}_{\boldsymbol{RL}}^{m-3}x_{gH}^{(m-3)}(t) \ominus_H \boldsymbol{I}_{\boldsymbol{RL}}^{m-3}x_{gH}^{(m-3)}(a) \ominus_H \boldsymbol{I}_{\boldsymbol{RL}}^{m-2}x_{gH}^{(m-2)}(a)\\ \ominus_H \boldsymbol{I}_{\boldsymbol{RL}}^{m-1}x_{gH}^{(m-1)}(a)\end{aligned}$$

Proceeding straight forward,

$$\begin{aligned}\boldsymbol{I}_{\boldsymbol{RL}}^{\alpha}\boldsymbol{D}_{C_{gH}}^{\alpha}x(t) = x(t) \ominus_H x(a) \ominus_H \boldsymbol{I}_{\boldsymbol{RL}}^{1}x'_{gH}(a) \ominus_H \boldsymbol{I}_{\boldsymbol{RL}}^{2}x''_{gH}(a)\\ \ominus_H \cdots \ominus_H \boldsymbol{I}_{\boldsymbol{RL}}^{m-1}x_{gH}^{(m-1)}(a)\end{aligned}$$

where

$$\boldsymbol{I}_{\boldsymbol{RL}}^{k} := \underbrace{\int_a^t\int_a^t\int_a^t\cdots\int_a^t d\tau d\tau\cdots d\tau}_{k-times} = \frac{(t-a)^k}{k!}$$

So we found out,

$$\boldsymbol{I}_{\boldsymbol{RL}}^{\alpha}\boldsymbol{D}_{\boldsymbol{C}_{gH}}^{\alpha}x(t) = x(t) \ominus_H x(a) \ominus_H (t-a)x'_{gH}(a) \ominus_H \frac{(t-a)^2}{2!}x''_{gH}(a) \\ \ominus_H \cdots \ominus_H \frac{(t-a)^{m-1}}{(m-1)!}x_{gH}^{(m-1)}(a)$$

And finally based on the definition of H-difference,

$$x(t) = x(a) \oplus \sum_{k=1}^{m-1} \frac{(t-a)^k}{k!} \odot x_{gH}^{(k)}(a) \oplus \boldsymbol{I}_{\boldsymbol{RL}}^{\alpha}\boldsymbol{D}_{\boldsymbol{C}_{gH}}^{\alpha}x(t)$$

Case 2 Proof. All derivatives $x_{gH}^{(k)}, k = 1, 2, \ldots, m$ are $ii - gH$ differentiable.

$$\boldsymbol{I}_{\boldsymbol{RL}}^{\alpha}\boldsymbol{D}_{\boldsymbol{C}_{gH}}^{\alpha}x(t) = \boldsymbol{I}_{\boldsymbol{RL}}^{m}x_{gH}^{(m)}(t) = \boldsymbol{I}_{\boldsymbol{RL}}^{m-1}x_{gH}^{(m-1)}(t) \ominus_H \boldsymbol{I}_{\boldsymbol{RL}}^{m-1}x_{gH}^{(m-1)}(a)$$

Because in the operator form it is,

$$FR\int_a^s x_{gH}^{(m)}(t)dt = \boldsymbol{I}_{\boldsymbol{RL}}^{1}x_{gH}^{(m)}(t) = \ominus_H(-1)x_{gH}^{(m-1)}(t) \ominus_H \ominus_H(-1)x_{gH}^{(m-1)}(a) \\ = \ominus_H(-1)x_{gH}^{(m-1)}(t) \oplus (-1)x_{gH}^{(m-1)}(a)$$

Similar to the case (1), finally we get,

$$\boldsymbol{I}_{\boldsymbol{RL}}^{\alpha}\boldsymbol{D}_{\boldsymbol{C}_{gH}}^{\alpha}x(t) = \boldsymbol{I}_{\boldsymbol{RL}}^{1}x'_{gH}(t) \ominus_H \boldsymbol{I}_{\boldsymbol{RL}}^{2}x''_{gH}(a) \ominus_H \cdots \ominus_H \boldsymbol{I}_{\boldsymbol{RL}}^{m-1}x_{gH}^{(m-1)}(a)$$

Proceeding straight forward,

$$\boldsymbol{I}_{\boldsymbol{RL}}^{\alpha}\boldsymbol{D}_{\boldsymbol{C}_{gH}}^{\alpha}x(t) \\ = \ominus_H(-1)x(t) \oplus (-1)x(a) \ominus_H \boldsymbol{I}_{\boldsymbol{RL}}^{1}x'_{gH}(a) \ominus_H \boldsymbol{I}_{\boldsymbol{RL}}^{2}x''_{gH}(a) \\ \ominus_H \cdots \ominus_H \boldsymbol{I}_{\boldsymbol{RL}}^{m-1}x_{gH}^{(m-1)}(a)$$

Because,

$$\boldsymbol{I}_{\boldsymbol{RL}}^{1}x'_{gH}(t) = \int_a^t x'_{gH}(s)ds = (\ominus_H(-1)x(t)) \ominus_H (\ominus_H(-1)x(a)) \\ = \ominus_H(-1)x(t) \oplus (-1)x(a)$$

where

$$I_{RL}^{k} := \underbrace{\int_a^t \int_a^t \int_a^t \cdots \int_a^t d\tau d\tau \cdots d\tau}_{k-times} = \frac{(t-a)^k}{k!}$$

Indeed we have,

$$I_{RL}^{\alpha} D_{C_{gH}}^{\alpha} x(t) = \ominus_H (-1)x(t) \oplus (-1)x(a) \ominus_H \sum_{k=1}^{m-1} \frac{(t-a)^k}{k!} \odot x_{gH}^{(k)}(a)$$

$$\ominus_H (-1)x(t) \oplus (-1)x(a) = I_{RL}^{\alpha} D_{C_{gH}}^{\alpha} x(t) \oplus \sum_{k=1}^{m-1} \frac{(t-a)^k}{k!} \odot x_{gH}^{(k)}(a)$$

And finally based on the definition of gH-difference in type (2),

$$x(t) = x(a) \ominus_H (-1) \sum_{k=1}^{m-1} \frac{(t-a)^k}{k!} \odot x_{gH}^{(m)}(a) \ominus_H (-1) I_{RL}^{\alpha} D_{C_{gH}}^{\alpha} x(t)$$

Case 3 If $x(t)$ is $i-gH$ differentiable and its higher-order derivatives are changing in every other type periodically. In the other word, $x_{gH}^{(2k)}(t), k = 0, 1, \ldots, \frac{m}{2}$ are $i-gH$-differentiable and, $x_{gH}^{(2k-1)}(t), k = 0, 1, \ldots, \frac{m}{2}$ are $ii-gH$ differentiable.

$$I_{RL}^{\alpha} D_{C_{gH}}^{\alpha} x(t) = I_{RL}^{m-1} I_{RL}^{1} x_{gH}^{(m)}(t) = I_{RL}^{m-1} x_{gH}^{(m-1)}(t) \ominus_H I_{RL}^{m-1} x_{gH}^{(m-1)}(a)$$

By assumption $x_{gH}^{(m)}(t)$ is in the $i-gH$ sense and we have,

$$FR \int_a^t x_{gH}^{(m)}(\tau) d\tau = I_{RL}^{1} x_{gH}^{(m)}(t) = x_{gH}^{(m-1)}(t) \ominus_H x_{gH}^{(m-1)}(a)$$

Thus

$$\begin{aligned} I_{RL}^{\alpha} D_{C_{gH}}^{\alpha} x(t) &= I_{RL}^{m-1} \left[x_{gH}^{(m-1)}(t) \ominus_H x_{gH}^{(m-1)}(a) \right] \\ &= I_{RL}^{m-1} x_{gH}^{(m-1)}(t) \ominus_H I_{RL}^{m-1} x_{gH}^{(m-1)}(a) \end{aligned}$$

Since $x^{(m-1)}(t)$ is $ii-gH$ differentiable,

$$\boldsymbol{I}_{RL}^{m-1}x_{gH}^{(m-1)}(t) = \ominus_H(-1)\boldsymbol{I}_{RL}^{m-2}x_{gH}^{(m-2)}(t) \oplus (-1)\boldsymbol{I}_{RL}^{m-2}x_{gH}^{(m-2)}(a)$$
$$\boldsymbol{I}_{RL}^{\alpha}\boldsymbol{D}_{C_{gH}}^{\alpha}x(t) = \ominus_H\boldsymbol{I}_{RL}^{m-1}x_{gH}^{(m-1)}(a) \ominus_H (-1)\boldsymbol{I}_{RL}^{m-2}x_{gH}^{(m-2)}(t) \oplus (-1)\boldsymbol{I}_{RL}^{m-2}x_{gH}^{(m-2)}(a)$$

Since $x^{(m-2)}(t)$ is $i - gH$ differentiable,

$$\boldsymbol{I}_{RL}^{m-2}x_{gH}^{(m-2)}(t) = \boldsymbol{I}_{RL}^{m-3}x_{gH}^{(m-3)}(t) \ominus_H \boldsymbol{I}_{RL}^{m-3}x_{gH}^{(m-3)}(a)$$

Then

$$\boldsymbol{I}_{RL}^{\alpha}\boldsymbol{D}_{C_{gH}}^{\alpha}x(t) = \ominus_H\boldsymbol{I}_{RL}^{m-1}x_{gH}^{(m-1)}(a) \oplus (-1)\boldsymbol{I}_{RL}^{m-2}x_{gH}^{(m-2)}(a) \ominus_H (-1)\boldsymbol{I}_{RL}^{m-3}x_{gH}^{(m-3)}(t) \oplus (-1)\boldsymbol{I}_{RL}^{m-3}x_{gH}^{(m-3)}(a)$$

Since $x^{(m-3)}(t)$ is $ii - gH$ differentiable,

$$\boldsymbol{I}_{RL}^{m-3}x_{gH}^{(m-3)}(t) = \ominus_H(-1)\boldsymbol{I}_{RL}^{m-4}x_{gH}^{(m-4)}(t) \oplus (-1)\boldsymbol{I}_{RL}^{m-4}x_{gH}^{(m-4)}(a)$$

By substituting

$$\boldsymbol{I}_{RL}^{\alpha}\boldsymbol{D}_{C_{gH}}^{\alpha}x(t) = \ominus_H\boldsymbol{I}_{RL}^{m-1}x_{gH}^{(m-1)}(a) \oplus (-1)\boldsymbol{I}_{RL}^{m-2}x_{gH}^{(m-2)}(a) \oplus (-1)\boldsymbol{I}_{RL}^{m-3}x_{gH}^{(m-3)}(a) \ominus_H (-1)\left[\ominus_H(-1)\boldsymbol{I}_{RL}^{m-4}x_{gH}^{(m-4)}(t) \oplus (-1)\boldsymbol{I}_{RL}^{m-4}x_{gH}^{(m-4)}(a)\right]$$
$$\boldsymbol{I}_{RL}^{\alpha}\boldsymbol{D}_{C_{gH}}^{\alpha}x(t) = \ominus_H\boldsymbol{I}_{RL}^{m-1}x_{gH}^{(m-1)}(a) \oplus (-1)\boldsymbol{I}_{RL}^{m-2}x_{gH}^{(m-2)}(a) \oplus (-1)\boldsymbol{I}_{RL}^{m-3}x_{gH}^{(m-3)}(a) \ominus_H \boldsymbol{I}_{RL}^{m-4}x_{gH}^{(m-4)}(a) \oplus \boldsymbol{I}_{RL}^{m-4}x_{gH}^{(m-4)}(t)$$

Proceed straightforward and,

$$\boldsymbol{I}_{RL}^{m-4}x_{gH}^{(m-4)}(t) = \boldsymbol{I}_{RL}^{m-5}x_{gH}^{(m-5)}(t)\ominus_H\boldsymbol{I}_{RL}^{m-5}x_{gH}^{(m-5)}(a)$$
$$\boldsymbol{I}_{RL}^{\alpha}\boldsymbol{D}_{C_{gH}}^{\alpha}x(t) = \ominus_H\boldsymbol{I}_{RL}^{m-1}x_{gH}^{(m-1)}(a) \oplus (-1)\boldsymbol{I}_{RL}^{m-2}x_{gH}^{(m-2)}(a) \oplus (-1)\boldsymbol{I}_{RL}^{m-3}x_{gH}^{(m-3)}(a) \ominus_H\boldsymbol{I}_{RL}^{m-4}x_{gH}^{(m-4)}(a)\ominus_H\boldsymbol{I}_{RL}^{m-5}x_{gH}^{(m-5)}(a) \oplus \boldsymbol{I}_{RL}^{m-5}x_{gH}^{(m-5)}(t)$$

Finally,

$$\boldsymbol{I}_{RL}^{\alpha}\boldsymbol{D}_{C_{gH}}^{\alpha}x(t) = \ominus_H\boldsymbol{I}_{RL}^{m-1}x_{gH}^{(m-1)}(a) \oplus (-1)\boldsymbol{I}_{RL}^{m-2}x_{gH}^{(m-2)}(a) \oplus (-1)\boldsymbol{I}_{RL}^{m-3}x_{gH}^{(m-3)}(a)$$

$$\boldsymbol{I}_{RL}^{\alpha}\boldsymbol{D}_{C_{gH}}^{\alpha}x(t) = \ominus_H\frac{(t-a)^{m-1}}{(m-1)!}\odot x_{gH}^{(m-1)}(a) \oplus (-1)\frac{(t-a)^{m-2}}{(m-2)!}\odot x_{gH}^{(m-2)}(a) \ominus_H \cdots \oplus (-1)\frac{(t-a)^2}{2!}\odot x_{gH}^{(2)}(a)\ominus_H(t-a)\odot x_{gH}^{(1)}(a) \oplus x(t)\ominus_H x(a)$$

Based on definition of H-difference,

$$x(t)\ominus_H x(a) = \boldsymbol{I}_{\boldsymbol{RL}}^{\alpha}\boldsymbol{D}_{\boldsymbol{C}_{gH}}^{\alpha}x(t) \oplus (t-a)\odot x_{gH}^{(1)}(a)\ominus_H(-1)\frac{(t-a)^2}{2!}\odot x_{gH}^{(2)}(a)$$
$$\oplus\cdots\ominus_H(-1)\frac{(t-a)^{m-2}}{(m-2)!}\odot x_{gH}^{(m-2)}(a)\oplus\frac{(t-a)^{m-1}}{(m-1)!}\odot x_{gH}^{(m-1)}(a)$$
$$x(t) = x(a)\ominus_H(-1)\sum_{k=1,even}^{m-1}\frac{(t-a)^k}{k!}\odot x_{gH}^{(k)}(a)\oplus\sum_{k=1,odd}^{m-1}\frac{(t-a)^k}{k!}\odot x_{gH}^{(k)}(a)$$
$$\oplus\,\boldsymbol{I}_{\boldsymbol{RL}}^{\alpha}\boldsymbol{D}_{\boldsymbol{C}_{gH}}^{\alpha}x(t)$$

Case 4 Now suppose that $x_{gH}^{(k)}(t), k=2,3,\ldots,m$ is $i-gH$ differentiable,

$$\boldsymbol{I}_{\boldsymbol{RL}}^{\alpha}\boldsymbol{D}_{\boldsymbol{C}_{gH}}^{\alpha}x(t) = \boldsymbol{I}_{\boldsymbol{RL}}^{m}x_{gH}^{(m)}(t) = \boldsymbol{I}_{\boldsymbol{RL}}^{m-1}x_{gH}^{(m-1)}(t)\ominus_H \boldsymbol{I}_{\boldsymbol{RL}}^{m-1}x_{gH}^{(m-1)}(a)$$

By the same process in case 1,

$$\boldsymbol{I}_{\boldsymbol{RL}}^{\alpha}\boldsymbol{D}_{\boldsymbol{C}_{gH}}^{\alpha}x(t) = \boldsymbol{I}_{\boldsymbol{RL}}^{1}x_{gH}^{(1)}(t)\ominus_H \boldsymbol{I}_{\boldsymbol{RL}}^{1}x_{gH}^{(1)}(a)\ominus_H \boldsymbol{I}_{\boldsymbol{RL}}^{2}x_{gH}^{(2)}(t)\ominus_H \boldsymbol{I}_{\boldsymbol{RL}}^{2}x_{gH}^{(2)}(a)$$
$$\ominus_H\cdots\ominus_H \boldsymbol{I}_{\boldsymbol{RL}}^{m-1}x_{gH}^{(m-1)}(a)$$

Since $x'(t)$ is $i-gH$ differentiable on $[a,\sigma]$ and $ii-gH$ differentiable after $[\sigma,b]$,

$$\boldsymbol{I}_{\boldsymbol{RL}}^{2}x_{gH}^{(2)}(t) = \begin{cases} \boldsymbol{I}_{\boldsymbol{RL}}^{1}x_{gH}^{(1)}(t)\ominus_H \boldsymbol{I}_{\boldsymbol{RL}}^{1}x_{gH}^{(1)}(a), & a\le t\le\sigma \\ (-1)\boldsymbol{I}_{\boldsymbol{RL}}^{1}x_{gH}^{1}(a)\ominus_H(-1)\boldsymbol{I}_{\boldsymbol{RL}}^{1}x_{gH}^{(1)}(t), & \sigma\le t\le b \end{cases}$$

Also $x(t)$ is $ii-gH$ differentiable on $[a,\xi]$ and $i-gH$ differentiable after $[\xi,b]$,

$$\boldsymbol{I}_{\boldsymbol{RL}}^{1}x_{gH}^{(1)}(t) = \begin{cases} (-1)x(a)\ominus_H(-1)x(t), & a\le t\le\xi \\ x(t)\ominus_H x(t), & \xi\le t\le\sigma \\ x(t)\ominus_H x(a), & \sigma\le t\le b \end{cases}$$

Substituting the recent equations in $\boldsymbol{I}_{\boldsymbol{RL}}^{\alpha}\boldsymbol{D}_{\boldsymbol{C}_{gH}}^{\alpha}x(t)$ the assertion of this case are obtained.

In this section we are going to obtain a generalized Taylor's expansion for fuzzy-valued functions by using the concept of Caputo generalized Hukuhara derivative. To this end, we need some results and definitions that are explained in the following.

Remark Consider the fuzzy continuous function $\boldsymbol{D}_{\boldsymbol{C}_{gH}}^{k\alpha}x(t)\in\mathbb{F}_R$ in (a,b) for $k=0,1,\ldots,n$, such that $0<\alpha\le 1$. Then we have the following items,

Case 1. If the derivatives $D^{n\alpha}_{C_{gH}}x(t), D^{(n+1)\alpha}_{C_{gH}}x(t)$ are in the same types of differentiability for $n \geq 1$, then

$$I^{\alpha}_{RL} D^{(n+1)\alpha}_{C_{gH}} x(t) = D^{n\alpha}_{C_{gH}} x(t) \ominus_H D^{n\alpha}_{C_{gH}} x(a)$$

Case 2. If the derivatives $D^{n\alpha}_{C_{gH}}x(t), D^{(n+1)\alpha}_{C_{gH}}x(t)$ are in different types of differentiability for $n \geq 1$, then

$$I^{\alpha}_{RL} D^{(n+1)\alpha}_{C_{gH}} x(t) = (-1) D^{n\alpha}_{C_{gH}} x(a) \ominus_H (-1) D^{n\alpha}_{C_{gH}} x(t)$$

To show two cases, we have the following relation,

$$\begin{aligned} I^{\alpha}_{RL} D^{(n+1)\alpha}_{C_{gH}} x(t) &= I^{\alpha}_{RL}\left(D^{n\alpha}_{C_{gH}} D^{\alpha}_{C_{gH}} x\right)(t) = \left(D^{n\alpha}_{C_{gH}} I^{\alpha}_{RL} D^{\alpha}_{C_{gH}} x\right)(t) = D^{n\alpha}_{C_{gH}} x(t) \\ &= \left(D^{(n-1)\alpha}_{C_{gH}} D^{\alpha}_{C_{gH}} x\right)(t) = \left(D^{n\alpha}_{C_{gH}} I^{\alpha}_{RL} D^{\alpha}_{C_{gH}} x\right)(t) \end{aligned}$$

Since

$$\begin{aligned} &I^{\alpha}_{RL} D^{\alpha}_{C_{gH}} x(t) = x(t) \ominus_{gH} x(a) \\ &I^{\alpha}_{RL} D^{(n+1)\alpha}_{C_{gH}} x(t) = D^{n\alpha}_{C_{gH}}\left(I^{\alpha}_{RL} D^{\alpha}_{C_{gH}} x(t)\right) = D^{n\alpha}_{C_{gH}}\left(x(t) \ominus_{gH} x(a)\right) \\ &I^{\alpha}_{RL} D^{(n+1)\alpha}_{C_{gH}} x(t) = D^{n\alpha}_{C_{gH}} x(t) \ominus_{gH} D^{n\alpha}_{C_{gH}} x(a) \end{aligned}$$

Two cases are obtained by the definition of gH-difference in two types.

Remark Consider the fuzzy continuous function $D^{k\alpha}_{C_{gH}} x(t) \in \mathbb{F}_R$ in (a, b) for $k = 0, 1, \ldots, n$, such that $0 < \alpha \leq 1$. Then we have the following items,

Case 1. If the derivatives $D^{n\alpha}_{C_{gH}}x(t), D^{(n+1)\alpha}_{C_{gH}}x(t)$ are in the same types of differentiability for $n \geq 1$, then

$$I^{n\alpha}_{RL} D^{n\alpha}_{C_{gH}} x(t) \ominus_H I^{(n+1)\alpha}_{RL} D^{(n+1)\alpha}_{C_{gH}} x(t) = \frac{(t-a)^{n\alpha}}{\Gamma(n\alpha+1)} \odot D^{n\alpha}_{C_{gH}} x(a)$$

Case 2. If the derivatives $D^{n\alpha}_{C_{gH}}x(t), D^{(n+1)\alpha}_{C_{gH}}x(t)$ are in different types of differentiability for $n \geq 1$, then

$$I^{n\alpha}_{RL} D^{n\alpha}_{C_{gH}} x(t) \oplus (-1) I^{(n+1)\alpha}_{RL} D^{(n+1)\alpha}_{C_{gH}} x(t) = \frac{(t-a)^{n\alpha}}{\Gamma(n\alpha+1)} \odot D^{n\alpha}_{C_{gH}} x(a)$$

Two cases can be proved as follows,

$$\boldsymbol{I}_{\boldsymbol{RL}}^{n\alpha}\boldsymbol{D}_{C_{gH}}^{n\alpha}x(t)\ominus_{gH}\boldsymbol{I}_{\boldsymbol{RL}}^{(n+1)\alpha}\boldsymbol{D}_{C_{gH}}^{(n+1)\alpha}x(t)=\boldsymbol{I}_{\boldsymbol{RL}}^{n\alpha}\left(\boldsymbol{D}_{C_{gH}}^{n\alpha}x(t)\ominus_{gH}\boldsymbol{I}_{\boldsymbol{RL}}^{n\alpha}\boldsymbol{D}_{C_{gH}}^{(n+1)\alpha}x(t)\right)$$

We proved in the previous remark, that is

$$\boldsymbol{I}_{\boldsymbol{RL}}^{\alpha}\boldsymbol{D}_{C_{gH}}^{(n+1)\alpha}x(t)=\boldsymbol{D}_{C_{gH}}^{n\alpha}x(t)\ominus_{gH}\boldsymbol{D}_{C_{gH}}^{n\alpha}x(a)$$

Substituting and

$$\begin{aligned}&\boldsymbol{I}_{\boldsymbol{RL}}^{n\alpha}\boldsymbol{D}_{C_{gH}}^{n\alpha}x(t)\ominus_{gH}\boldsymbol{I}_{\boldsymbol{RL}}^{(n+1)\alpha}\boldsymbol{D}_{C_{gH}}^{(n+1)\alpha}x(t)\\&=\boldsymbol{I}_{\boldsymbol{RL}}^{n\alpha}\left(\boldsymbol{D}_{C_{gH}}^{n\alpha}x(t)\ominus_{gH}\left(\boldsymbol{D}_{C_{gH}}^{n\alpha}x(t)\ominus_{gH}\boldsymbol{D}_{C_{gH}}^{n\alpha}x(a)\right)\right)=\boldsymbol{I}_{\boldsymbol{RL}}^{n\alpha}\boldsymbol{D}_{C_{gH}}^{n\alpha}x(a)\\&=\frac{(t-a)^{n\alpha}}{\Gamma(n\alpha+1)}\odot\boldsymbol{D}_{C_{gH}}^{n\alpha}x(a)\end{aligned}$$

So we proved,

$$\boldsymbol{I}_{\boldsymbol{RL}}^{n\alpha}\boldsymbol{D}_{C_{gH}}^{n\alpha}x(t)\ominus_{gH}\boldsymbol{I}_{\boldsymbol{RL}}^{(n+1)\alpha}\boldsymbol{D}_{C_{gH}}^{(n+1)\alpha}x(t)=\frac{(t-a)^{n\alpha}}{\Gamma(n\alpha+1)}\odot\boldsymbol{D}_{C_{gH}}^{n\alpha}x(a)$$

In accordance with the definition of the gH-difference two cases are gotten.

5.3.2 *Fuzzy Generalized Taylor's Expansion*

Let us consider the fuzzy continuous function $\boldsymbol{D}_{C_{gH}}^{k\alpha}x(t)\in\mathbb{F}_R$ in (a,b) for $k=0,1,\ldots,n$, such that $0<\alpha\leq 1$. Then we have the following items,

Case 1 If $x(t)$ is $i-C_{gH}$ differentiable of order $k\alpha$ then there exists $\xi\in(a,b)$ such that,

$$x(t)=x(a)\oplus\sum_{k=1}^{n}\frac{(t-a)^{k\alpha}}{\Gamma(k\alpha+1)}\odot\boldsymbol{D}_{C_{gH}}^{k\alpha}x(a)\oplus\frac{\boldsymbol{D}_{C_{gH}}^{(n+1)\alpha}x(\xi)}{\Gamma((n+1)\alpha+1)}\odot(t-a)^{(n+1)\alpha}$$

Case 2 If $x(t)$ is $ii-C_{gH}$ differentiable of order $k\alpha$ then there exists $\xi\in(a,b)$ such that,

$$\begin{aligned}x(t)=x(a)\ominus_H(-1)\sum_{k=1}^{n}\frac{(t-a)^{k\alpha}}{\Gamma(k\alpha+1)}\odot\boldsymbol{D}_{C_{gH}}^{k\alpha}x(a)\ominus_H(-1)\frac{\boldsymbol{D}_{C_{gH}}^{(n+1)\alpha}x(\xi)}{\Gamma((n+1)\alpha+1)}\\\odot(t-a)^{(n+1)\alpha}\end{aligned}$$

Case 3 If $x(t)$ is $i-C_{gH}$ differentiable of order $2k\alpha, k=0,1,\ldots,\left[\frac{n}{2}\right]$ and also it is $ii-C_{gH}$ differentiable of order $(2k-1)\alpha, k=0,1,\ldots,\left[\frac{n}{2}\right]$ then there exists $\xi \in (a,b)$ such that,

$$x(t)=x(a)\ominus_H(-1)\sum_{k=1,odd}^{n}\frac{(t-a)^{k\alpha}}{\Gamma(k\alpha+1)}\odot \boldsymbol{D}^{k\alpha}_{C_{gH}}x(a)\oplus\sum_{k=1,even}^{n}\frac{(t-a)^{k\alpha}}{\Gamma(k\alpha+1)}\odot \boldsymbol{D}^{k\alpha}_{C_{gH}}x(a)$$
$$\ominus_H(-1)\frac{\boldsymbol{D}^{(n+1)\alpha}_{C_{gH}}x(\xi)}{\Gamma((n+1)\alpha+1)}\odot(t-a)^{(n+1)\alpha}$$

Case 4 If $\zeta \in [a,b]$, and $x(t)$ is $ii-C_{gH}$ differentiable before ζ and $i-C_{gH}$ after it, and type of differentiability for $\boldsymbol{D}^{k\alpha}_{C_{gH}}x(t), k=1,2,\ldots,n$ are $i-C_{gH}$ differentiable then there exists $\xi \in (a,b)$ such that,

$$x(t)=\begin{cases} x(a)\ominus_H(-1)\frac{(t-a)^{\alpha}}{\Gamma(\alpha+1)}\odot \boldsymbol{D}^{\alpha}_{C_{gH}}x(a)\oplus\sum_{k=2}^{n}\frac{(t-a)^{k\alpha}}{\Gamma(k\alpha+1)}\odot \boldsymbol{D}^{k\alpha}_{C_{gH}}x(a) \\ \quad\oplus\frac{\boldsymbol{D}^{(n+1)\alpha}_{C_{gH}}x(\xi)}{\Gamma((n+1)\alpha+1)}\odot(t-a)^{(n+1)\alpha}, \quad a\le t\le\zeta \\ x(a)\oplus\sum_{k=1}^{n}\frac{(t-a)^{k\alpha}}{\Gamma(k\alpha+1)}\odot \boldsymbol{D}^{k\alpha}_{C_{gH}}x(a)\oplus\frac{\boldsymbol{D}^{(n+1)\alpha}_{C_{gH}}x(\xi)}{\Gamma((n+1)\alpha+1)}\odot(t-a)^{(n+1)\alpha}, \zeta\le t\le b \end{cases}$$

Case 5 If $\zeta \in [a,b]$, and $x(t)$ is $i-C_{gH}$ differentiable before ζ and $ii-C_{gH}$ after it, and type of differentiability for $\boldsymbol{D}^{k\alpha}_{C_{gH}}x(t), k=1,2,\ldots,n$ are $ii-C_{gH}$ differentiable then there exists $\xi \in (a,b)$ such that,

$$x(t)=\begin{cases} x(a)\oplus\frac{(t-a)^{\alpha}}{\Gamma(\alpha+1)}\odot \boldsymbol{D}^{\alpha}_{C_{gH}}x(a)\ominus_H(-1)\sum_{k=2}^{n}\frac{(t-a)^{k\alpha}}{\Gamma(k\alpha+1)}\odot \boldsymbol{D}^{k\alpha}_{C_{gH}}x(a) \\ \quad\ominus_H(-1)\frac{\boldsymbol{D}^{(n+1)\alpha}_{C_{gH}}x(\xi)}{\Gamma((n+1)\alpha+1)}\odot(t-a)^{(n+1)\alpha}, \quad a\le t\le\zeta \\ x(a)\ominus_H(-1)\sum_{k=1}^{n}\frac{(t-a)^{k\alpha}}{\Gamma(k\alpha+1)}\odot \boldsymbol{D}^{k\alpha}_{C_{gH}}x(a)\ominus_H \\ \quad(-1)\frac{\boldsymbol{D}^{(n+1)\alpha}_{C_{gH}}x(\xi)}{\Gamma((n+1)\alpha+1)}\odot(t-a)^{(n+1)\alpha}, \quad \zeta\le t\le b \end{cases}$$

In general, in accordance with gH-difference, we have the following relations. Since

$$\boldsymbol{I}_{\boldsymbol{RL}}^{n\alpha}\boldsymbol{D}_{C_{gH}}^{n\alpha}x(t)\ominus_{gH}\boldsymbol{I}_{\boldsymbol{RL}}^{(n+1)\alpha}\boldsymbol{D}_{C_{gH}}^{(n+1)\alpha}x(t)=\frac{(t-a)^{n\alpha}}{\Gamma(n\alpha+1)}\odot\boldsymbol{D}_{C_{gH}}^{n\alpha}x(a)$$

$$\sum_{k=0}^{n}\left(\boldsymbol{I}_{\boldsymbol{RL}}^{k\alpha}\boldsymbol{D}_{C_{gH}}^{k\alpha}x(t)\ominus_{gH}\boldsymbol{I}_{\boldsymbol{RL}}^{(k+1)\alpha}\boldsymbol{D}_{C_{gH}}^{(k+1)\alpha}x(t)\right)=x(t)\ominus_{gH}\boldsymbol{I}_{\boldsymbol{RL}}^{(n+1)\alpha}\boldsymbol{D}_{C_{gH}}^{(n+1)\alpha}x(t)$$

$$\sum_{k=0}^{n}\left(\boldsymbol{I}_{\boldsymbol{RL}}^{k\alpha}\boldsymbol{D}_{C_{gH}}^{k\alpha}x(t)\ominus_{gH}\boldsymbol{I}_{\boldsymbol{RL}}^{(k+1)\alpha}\boldsymbol{D}_{C_{gH}}^{(k+1)\alpha}x(t)\right)=\sum_{k=0}^{n}\frac{(t-a)^{k\alpha}}{\Gamma(k\alpha+1)}\odot\boldsymbol{D}_{C_{gH}}^{k\alpha}x(a)$$

$$x(t)\ominus_{gH}\boldsymbol{I}_{\boldsymbol{RL}}^{(n+1)\alpha}\boldsymbol{D}_{C_{gH}}^{(n+1)\alpha}x(t)=\sum_{k=0}^{n}\frac{(t-a)^{k\alpha}}{\Gamma(k\alpha+1)}\odot\boldsymbol{D}_{C_{gH}}^{k\alpha}x(a)$$

$$x(t)=\sum_{k=0}^{n}\frac{(t-a)^{k\alpha}}{\Gamma(k\alpha+1)}\odot\boldsymbol{D}_{C_{gH}}^{k\alpha}x(a)\oplus\boldsymbol{I}_{\boldsymbol{RL}}^{(n+1)\alpha}\boldsymbol{D}_{C_{gH}}^{(n+1)\alpha}x(t)$$

$$x(t)=x(a)\oplus\sum_{k=1}^{n}\frac{(t-a)^{k\alpha}}{\Gamma(k\alpha+1)}\odot\boldsymbol{D}_{C_{gH}}^{k\alpha}x(a)\oplus\boldsymbol{I}_{\boldsymbol{RL}}^{(n+1)\alpha}\boldsymbol{D}_{C_{gH}}^{(n+1)\alpha}x(t)$$

Using fractional mean value theorem,

$$\begin{aligned}\boldsymbol{I}_{\boldsymbol{RL}}^{(n+1)\alpha}\boldsymbol{D}_{C_{gH}}^{(n+1)\alpha}x(t)&=\frac{1}{\Gamma((n+1)\alpha)}\odot\int_{a}^{t}(t-\tau)^{(n+1)\alpha-1}\odot\boldsymbol{D}_{C_{gH}}^{(n+1)\alpha}x(\tau)d\tau\\&=\frac{\boldsymbol{D}_{C_{gH}}^{(n+1)\alpha}x(\xi)}{\Gamma((n+1)\alpha+1)}\odot(t-a)^{(n+1)\alpha}\end{aligned}$$

Proof of Case 1 Since $x(t)$ is $i-C_{gH}$ differentiable of order $k\alpha$ then all gH-differences are defined in the sense of $i-gH$ difference and we have,

$$x(t)=x(a)\oplus\sum_{k=1}^{n}\frac{(t-a)^{k\alpha}}{\Gamma(k\alpha+1)}\odot\boldsymbol{D}_{C_{gH}}^{k\alpha}x(a)\oplus\boldsymbol{I}_{\boldsymbol{RL}}^{(n+1)\alpha}\boldsymbol{D}_{C_{gH}}^{(n+1)\alpha}x(\zeta)$$

Proof of case 2 Since $x(t)$ is $ii-C_{gH}$ differentiable of order $k\alpha$ then all gH-differences are defined in the sense of $ii-gH$ difference and also fractional mean value theorem we have,

$$\boldsymbol{I}_{\boldsymbol{RL}}^{n\alpha}\boldsymbol{D}_{C_{gH}}^{n\alpha}x(t)\oplus(-1)\boldsymbol{I}_{\boldsymbol{RL}}^{(n+1)\alpha}\boldsymbol{D}_{C_{gH}}^{(n+1)\alpha}x(t)=\frac{(t-a)^{n\alpha}}{\Gamma(n\alpha+1)}\odot\boldsymbol{D}_{C_{gH}}^{n\alpha}x(a)$$

$$x(t)=x(a)\ominus_{H}(-1)\sum_{k=1}^{n}\frac{(t-a)^{k\alpha}}{\Gamma(k\alpha+1)}\odot\boldsymbol{D}_{C_{gH}}^{k\alpha}x(a)\ominus_{H}(-1)\frac{\boldsymbol{D}_{C_{gH}}^{(n+1)\alpha}x(\xi)}{\Gamma((n+1)\alpha+1)}\odot(t-a)^{(n+1)\alpha}$$

Proof of Case 3 If $x(t)$ is $i-C_{gH}$ differentiable of order $2k\alpha, k=0,1,\ldots,\left[\frac{n}{2}\right]$ and also it is $ii-C_{gH}$ differentiable of order $(2k-1)\alpha, k=0,1,\ldots,\left[\frac{n}{2}\right]$ then there exists $\xi\in(a,b)$.

$$\sum_{k=0}^{n}\left(\boldsymbol{I}_{RL}^{k\alpha}\boldsymbol{D}_{C_{gH}}^{k\alpha}x(t)\ominus_{gH}\boldsymbol{I}_{RL}^{(k+1)\alpha}\boldsymbol{D}_{C_{gH}}^{(k+1)\alpha}x(t)\right)=x(t)\oplus(-1)\boldsymbol{I}_{RL}^{\alpha}\boldsymbol{D}_{C_{gH}}^{\alpha}x(t)$$
$$\ominus_{H}(-1)\boldsymbol{I}_{RL}^{\alpha}\boldsymbol{D}_{C_{gH}}^{\alpha}x(t)\ominus_{H}\boldsymbol{I}_{RL}^{2\alpha}\boldsymbol{D}_{C_{gH}}^{2\alpha}x(t)$$
$$\oplus\cdots\oplus\boldsymbol{I}_{RL}^{n\alpha}\boldsymbol{D}_{C_{gH}}^{n\alpha}x(t)\ominus_{H}(-1)\boldsymbol{I}_{RL}^{(n+1)\alpha}\boldsymbol{D}_{C_{gH}}^{(n+1)\alpha}x(t)$$

By similar reasoning in case 1,

$$x(t)\oplus(-1)\boldsymbol{I}_{RL}^{(n+1)\alpha}\boldsymbol{D}_{C_{gH}}^{(n+1)\alpha}x(t)=x(a)\ominus_{H}\sum_{k=1,odd}^{n}\boldsymbol{I}_{RL}^{k\alpha}\boldsymbol{D}_{C_{gH}}^{k\alpha}x(a)\oplus\sum_{k=1,even}^{n}\boldsymbol{I}_{RL}^{k\alpha}\boldsymbol{D}_{C_{gH}}^{k\alpha}x(a)$$

$$=x(a)\ominus_{H}\sum_{k=1,odd}^{n}\frac{(t-a)^{k\alpha}}{\Gamma(k\alpha+1)}\odot\boldsymbol{D}_{C_{gH}}^{k\alpha}x(a)\oplus\sum_{k=1,even}^{n}\frac{(t-a)^{k\alpha}}{\Gamma(k\alpha+1)}\odot\boldsymbol{D}_{C_{gH}}^{k\alpha}x(a)$$

By using definition of H-difference,

$$x(t)=x(a)\ominus_{H}(-1)\sum_{k=1,odd}^{n}\frac{(t-a)^{k\alpha}}{\Gamma(k\alpha+1)}\odot\boldsymbol{D}_{C_{gH}}^{k\alpha}x(a)\oplus\sum_{k=1,even}^{n}\frac{(t-a)^{k\alpha}}{\Gamma(k\alpha+1)}\odot\boldsymbol{D}_{C_{gH}}^{k\alpha}x(a)$$
$$\ominus_{H}(-1)\frac{\boldsymbol{D}_{C_{gH}}^{(n+1)\alpha}x(\xi)}{\Gamma((n+1)\alpha+1)}\odot(t-a)^{(n+1)\alpha}$$

The proof of the cases 4 and 5 are very similar to cases 1, 2 and 3.

One of the applications of Fuzzy fractional Taylor, is Fuzzy Euler fractional method that is immediate consequence of the Taylor expansion. Now ware going to cover the Fuzzy fractional Euler method.

5.4 Fuzzy Fractional Euler with Caputo gH-Derivative

Consider the following fuzzy fractional differential equation,

$$\boldsymbol{D}_{C_{gH}}^{\alpha}x(t)=f(t,x(t)),\quad x(t_0)=x_0,\quad t\in[t_0,T],\quad 0<\alpha\leq 1$$

where $f\colon[t_0,T]\times\mathbb{F}_R\rightarrow\mathbb{F}_R$ is a fuzzy number valued function and $x(t)$ is a continuous fuzzy set valued solution and $\boldsymbol{D}_{C_{gH}}^{\alpha}$ is the fractional Caputo fractional operator. In this method, Indeed, a sequence of approximations to the solution $x(t)$ will be obtained at several points, called gride points. To derive Euler's Method, the interval $[t_0,T]$ is divided to into N equal subintervals, each of length h, by the grid points $t_i=t_0+ih, i=0,1,2,\ldots,N$. The distance between points, $h=\frac{T-t_0}{N}$ is called the grid size. To explain the method, we should discuss the type of differentiability of the solution and to this end we will have several cases like as Fuzzy Taylor expansion.

Case 1 Suppose that the fuzzy solution $x(t)$ is $i - C_{gH}$ differentiable on $[t_0, T]$. The Taylor expansion about t_k on any subinterval $[t_k, t_{k+1}]$ for $k = 0, 1, \ldots, N-1$ can be expressed as the following form, where $h = t_{k+1} - t_k, \exists \zeta_k \in [t_k, t_{k+1}]$.

$$x(t_{k+1}) = x(t_k) \oplus \frac{(t_{k+1} - t_k)^{\alpha}}{\Gamma(\alpha+1)} \odot \boldsymbol{D}^{\alpha}_{C_{gH}} x(t_k) \oplus \frac{(t_{k+1} - t_k)^{2\alpha}}{\Gamma(2\alpha+1)} \odot \boldsymbol{D}^{2\alpha}_{C_{gH}} x(\zeta_k)$$

$$x(t_{k+1}) = x(t_k) \oplus \frac{h^{\alpha}}{\Gamma(\alpha+1)} \odot \boldsymbol{D}^{k\alpha}_{C_{gH}} x(t_k) \oplus \frac{h^{2\alpha}}{\Gamma(2\alpha+1)} \odot \boldsymbol{D}^{2\alpha}_{C_{gH}} x(\zeta_k)$$

Since $\boldsymbol{D}^{\alpha}_{C_{gH}} x(t_k) = f(t, x(t_k))$ then by substituting, we have

$$x(t_{k+1}) = x(t_k) \oplus \frac{h^{\alpha}}{\Gamma(\alpha+1)} \odot f(t_k, x(t_k)) \oplus \frac{h^{2\alpha}}{\Gamma(2\alpha+1)} \odot \boldsymbol{D}^{2\alpha}_{C_{gH}} x(\zeta_k)$$

where the term $\frac{h^{2\alpha}}{\Gamma(2\alpha+1)} \odot \boldsymbol{D}^{2\alpha}_{C_{gH}} x(\zeta_k)$ can be denoted as the error for the numerical method. As a result, the Fuzzy fractional Euler method on $[t_k, t_{k+1}]$ can be introduced as,

$$x(t_{k+1}) = x(t_k) \oplus \frac{h^{\alpha}}{\Gamma(\alpha+1)} \odot f(t_k, x(t_k)),\ h = t_{k+1} - t_k,\ k = 0, 1, \ldots, N-1$$

Case 2 Suppose that the fuzzy solution $x(t)$ is $ii - C_{gH}$ differentiable on $[t_0, T]$. The Taylor expansion about t_k on any subinterval $[t_k, t_{k+1}]$ for $k = 0, 1, \ldots, N-1$ can be expressed as the following form, where $h = t_{k+1} - t_k, \exists \zeta_k \in [t_k, t_{k+1}]$.

$$x(t_{k+1}) = x(t_k) \ominus_H (-1) \frac{h^{\alpha}}{\Gamma(\alpha+1)} \odot \boldsymbol{D}^{\alpha}_{C_{gH}} x(t_k) \ominus_H (-1) \frac{h^{\alpha}}{\Gamma(2\alpha+1)} \odot \boldsymbol{D}^{2\alpha}_{C_{gH}} x(\zeta_k)$$

Since $\boldsymbol{D}^{\alpha}_{C_{gH}} x(t_k) = f(t, x(t_k))$ then by substituting, we have

$$x(t_{k+1}) = x(t_k) \ominus_H (-1) \frac{h^{\alpha}}{\Gamma(\alpha+1)} \odot f(t_k, x(t_k)) \ominus_H (-1) \frac{h^{2\alpha}}{\Gamma(2\alpha+1)} \odot \boldsymbol{D}^{2\alpha}_{C_{gH}} x(\zeta_k)$$

where the term $\frac{h^{2\alpha}}{\Gamma(2\alpha+1)} \odot \boldsymbol{D}^{2\alpha}_{C_{gH}} x(\zeta_k)$ can be denoted as the error for the numerical method. As a result, the Fuzzy fractional Euler method on $[t_k, t_{k+1}]$ can be introduced as,

$$\begin{aligned} x(t_{k+1}) &= x(t_k) \ominus_H (-1) \frac{h^{\alpha}}{\Gamma(\alpha+1)} \odot f(t_k, x(t_k)),\ h = t_{k+1} - t_k,\ k \\ &= 0, 1, \ldots, N-1 \end{aligned}$$

Case 3 Suppose that the fuzzy solution $x(t)$ has a switching point at $\xi \in [t_0, T]$ and it is $i - C_{gH}$ differentiable at the points $t_0, t_1, \ldots, t_j$ and $ii - C_{gH}$ differentiable at the points $t_{j+1}, t_{j+2}, \ldots, t_N$ on $[t_0, T]$. The Taylor expansion about t_k on any subinterval $[t_k, t_{k+1}]$ for $k = 0, 1, \ldots, N-1$ can be expressed as the following form,

$$\begin{cases} x(t_{k+1}) = x(t_k) \oplus \frac{h^\alpha}{\Gamma(\alpha+1)} \odot f(t_k, x(t_k)), & k = 1, 2, \ldots, j \\ x(t_{k+1}) = x(t_k) \ominus_H (-1) \frac{h^\alpha}{\Gamma(\alpha+1)} \odot f(t_k, x(t_k)), & k = j+1, j+2, \ldots, N \end{cases}$$

Case 4 Suppose that the fuzzy solution $x(t)$ has a switching point at $\xi \in [t_0, T]$ and it is $ii - C_{gH}$ differentiable at the points $t_0, t_1, \ldots, t_j$ and $i - C_{gH}$ differentiable at the points $t_{j+1}, t_{j+2}, \ldots, t_N$ on $[t_0, T]$. The Taylor expansion about t_k on any subinterval $[t_k, t_{k+1}]$ for $k = 0, 1, \ldots, N-1$ can be expressed as the following form,

$$\begin{cases} x(t_{k+1}) = x(t_k) \ominus_H (-1) \frac{h^\alpha}{\Gamma(\alpha+1)} \odot f(t_k, x(t_k)), & k = 1, 2, \ldots, j \\ x(t_{k+1}) = x(t_k) \oplus \frac{h^\alpha}{\Gamma(\alpha+1)} \odot f(t_k, x(t_k)), & k = j+1, j+2, \ldots, N \end{cases}$$

Example Consider the following fuzzy fractional differential equation,

$$\boldsymbol{D}^\alpha_{C_{gH}} x(t) = (0, 1, 1.5) \odot \Gamma(\alpha + 1), \quad x(0) = 0, \quad t \in [0, 1], \quad 0 < \alpha \leq 1$$

The exact solution is,

$$x(t) = (0, 1, 1.5) \odot t^\alpha := c \odot t^\alpha$$

This function is $i - gH$ differentiable, because the length of $x(t)$ is increasing, Note. The Caputo gH-derivative of $x(t) = (0, 1, 1.5) \odot t^\alpha$ is,

$$\begin{aligned} \boldsymbol{D}^\alpha_{C_{gH}}((0, 1, 1.5) \odot t^\alpha) &= \frac{1}{\Gamma(1-\alpha)} \odot \int_0^t \frac{((0, 1, 1.5) \odot \tau^\alpha)'}{(t-\tau)^\alpha} d\tau \\ &= \frac{(0, 1, 1.5)\alpha}{\Gamma(1-\alpha)} \odot \int_0^t \frac{\tau^{\alpha-1}}{(t-\tau)^\alpha} d\tau \end{aligned}$$

Since,

$$\int_0^t \frac{\tau^{\alpha-1}}{(t-\tau)^\alpha} d\tau = \frac{4^{-\alpha}\sqrt{\pi}\Gamma(\alpha)t^{2\alpha}}{\Gamma\left(\alpha + \frac{1}{2}\right)}$$

Then

$$D^{\alpha}_{C_{gH}}((0,1,1.5)\odot t^{\alpha}) = (0,1,1.5)\odot \frac{\alpha 4^{-\alpha}\sqrt{\pi}\Gamma(\alpha)t^{2\alpha}}{\Gamma(\alpha+\frac{1}{2})\Gamma(1-\alpha)}$$

In the level-wise form,

$$D^{\alpha}_{C}(rt^{\alpha}) = r\cdot \frac{\alpha 4^{-\alpha}\sqrt{\pi}\Gamma(\alpha)t^{2\alpha}}{\Gamma(\alpha+\frac{1}{2})\Gamma(1-\alpha)}$$
$$D^{\alpha}_{C}((1.5-1.5r)t^{\alpha}) = (1.5-1.5r)\cdot \frac{\alpha 4^{-\alpha}\sqrt{\pi}\Gamma(\alpha)t^{2\alpha}}{\Gamma(\alpha+\frac{1}{2})\Gamma(1-\alpha)}$$

So, the Euler method is in the form of case 1,

$$x(t_{k+1}) = x(t_k)\oplus\frac{h^{\alpha}}{\Gamma(\alpha+1)}\odot(0,1,1.5)\odot\Gamma(\alpha+1),\quad k=0,1,\ldots,N-1$$

For instance, consider the step size $h=0.1$ and $\alpha=0.5$. The level-wise form of the Euler method is,

$$x(t_{k+1},r) = x(t_k,r)\oplus(0.1)^{0.5}\odot(r,1.5-1.5r),\quad k=0,1,\ldots,N-1$$

$$x(t_{k+1},r) = x(t_k,r)\oplus\left((0.1)^{0.5}r,(0.1)^{0.5}(1.5-1.5r)\right),\quad k=0,1,\ldots,N-1$$

In the level-wise form,

$$x_l(t_{k+1},r) = x_l(t_k,r)+(0.1)^{0.5}r,\ k=0,1,\ldots,N-1, 0\le r\le 1$$

$$x_u(t_{k+1},r) = x_u(t_k,r)+(0.1)^{0.5}(1.5-1.5r),\quad k=0,1,\ldots,N-1, 0\le r\le 1$$

Using the recursive method and finding the values for $x(t_k), k=0,1,\ldots,9$ finally we get,

$$x_l(t_{10},r) = 10(0.1)^{0.5}r,\quad x_u(t_{10},r) = 10(0.1)^{0.5}(1.5-1.5r),\quad 0\le r\le 1$$

In Fig. 5.1 the exact solution and its Caputo gH-derivative with $\alpha=0.5$ are presented at the grid points.

In Table 5.1 the numerical solutions with order $\alpha=0.5$ and $h=0.1$ are expressed.

Also, in Fig. 5.2 the exact solution and its Caputo gH-derivative with $\alpha=0.75$ are presented at the grid points.

In Table 5.2 the numerical solutions with order $\alpha=0.75$ and $h=0.1$ are expressed.

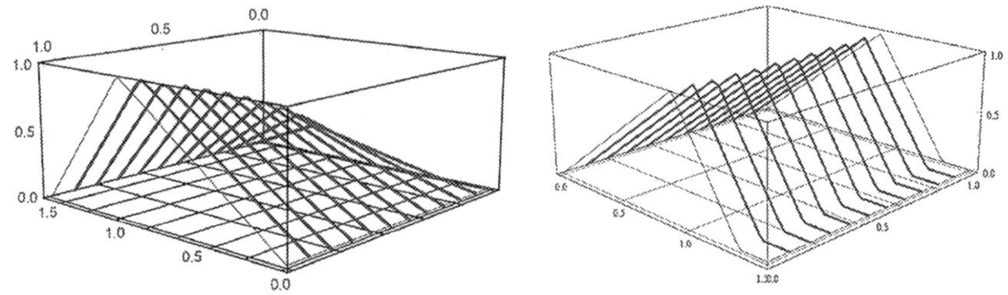

Fig. 5.1 Exact solution (left) and its C-gH-derivative (right) at the grid points with $\alpha = 0.5$

Table 5.1 The numerical results for step size $h = 0.1$

t	h=0.1
0.1	(0, 0.31, 0.47)
0.2	(0, 0.63, 0.94)
0.3	(0, 0.94, 1.42)
0.4	(0, 1.26, 1.89)
0.5	(0, 1.58, 2.37)
0.6	(0, 1.89, 2.84)
0.7	(0, 2.21, 3.32)
0.8	(0, 2.52, 3.79)
0.9	(0, 2.84, 4.26)
1.0	(0, 3.16, 4.74)

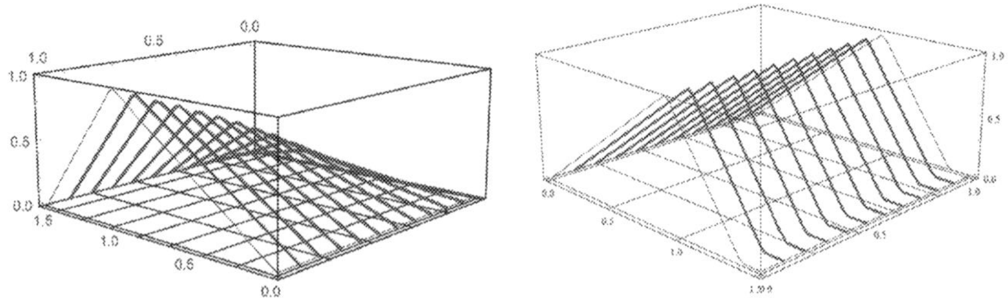

Fig. 5.2 Exact solution (left) and its C-gH-derivative (right) at the grid points with $\alpha = 0.75$

Table 5.2 The numerical results with order $\alpha = 0.75$ for step size $h = 0.1$

t	h=0.1
0.1	(0, 0.17, 0.26)
0.2	(0, 0.35, 0.53)
0.3	(0, 0.53, 0.80)
0.4	(0, 0.71, 1.06)
0.5	(0, 0.88, 1.33)
0.6	(0, 1.06, 1.60)
0.7	(0, 1.24, 1.86)
0.8	(0, 1.42, 2.13)
0.9	(0, 1.60, 2.40)
1.0	(0, 1.77, 2.66)

Example Consider the following fuzzy fractional differential equation,

$$\boldsymbol{D}^{\alpha}_{C_{gH}}x(t) = (-1) \odot x(t), \quad x(0) = (0, 1, 2), \quad t \in [0, 1], \quad 0 < \alpha \leq 1$$

The exact solution is,

$$x(t) = (0, 1, 2) \odot E_{\alpha}(-t^{\alpha}) := c \odot E_{\alpha}(-t^{\alpha})$$

where

$$E_{\alpha}(-t^{\alpha}) = \sum_{i=0}^{\infty} \frac{(-t^{\alpha})^i}{\Gamma(i\alpha + 1)}$$

This function is $ii - gH$ differentiable, because the length of $x(t)$ is decreasing, So, the Euler method is in the form of case 2,

$$x(t_{k+1}) = x(t_k) \ominus_H (-1)\frac{h^{\alpha}}{\Gamma(\alpha+1)} \odot (0, 1, 2) \odot E_{\alpha}\left(-t_k^{\alpha}\right), \quad k = 0, 1, \ldots, N - 1$$

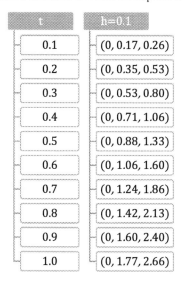

For instance, consider the step size $h = 0.1$ and $\alpha = 0.5$. The level-wise form of the Euler method is,

$$x(t_{k+1}, r) = x(t_k, r) \ominus_H (-1)\frac{0.5^{\alpha} E_{0.5}\left(-t_k^{0.5}\right)}{\Gamma(0.5+1)} \odot (r-1, 2-r),$$

For $k = 0, 1, \ldots, N-1$.

$$\frac{0.687498}{(\sqrt{t_k}+1)\Gamma(0.5+1)} = \frac{0.775758}{(\sqrt{t_k}+1)}$$

Since the value $\frac{0.5^{\alpha} E_{0.5}\left(-t^{0.5}\right)}{\Gamma(0.5+1)} > 0$ then, in the level-wise form,

$$x_l(t_{k+1}, r) = x_l(t_k, r) + \frac{0.775758}{(\sqrt{t_k}+1)}(2-r), \quad k = 0, 1, \ldots, N-1, 0 \leq r \leq 1$$

$$x_u(t_{k+1}, r) = x_u(t_k, r) + \frac{0.775758}{(\sqrt{t_k}+1)}(r-1), \quad k = 0, 1, \ldots, N-1, 0 \leq r \leq 1$$

In Fig. 5.3 the exact solution and its Caputo gH-derivative with $\alpha = 0.5$ are presented at the grid points.

In Table 5.3 the numerical solutions with order $\alpha = 0.5$ and $h = 0.1$ are expressed.

Also, in Fig. 5.4 the exact solution and its Caputo gH-derivative with $\alpha = 0.75$ are presented at the grid points.

In Table 5.4 the numerical solutions with order $\alpha = 0.75$ and $h = 0.1$ are expressed.

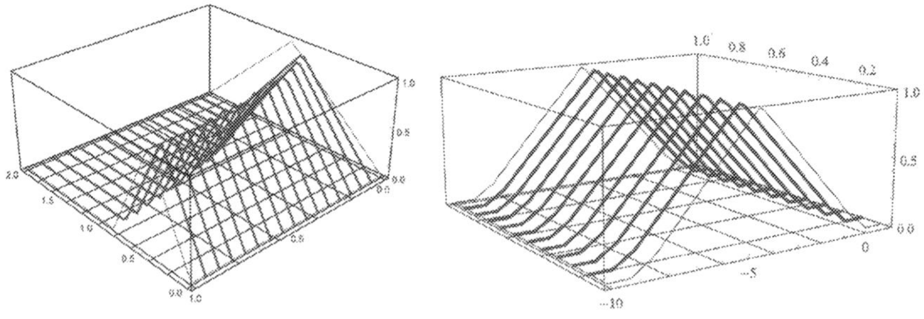

Fig. 5.3 Exact solution (left) and its C-gH-derivative (right) at the grid points with $\alpha = 0.5$

Table 5.3 The numerical results for step size $h = 0.1$

t	h=0.1
0.1	(0, 0.64, 1.28)
0.2	(0, 0.41, 0.82)
0.3	(0, 0.26, 0.53)
0.4	(0, 0.26, 0.34)
0.5	(0, 0.17, 0.22)
0.6	(0, 0.11, 0.22)
0.7	(0, 0.07, 0.14)
0.8	(0, 0.04, 0.09)
0.9	(0, 0.02, 0.05)
1.0	(0, 0.01, 0.03)

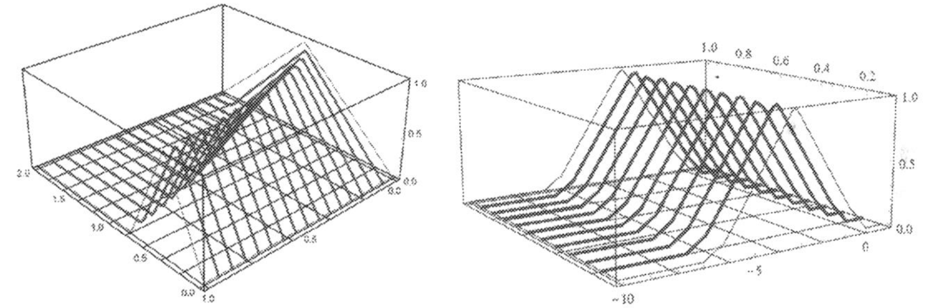

Fig. 5.4 Exact solution (left) and its C-gH-derivative (right) at the grid points with $\alpha = 0.75$

Example Consider another fuzzy fractional differential equation,

$$\begin{aligned} \boldsymbol{D}^{\alpha}_{C_{gH}} x(t) &= \frac{\pi t^{1-\alpha}\alpha}{\Gamma(2-\alpha)} \odot \left(0, \frac{1}{2}, 1\right) \\ &\odot {}_1F_1\left(1; \left[1 - \frac{\alpha}{2}, \frac{3}{2} - \frac{\alpha}{2}\right]; -\frac{1}{4}\pi^2 t^2 \alpha^2\right), 1 \leq t \leq 2 \end{aligned}$$

Table 5.4 The numerical results with order $\alpha = 0.75$ for step size $h = 0.1$

t	h=0.1
0.1	(0, 0.80, 1.61)
0.2	(0, 0.65, 1.30)
0.3	(0, 0.52, 1.04)
0.4	(0, 0.42, 0.84)
0.5	(0, 0.34, 0.68)
0.6	(0, 0.27, 0.55)
0.7	(0, 0.22, 0.44)
0.8	(0, 0.17, 0.35)
0.9	(0, 0.14, 0.28)
1.0	(0, 0.11, 0.23)

With fuzzy initial value $x(1) = (0, 0.5, 1) \odot \sin \alpha\pi$. where ${}_1F_1(a; b; z)$ is a generalized hypergeometric function and defined as the following form,

$$ {}_pF_q(a_1, a_2, \ldots, a_p; b_1, b_2, \ldots, b_q; z) = \sum_{n=0}^{\infty} \frac{(a_1)_n \cdots (a_p)_n}{(b_1)_n \cdots (b_q)_n} \cdot \frac{z^n}{n!} $$

Here

$$ {}_1F_1(a; b; z) = \sum_{n=0}^{\infty} \frac{(a)_n}{(b)_n} \cdot \frac{z^n}{n!} = \sum_{n=0}^{\infty} \frac{\Gamma(a+n)}{\Gamma(a)} \cdot \frac{\Gamma(b)}{\Gamma(b+n)} \cdot \frac{z^n}{n!} $$

where

$$ (\cdot)_n = \frac{\Gamma(\cdot + n)}{\Gamma(\cdot)} $$

Note that by the ratio test the series is convergent. In this example, using $\alpha = 0.5$ the exact solution is

$$ x(t) = (0, 0.5, 1) \odot \sin(\alpha\pi t) $$

And the solution has a switching point at the point $t = 1.463$ and the switching point is type I, before the point it is $i - gH$ differentiable and after it is $ii - gH$ differentiable. In the subintervals $[t_k, t_{k+1}], k = 0, 1, \ldots, N-1$ with assumption that switching point is in the interval $[t_j, t_{j+1}]$, the fuzzy Euler's method is denoted as,

$$\begin{cases} x(t_{k+1}) = x(t_k) \oplus \frac{h^\alpha}{\Gamma(\alpha+1)} \odot f(t_k, x(t_k)), & k = 1, 2, \ldots, j \\ x(t_{k+1}) = x(t_k) \ominus_H (-1) \frac{h^\alpha}{\Gamma(\alpha+1)} \odot f(t_k, x(t_k)), & k = j+1, j+2, \ldots, N \end{cases}$$

where

$$f(t_k, x(t_k)) = \frac{\pi t_k^{1-\alpha} \alpha}{\Gamma(2-\alpha)} \odot \left(0, \frac{1}{2}, 1\right) \odot {}_1F_1\left(1; \left[1 - \frac{\alpha}{2}, \frac{3}{2} - \frac{\alpha}{2}\right]; -\frac{1}{4}\pi^2 t_k^2 \alpha^2\right)$$

Suppose $h = 0.1, k = 0, \alpha = 0.5, t_0 = 1, x(1) = (0, 0.5, 1) \odot \sin \alpha\pi$. Then

$$x(t_1) = (0, 0.5, 1) \oplus \frac{h^{0.5}}{\Gamma(0.5+1)} \odot f(1, (0, 0.5, 1))$$

where

$$f(1, (0, 0.5, 1)) = \frac{\pi 0.5}{\Gamma\left(\frac{3}{2}\right)} \odot \left(0, \tfrac{1}{2}, 1\right) \odot_1 F_1\left(1; \left[\tfrac{3}{4}, \tfrac{5}{4}\right]; -\tfrac{1}{16}\pi^2\right)$$

$${}_1F_1\left(1; \left[\tfrac{3}{4}, \tfrac{5}{4}\right]; -\tfrac{1}{16}\pi^2\right) \sum_{n=0}^{\infty} \frac{\Gamma(1+n)}{\Gamma(1)} \cdot \frac{\Gamma\left(\left[\frac{3}{4}, \frac{5}{4}\right]\right)}{\Gamma\left(\left[\frac{3}{4}+n, \frac{5}{4}+n\right]\right)} \cdot \frac{\left(-\frac{1}{16}\pi^2\right)^n}{n!}$$

$$= \left[\sum_{n=0}^{\infty} \frac{\Gamma(1+n)}{\Gamma(1)} \cdot \frac{\Gamma\left(\frac{3}{4}\right)}{\Gamma\left(\frac{3}{4}\right)} \cdot \frac{\left(-\frac{1}{16}\pi^2\right)^n}{n!}, \sum_{n=0}^{\infty} \frac{\Gamma(1+n)}{\Gamma(1)} \cdot \frac{\Gamma\left(\frac{5}{4}\right)}{\Gamma\left(\frac{5}{4}\right)} \cdot \frac{\left(-\frac{1}{16}\pi^2\right)^n}{n!}\right]$$

$$= \left[\frac{16}{16+\pi^2}, \frac{16}{16+\pi^2}\right] = \frac{16}{16+\pi^2}$$

Thus,

$$f(t_0, x(t_0)) = f(1, (0, 0.5, 1)) = \frac{\pi 0.5}{\Gamma\left(\frac{3}{2}\right)} \odot \left(0, \frac{1}{2}, 1\right) \odot \frac{16}{16+\pi^2}$$

Since $\frac{\pi 0.5}{\Gamma\left(\frac{3}{2}\right)} \frac{16}{16+\pi^2} \approx 1.1$ and positive then, $f(t_0, x(t_0)) = 1.1 \odot (0, 0.5, 1)$ and in the level-wise form,

$$f_l(t_0, x(t_0), r) = \frac{1.1}{2} r, f_u(t_0, x(t_0), r) = 1.1 - \frac{1.1}{2} r$$

and the method for approximate the solution at the second point $x(t_1)$ is,

$$x_l(t_1, r) = \frac{1}{2}r + \frac{h^{0.5}}{\Gamma(0.5+1)} \cdot \frac{1.1}{2}r, x_u(t_1, r)$$
$$= 1 - \frac{1}{2}r + \frac{h^{0.5}}{\Gamma(0.5+1)} \cdot \left(1.1 - \frac{1.1}{2}r\right)$$

Putting $h = 0.1$ and in the level for instance $r = 0.5$, we have

$$x_l(t_1, 0.5) = \frac{1}{4} + \frac{0.5^{0.5}}{\Gamma(1.5)} \cdot \frac{1.1}{4} \approx 0.47,\ x_u(t_1, r) = \frac{3}{4} + \frac{0.5^{0.5}}{\Gamma(1.5)} \cdot \frac{3}{4} \cdot 1.1 \approx 1.41$$

In general, for $\alpha = 0.5$ we have,

$$f(t_k, x(t_k)) = \frac{\pi t_k^{o.5} 0.5}{\Gamma(1.5)} \odot \left(0, \frac{1}{2}, 1\right) \odot_1 F_1\left(1; [0.5, 1]; -\frac{1}{16}\pi^2 t_k^2\right)$$

And

$$\begin{cases} x(t_{k+1}) = x(t_k) \oplus \frac{h^{0.5}}{\Gamma(1.5)} \odot f(t_k, x(t_k)), & k = 1, 2, \ldots, j \\ x(t_{k+1}) = x(t_k) \ominus_H (-1) \frac{h^{0.5}}{\Gamma(1.5)} \odot f(t_k, x(t_k)), & k = j+1, j+2, \ldots, N \end{cases}$$

Table 5.5 The numerical results with order $\alpha = 0.75$ for step size $h = 0.1$

t	h=0.1
1.1	(0, 0.49, 0.98)
1.2	(0, 0.46, 0.93)
1.3	(0, 0.42, 0.86
1.4	(0, 0.41, 0.80)
1.5	(0, 0.35, 0.71)
1.6	(0, 0.30, 0.59)
1.7	(0, 0.20, 0.43)
1.8	(0, 0.16, 0.31)
1.9	(0, 0.04, 0.13)
1.0	(0, 0, 0)

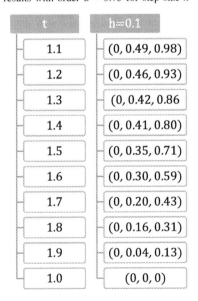

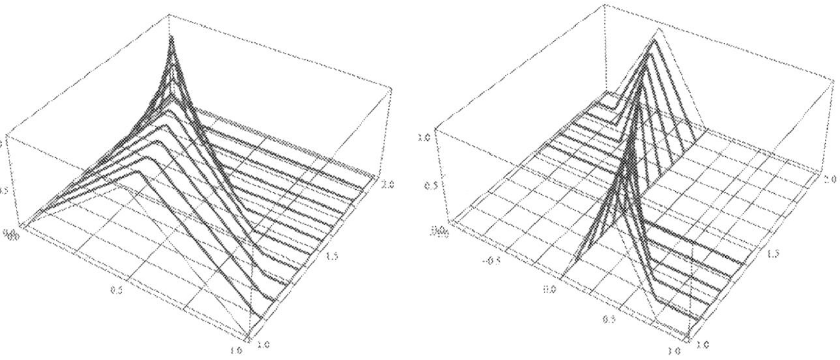

Fig. 5.5 Exact solution (left) and its C-gH-derivative (right) at the grid points with $\alpha = 0.5$

The results for $h = 0.1$ are listed as in Table 5.5.

Figure 5.5 show the fuzzy solution and its gH-derivative with $\alpha = 0.5$.

5.5 ABC-PI Numerical Method with ABC gH-Derivative

In this section, first, the ABC fractional derivative on fuzzy number-valued functions in parametric interval form is defined. Then it is applied for proving the existence and uniqueness of the solution of fuzzy fractional differential equation with ABC fractional derivative. In general, it is shown that the last interval model is as a coupled system of nonlinear equations in interval form. To solve the final system an efficient numerical method called ABC-PI is used.

5.5.1 *Definition—ABC_{GH} Fractional Derivative in the Sense of Caputo Derivative*

The generalized ABC_{gH} derivative of a fuzzy number valued function $x(t)$ on interval $[t_0, T]$ starting at t_0 with kernel $E^{\gamma}_{\alpha,\mu}(\lambda, t)$ where $0<\alpha\langle 1, Re(\mu)\rangle 0, \gamma \in R$, $\lambda = \frac{-\alpha}{1-\alpha}$ is defined in the following form,

$$\boldsymbol{D}^{\alpha,\mu,\gamma}_{\boldsymbol{ABC_{gH}}} x(t) = \frac{B(\alpha)}{1-\alpha} \odot \int_{t_0}^{t} x'_{gH}(\tau) \odot E^{\gamma}_{\alpha,\mu}(\lambda, t-\tau)d\tau$$

where $B(\alpha) > 0$ is a normalizing function and is defined as,

$$B(\alpha) = 1 - \alpha + \frac{\alpha}{\Gamma(\alpha)}$$

With properties $B(0) = B(1) = 1$. Its corresponding AB fractional integral operator is defined as,

$$\boldsymbol{I}_{AB}^{\alpha,\mu.\gamma}x(t) = \sum_{i=0}^{\gamma}\binom{\gamma}{i}\frac{\alpha^{i}}{B(\alpha)(1-\alpha)^{i-1}}\boldsymbol{I}^{\alpha,\mu,\gamma}x(t)$$

Subject to,

$$\boldsymbol{I}_{AB}^{\alpha,\mu.\gamma}\boldsymbol{D}_{ABC_{gH}}^{\alpha,\mu.\gamma}x(t) = x(t) \ominus_{gH} x(t_0)$$

In case $\mu = \gamma = 1$ we get the fractional derivative and integral with order $0<\alpha<1$ then the ABC_{gH} derivative that is denoted by $\boldsymbol{D}_{ABC_{gH}}^{\alpha}$ is defined on a fuzzy number valued function $x(t)$ on interval $[t_0, T]$ in the following form,

$$\boldsymbol{D}_{ABC_{gH}}^{\alpha}x(t) = \frac{B(\alpha)}{1-\alpha} \odot \int_{t_0}^{t} x'_{gH}(\tau) \odot E_{\alpha}\left(-\frac{\alpha}{1-\alpha}(t-\tau)^{\alpha}\right)d\tau$$

Its corresponding fractional integral is defined as,

$$\boldsymbol{I}_{AB}^{\alpha}x(t) = \frac{1-\alpha}{B(\alpha)} \odot x(t) \oplus \frac{\alpha}{B(\alpha)\Gamma(\alpha)} \odot \int_{t_0}^{t} x(\tau) \odot (t-\tau)^{\alpha-1}d\tau$$

In case $\alpha = 0$ it recovers initial function and if $\alpha = 1$ it is ordinary integral. Here we are going to explain its level-wise form regarding the type of differentiability of fuzzy valued function $x(t)$.

Case 1 $x(t)$ is $i - gH$ differentiable,

$$\boldsymbol{D}_{ABC_{i-gH}}^{\alpha}x(t,r) = \left[\boldsymbol{D}_{ABC}^{\alpha}x_l(t,r), \boldsymbol{D}_{ABC}^{\alpha}x_u(t,r)\right]$$

Case 2 $x(t)$ is $ii - gH$ differentiable,

$$\boldsymbol{D}_{ABC_{ii-gH}}^{\alpha}x(t,r) = \left[\boldsymbol{D}_{ABC}^{\alpha}x_u(t,r), \boldsymbol{D}_{ABC}^{\alpha}x_l(t,r)\right]$$

where

$$\boldsymbol{D}^{\alpha}_{\boldsymbol{ABC}} x_l(t,r) = \frac{B(\alpha)}{1-\alpha}\int_{t_0}^{t} x_l'(\tau,r)E_{\alpha}\left(-\frac{\alpha}{1-\alpha}(t-\tau)^{\alpha}\right)d\tau$$

$$\boldsymbol{D}^{\alpha}_{\boldsymbol{ABC}} x_u(t,r) = \frac{B(\alpha)}{1-\alpha}\int_{t_0}^{t} x_u'(\tau,r)E_{\alpha}\left(-\frac{\alpha}{1-\alpha}(t-\tau)^{\alpha}\right)d\tau$$

And its fractional integral,

$$\boldsymbol{I}^{\alpha}_{\boldsymbol{AB}} x_l(t) = \frac{1-\alpha}{B(\alpha)} \odot x_l(t) \oplus \frac{\alpha}{B(\alpha)\Gamma(\alpha)} \odot \int_{t_0}^{t} x_l(\tau) \odot (t-\tau)^{\alpha-1} d\tau$$

$$\boldsymbol{I}^{\alpha}_{\boldsymbol{AB}} x_u(t) = \frac{1-\alpha}{B(\alpha)} \odot x_u(t) \oplus \frac{\alpha}{B(\alpha)\Gamma(\alpha)} \odot \int_{t_0}^{t} x_u(\tau) \odot (t-\tau)^{\alpha-1} d\tau$$

Laplace transform of the $\boldsymbol{D}^{\alpha}_{\boldsymbol{ABC}_{gH}}$ can be explained as,

$$\boldsymbol{L}\left(\boldsymbol{D}^{\alpha}_{\boldsymbol{ABC}_{gH}} x(t)\right) = \frac{B(\alpha)}{1-\alpha} \frac{s^{\alpha} \odot \boldsymbol{L}(x(t)) \ominus_{gH} s^{\alpha-1} \odot x(0)}{s^{\alpha} + \frac{\alpha}{1-\alpha}}$$

where

$$\boldsymbol{L}(x(t)) = \int_{0}^{\infty} e^{-st} \odot x(t)dt, \quad s > 0$$

Based on the definition of gH-difference there are two cases for the Laplace transform,

Case 1 $i - gH$ difference

$$\boldsymbol{L}\left(\boldsymbol{D}^{\alpha}_{\boldsymbol{ABC}_{gH}} x(t)\right) = \frac{B(\alpha)}{1-\alpha} \frac{s^{\alpha} \odot \boldsymbol{L}(x(t)) \ominus_{H} s^{\alpha-1} \odot x(0)}{s^{\alpha} + \frac{\alpha}{1-\alpha}}$$

$$L\left(\boldsymbol{D}^{\alpha}_{\boldsymbol{ABC}} x(t,r)\right) = \left[L\left(\boldsymbol{D}^{\alpha}_{\boldsymbol{ABC}} x_l(t,r)\right), L\left(\boldsymbol{D}^{\alpha}_{\boldsymbol{ABC}} x_u(t,r)\right)\right]$$

Case 2 $ii - gH$ difference

$$\boldsymbol{L}\left(\boldsymbol{D}^{\alpha}_{ABC_{gH}}x(t)\right) = \frac{B(\alpha)}{1-\alpha}\frac{\ominus_H(-1)s^{\alpha}\odot \boldsymbol{L}(x(t))\oplus(-1)s^{\alpha-1}\odot x(0)}{s^{\alpha}+\frac{\alpha}{1-\alpha}}$$
$$= \frac{B(\alpha)}{\alpha-1}\frac{s^{\alpha-1}\odot x(0)\ominus_H s^{\alpha}\odot \boldsymbol{L}(x(t))}{s^{\alpha}+\frac{\alpha}{1-\alpha}}$$

$$L\left(\boldsymbol{D}^{\alpha}_{ABC}x(t,r)\right) = \left[L\left(\boldsymbol{D}^{\alpha}_{ABC}x_u(t,r)\right), L\left(\boldsymbol{D}^{\alpha}_{ABC}x_l(t,r)\right)\right]$$

where

$$L\left(\boldsymbol{D}^{\alpha}_{ABC}x_l(t,r)\right) = \frac{B(\alpha)}{1-\alpha}\frac{s^{\alpha}L(x_l(t,r)) - s^{\alpha-1}x_l(0,r)}{s^{\alpha}+\frac{\alpha}{1-\alpha}}$$

$$L\left(\boldsymbol{D}^{\alpha}_{ABC}x_u(t,r)\right) = \frac{B(\alpha)}{1-\alpha}\frac{s^{\alpha}L(x_u(t,r)) - s^{\alpha-1}x_u(0,r)}{s^{\alpha}+\frac{\alpha}{1-\alpha}}$$

5.5.2 *Fuzzy Time Fractional Ordinary Differential Equation*

Consider the following fuzzy fractional differential equation,

$$\boldsymbol{D}^{\alpha}_{ABC_{gH}}x(t) = f(t,x(t)), \quad x(t_0) = x_0 \in \mathbb{F}_R, \quad t \in [t_0, T], \quad 0<\alpha<1$$

By taking the fractional integral $\boldsymbol{I}^{\alpha}_{AB}$ from both sides of the above fuzzy time fractional equation, the fuzzy solution of this equation satisfies its corresponding fuzzy integral equations in the following form,

$$\boldsymbol{I}^{\alpha}_{AB}\boldsymbol{D}^{\alpha}_{ABC_{gH}}x(t) = \boldsymbol{I}^{\alpha}_{AB}f(t,x(t))$$

$$\boldsymbol{I}^{\alpha}_{AB}\boldsymbol{D}^{\alpha}_{ABC_{gH}}x(t) = x(t)\ominus_{gH}x(t_0) = \boldsymbol{I}^{\alpha}_{AB}f(t,x(t))$$

$$x(t) = x(t_0)\oplus\frac{1-\alpha}{B(\alpha)}\odot f(t,x(t))\oplus\frac{\alpha}{B(\alpha)\Gamma(\alpha)}\odot\int_{t_0}^{t} f(\tau,x(\tau))\odot(t-\tau)^{\alpha-1}d\tau$$

Case 1 $x(t)$ is $i-gH$ differentiable,

$$\boldsymbol{D}^{\alpha}_{\boldsymbol{ABC}_{i-gH}}x(t) = f(t, x(t))$$

$$x(t) = x(t_0) \oplus \frac{1-\alpha}{B(\alpha)} \odot f(t, x(t)) \oplus \frac{\alpha}{B(\alpha)\Gamma(\alpha)} \odot \int_{t_0}^{t} f(\tau, x(\tau)) \odot (t-\tau)^{\alpha-1} d\tau$$

$$\begin{aligned}\boldsymbol{D}^{\alpha}_{\boldsymbol{ABC}_{i-gH}}x(t,r) &= \left[\boldsymbol{D}^{\alpha}_{\boldsymbol{ABC}}x_l(t,r), \boldsymbol{D}^{\alpha}_{\boldsymbol{ABC}}x_u(t,r)\right] \\ &= [f_l(t, x_l(t,r), x_u(t,r)), f_u(t, x_l(t,r), x_u(t,r))]\end{aligned}$$

$$\boldsymbol{D}^{\alpha}_{\boldsymbol{ABC}}x_l(t,r) = f_l(t, x_l(t,r), x_u(t,r)), \boldsymbol{D}^{\alpha}_{\boldsymbol{ABC}}x_u(t,r) = f_u(t, x_l(t,r), x_u(t,r))$$

Case 2 $x(t)$ is $ii-gH$ differentiable,

$$\begin{aligned}&\boldsymbol{D}^{\alpha}_{\boldsymbol{ABC}_{ii-gH}}x(t) = f(t, x(t)) \\ &x(t) = x(t_0) \ominus_H (-1)\frac{1-\alpha}{B(\alpha)} \odot f(t, x(t)) \ominus_H (-1)\frac{\alpha}{B(\alpha)\Gamma(\alpha)} \\ &\quad \odot \int_{t_0}^{t} f(\tau, x(\tau)) \odot (t-\tau)^{\alpha-1} d\tau\end{aligned}$$

$$\begin{aligned}\boldsymbol{D}^{\alpha}_{\boldsymbol{ABC}_{ii-gH}}x(t,r) &= \left[\boldsymbol{D}^{\alpha}_{\boldsymbol{ABC}}x_u(t,r), \boldsymbol{D}^{\alpha}_{\boldsymbol{ABC}}x_l(t,r)\right] \\ &= [f_l(t, x_l(t,r), x_u(t,r)), f_u(t, x_l(t,r), x_u(t,r))]\end{aligned}$$

$$\boldsymbol{D}^{\alpha}_{\boldsymbol{ABC}}x_l(t,r) = f_u(t, x_l(t,r), x_u(t,r)), \boldsymbol{D}^{\alpha}_{\boldsymbol{ABC}}x_l(t,r) = f_u(t, x_l(t,r), x_u(t,r))$$

where

$$\begin{aligned}x_l(t,r) &= x_l(0,r) + \frac{1-\alpha}{B(\alpha)} f_l(t, x_l(t,r), x_u(t,r)) \\ &\quad + \frac{\alpha}{B(\alpha)\Gamma(\alpha)} \int_{t_0}^{t} f_l(\tau, x_l(\tau,r), x_u(\tau,r))(t-\tau)^{\alpha-1} d\tau \\ x_u(t,r) &= x_u(0,r) + \frac{1-\alpha}{B(\alpha)} f_u(t, x_l(t,r), x_u(t,r)) \\ &\quad + \frac{\alpha}{B(\alpha)\Gamma(\alpha)} \int_{t_0}^{t} f_u(\tau, x_l(\tau,r), x_u(\tau,r))(t-\tau)^{\alpha-1} d\tau\end{aligned}$$

Indeed, in case 2, we have a system of ABC fractional differential equations and usually solving this system is not straight forward and the solution should be obtained by numerical methods. Now we are going to investigate the conditions for the existence of a unique solution of the fuzzy ABC fractional differential equation. To this end, consider the following equations,

$$\boldsymbol{D}^{\alpha}_{\boldsymbol{ABC}_{gH}}x(t) = f(t, x(t)), \quad x(t_0) = x_0 \in \mathbb{F}_R, \quad t \in [t_0, T], \quad 0 < \alpha < 1$$

where $x(t)$ is a fuzzy continuous function on $[t_0, T]$.

5.5.3 Remark—Uniqueness

Suppose that the fuzzy function $f(t, x(t))$ is a continuous function and satisfies the Lipschitz condition, for $\forall x \forall y \exists M > 0$ subject to,

$$D_H(f(t, x(t)), f(t, y(t))) \leq M D_H(x(t), y(t))$$

Then the above-mentioned differential equation has a unique solution if the following condition is satisfied,

$$\frac{1-\alpha}{B(\alpha)} M + \frac{\alpha}{B(\alpha)\Gamma(\alpha)} M T^{\alpha+1} < 1$$

The proof is easy to show and it is expressed by using the contraction operator. Let us consider the operator,

$$F(x(t)) = x(t_0) \oplus \frac{1-\alpha}{B(\alpha)} \odot f(t, x(t)) \oplus \frac{\alpha}{B(\alpha)\Gamma(\alpha)} \odot \int_{t_0}^{t} f(t, x(t)) \odot (t-\tau)^{\alpha-1} d\tau$$

$$D_H(F(x(t)), F(y(t))) \leq D_H(x(t_0), y(t_0)) + \frac{1-\alpha}{B(\alpha)} D_H(f(t, x(t)), f(t, y(t)))$$
$$+ \frac{\alpha}{B(\alpha)\Gamma(\alpha)} \int_{t_0}^{t} D_H(f(\tau, x(\tau)), f(\tau, y(\tau)))(t-\tau)^{\alpha-1} d\tau$$

Since $D_H(x(t_0), y(t_0)) = 0$ then

$$D_H(F(x(t)), F(y(t))) \leq \frac{1-\alpha}{B(\alpha)} D_H(f(t, x(t)), f(t, y(t)))$$
$$+ \frac{\alpha}{B(\alpha)\Gamma(\alpha)} \int_{t_0}^{t} D_H(f(\tau, x(\tau)), f(\tau, y(\tau)))(t-\tau)^{\alpha-1} d\tau$$

Based on Lipschitz condition,

$$D_H(f(t, x(t)), f(t, y(t))) \leq M D_H(x(t), y(t))$$

We have,

$$D_H(F(x(t)), F(y(t))) \leq \frac{1-\alpha}{B(\alpha)} M D_H(x(t), y(t))$$
$$+ \frac{M\alpha D_H(x(t), y(t))}{B(\alpha)\Gamma(\alpha)} \int_{t_0}^{t} (t-\tau)^{\alpha-1} d\tau$$

Since, $\int_{t_0}^{t} (t-\tau)^{\alpha-1} d\tau < T^\alpha$, we have,

$$D_H(F(x(t)), F(y(t))) \leq \frac{1-\alpha}{B(\alpha)} M D_H(x(t), y(t)) + \frac{M\alpha D_H(x(t), y(t))}{B(\alpha)\Gamma(\alpha)} T^\alpha$$
$$= \left[\frac{1-\alpha}{B(\alpha)} M + \frac{M\alpha}{B(\alpha)\Gamma(\alpha)} T^\alpha\right] D_H(x(t), y(t))$$

Applying the assumption of the remark,

$$D_H(F(x(t)), F(y(t))) \leq D_H(x(t), y(t))$$

It means the operator F is a contraction operator and therefor based on the Banach fixed point theorem our problem does have a unique solution in the mentioned form.

5.5.4 An Efficient Numerical Method for ABC Fractional Problems

As we mentioned, the system of fractional differential equations in the level-wise form are as,

$$\boldsymbol{D}^{\alpha}_{\boldsymbol{ABC}} x_l(t, r) = f_l(t, x_l(t, r), x_u(t, r)), \boldsymbol{D}^{\alpha}_{\boldsymbol{ABC}} x_u(t, r) = f_u(t, x_l(t, r), x_u(t, r))$$

$$\boldsymbol{D}^{\alpha}_{\boldsymbol{ABC}} x_l(t, r) = f_u(t, x_l(t, r), x_u(t, r)), \boldsymbol{D}^{\alpha}_{\boldsymbol{ABC}} x_u(t, r) = f_l(t, x_l(t, r), x_u(t, r))$$

In general, these equations can be shown as the following form,

$$\begin{cases} \boldsymbol{D}^{\alpha}_{\boldsymbol{ABC}} x_l(t, r) = f_l(t, x_l(t, r), x_u(t, r)) \\ \boldsymbol{D}^{\alpha}_{\boldsymbol{ABC}} x_u(t, r) = f_u(t, x_l(t, r), x_u(t, r)) \end{cases}$$

where R_1, R_2 are two continuous functions. Applying AB fractional integral $\boldsymbol{I}_{\boldsymbol{AB}}^{\alpha}$ both sides, we get the following equations,

$$x_l(t,r) = x_l(t_0,r) + \frac{1-\alpha}{B(\alpha)} f_l(t, x_l(t,r), x_u(t,r)) + \frac{\alpha}{B(\alpha)\Gamma(\alpha)} \int_{t_0}^{t} f_l(\tau, x_l(\tau,r), x_u(\tau,r))(t-\tau)^{\alpha-1} d\tau$$

$$x_u(t,r) = x_u(t_0,r) + \frac{1-\alpha}{B(\alpha)} f_u(t, x_l(t,r), x_u(t,r)) + \frac{\alpha}{B(\alpha)\Gamma(\alpha)} \int_{t_0}^{t} f_u(\tau, x_l(\tau,r), x_u(\tau,r))(t-\tau)^{\alpha-1} d\tau$$

Taking $t = t_n = t_0 + nh, h = \frac{T-t_0}{n}$ the discretized recursive implicit equations are obtained,

$$x_l(t_n,r) = x_l(t_0,r) + \frac{1-\alpha}{B(\alpha)} f_l(t_n, x_l(t_n,r), x_u(t_n,r)) + \frac{\alpha}{B(\alpha)\Gamma(\alpha)} \sum_{i=0}^{n-1} \int_{t_i}^{t_{i+1}} f_l(t_n, x_l(\tau,r), x_u(\tau,r))(t_n-\tau)^{\alpha-1} d\tau$$

$$x_u(t_n,r) = x_u(t_0,r) + \frac{1-\alpha}{B(\alpha)} f_u(t_n, x_l(t_n,r), x_u(t_n,r)) + \frac{\alpha}{B(\alpha)\Gamma(\alpha)} \sum_{i=0}^{n-1} \int_{t_i}^{t_{i+1}} f_u(t_n, x_l(\tau,r), x_u(\tau,r))(t_n-\tau)^{\alpha-1} d\tau$$

For $0 \le r \le 1$.

The functions $f_*(t_n, x_l(\tau,r), x_u(\tau,r))$, $* \in \{l,u\}$ can be approximated by Lagrange interpolation on $[t_i, t_{i+1}]$ and it is,

$$f_*(t_n, x_l(\tau,r), x_u(\tau,r)) \approx f_*(t_{i+1}, x_l(t_{i+1},r), x_u(t_{i+1},r)) + \frac{\tau - t_{i+1}}{h}\left(f_*(t_{i+1}, x_l(t_{i+1},r), x_u(t_{i+1},r)) - f_*(t_i, x_l(t_i,r), x_u(t_i,r))\right)$$

By substituting and algebraic manipulating,

$$x_l(t_n,r) = x_l(t_0,r) + \frac{\alpha h^{\alpha}}{B(\alpha)}\left(\xi_n f_l(t_0, x_l(t_0,r), x_u(t_n,r)) + \sum_{i=1}^{n} \mu_{n-i} f_l(t_i, x_l(t_i,r), x_u(t_i,r))\right)$$

$$x_u(t_n, r) = x_u(t_0, r) + \frac{\alpha h^\alpha}{B(\alpha)} \left(\xi_n f_u(t_0, x_l(t_0, r), x_u(t_n, r)) + \sum_{i=1}^{n} \mu_{n-i} f_u(t_i, x_l(t_i, r), x_u(t_i, r)) \right)$$

where

$$\xi_n = \frac{(n-1)^{\alpha+1} - n^\alpha (n - \alpha - 1)}{\Gamma(\alpha + 2)},$$

$$\mu_j = \begin{cases} \frac{1}{\Gamma(\alpha+2)} + \frac{1-\alpha}{\alpha h^\alpha}, & j = 0 \\ \frac{(j-1)^{\alpha+1} - 2j^{\alpha+1} + (j+1)^{\alpha+1}}{\Gamma(\alpha+2)}, & j = 1, 2, \ldots, n-1 \end{cases}$$

Numerical example
Suppose

$$\boldsymbol{D}^{\alpha}_{\boldsymbol{ABC}_{gH}} x(t) = \lambda \odot x(t), \quad x(0, r) = (r, 3 - 2r), \quad t \in [0, 1], \quad 0 < \alpha < 1$$

- $\lambda \geq 0$ we have $i - gH$ solution

$$x(t) = (r, 3 - 2r) \oplus \frac{\lambda(1-\alpha)}{B(\alpha)} x(t) \oplus \frac{\lambda \alpha}{B(\alpha)\Gamma(\alpha)} \odot \int_{t_0}^{t} x(\tau) \odot (t - \tau)^{\alpha - 1} d\tau$$

In the level-wise form we have,

$$\boldsymbol{D}^{\alpha}_{\boldsymbol{ABC}_{i-gH}} x(t, r) = \left[\boldsymbol{D}^{\alpha}_{\boldsymbol{ABC}} x_l(t, r), \boldsymbol{D}^{\alpha}_{\boldsymbol{ABC}} x_u(t, r) \right] = [\lambda x_l(t, r), \lambda x_u(t, r)]$$

$$\boldsymbol{D}^{\alpha}_{\boldsymbol{ABC}} x_l(t, r) = \lambda x_l(t, r), \quad \boldsymbol{D}^{\alpha}_{\boldsymbol{ABC}} x_u(t, r) = \lambda x_u(t, r)$$

And the numerical scheme is as,

$$x_l(t_n, r) = r + \frac{\lambda \alpha h^\alpha}{B(\alpha)} \left(\xi_n r + \sum_{i=1}^{n} \mu_{n-i} x_l(t_i, r) \right)$$

$$x_u(t_n, r) = 3 - 2r + \frac{\lambda \alpha h^\alpha}{B(\alpha)} \left(\xi_n (3 - 2r) + \sum_{i=1}^{n} \mu_{n-i} x_u(t_i, r) \right)$$

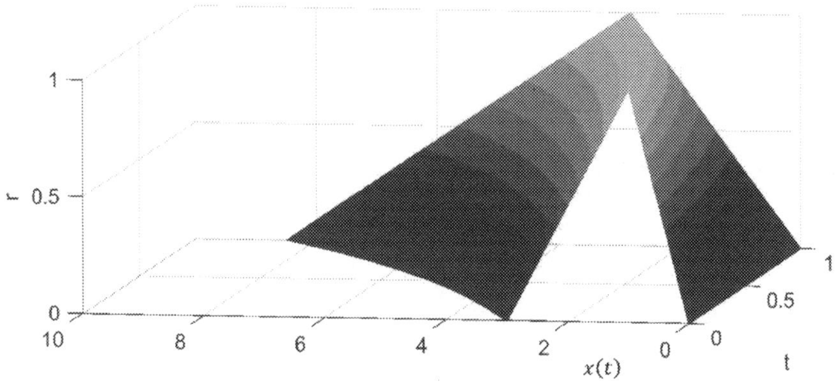

Fig. 5.6 $i - gH$ solution for $\lambda = 1, \alpha = 0.9$

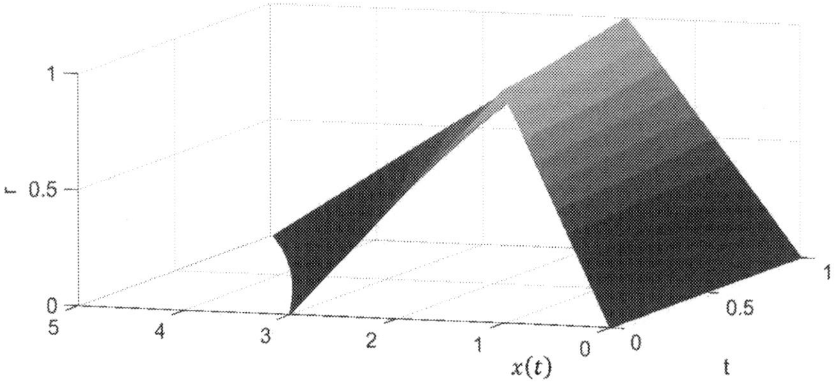

Fig. 5.7 $i - gH$ solution for $\lambda = 0.5, \alpha = 0.95$

In Figures 5.6 and 5.7, the numerical method has been used for approximating the fuzzy solution of our problem. To this end, first the level $0 \leq r \leq 1$ is fixed and then the algorithm is applied for two values of λ and α. It is seen that the solutions are fuzzy number valued functions because in all figures the triangular fuzzy number at each point is observed.

- $\lambda < 0$ we have $ii - gH$ solution

$$x(t) = (3 - 2r, r) \ominus_H (-1)\frac{\lambda(1-\alpha)}{B(\alpha)}x(t) \ominus_H (-1)\frac{\lambda\alpha}{B(\alpha)\Gamma(\alpha)} \odot \int_{t_0}^{t} x(\tau)$$
$$\odot\, (t-\tau)^{\alpha-1} d\tau$$

And in the level-wise for,

$$D^{\alpha}_{ABC_{ii-gH}}x(t,r) = \left[D^{\alpha}_{ABC}x_u(t,r), D^{\alpha}_{ABC}x_l(t,r)\right] = \left[\lambda x_u(t,r), \lambda x_l(t,r)\right]$$

$$D^{\alpha}_{ABC}x_l(t,r) = \lambda x_u(t,r), D^{\alpha}_{ABC}x_u(t,r) = \lambda x_l(t,r)$$

And the numerical solution is also obtained as,

$$x_l(t_n,r) = 3 - 2r + \frac{\lambda\alpha h^{\alpha}}{B(\alpha)}\left(\xi_n(3-2r) + \sum_{i=1}^{n}\mu_{n-i}x_u(t_i,r)\right)$$

$$x_u(t_n,r) = r + \frac{\lambda\alpha h^{\alpha}}{B(\alpha)}\left(\xi_n r + \sum_{i=1}^{n}\mu_{n-i}x_l(t_i,r)\right)$$

Here are $ii - gH$ solutions for some values of λ, α (Figs. 5.8 and 5.9).

Numerical example Consider the following fuzzy fractional differential equation,

$$D^{\alpha}_{ABC_{gH}}x(t) = t \odot x(t), x(-1,r) = \left(r^2+4, 5.5-0.5r\right),\ t \in [-1,1]$$

Here we also have two cases for $t \in [-1,1]$,

- $t \in [-1,0]$ we have $ii - gH$ solution

$$x(t) = \left(5.5-0.5r, r^2+4\right) \ominus_H (-1)\frac{t(1-\alpha)}{B(\alpha)}x(t)$$

$$\ominus_H (-1)\frac{\alpha}{B(\alpha)\Gamma(\alpha)} \odot \int_{t_0}^{t} \tau \odot x(\tau) \odot (t-\tau)^{\alpha-1}d\tau$$

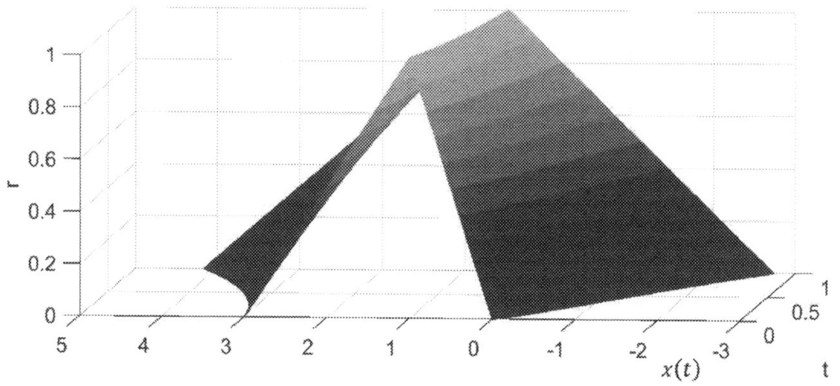

Fig. 5.8 $ii - gH$ solution for $\lambda = -0.8, \alpha = 0.85$

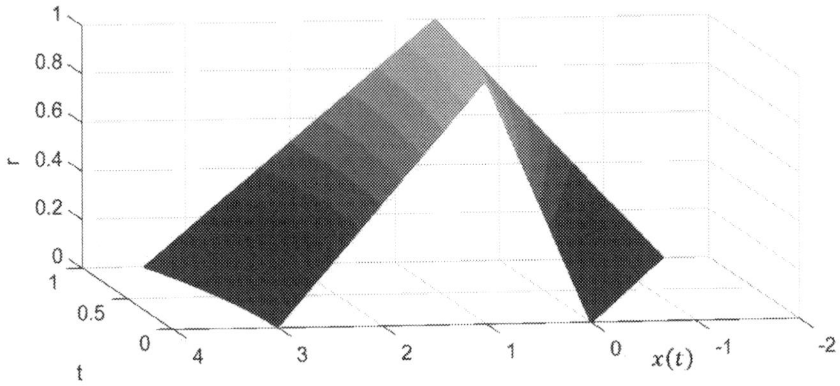

Fig. 5.9 $ii - gH$ solution for $\lambda = -0.5, \alpha = 0.95$

In the level-wise form we have,

$$\boldsymbol{D}^{\alpha}_{\boldsymbol{ABC}} x_l(t,r) = t x_u(t,r), \quad \boldsymbol{D}^{\alpha}_{\boldsymbol{ABC}} x_u(t,r) = t x_l(t,r)$$

The numerical scheme is as,

$$x_l(t_n,r) = 5.5 - 0.5r + \frac{\alpha h^{\alpha}}{B(\alpha)}\left(\xi_n(5.5 - 0.5r) + \sum_{i=1}^{n} \mu_{n-i} t_i x_u(t_i,r)\right)$$

$$x_u(t_n,r) = r^2 + 4 + \frac{\alpha h^{\alpha}}{B(\alpha)}\left(\xi_n\left(r^2 + 4\right) + \sum_{i=1}^{n} \mu_{n-i} t_i x_l(t_i,r)\right)$$

- $t \in [0,1]$ we have $i - gH$ solution

$$x(t) = \left(r^2 + 4, 5.5 - 0.5r\right) \oplus \frac{t(1-\alpha)}{B(\alpha)} x(t) \oplus \frac{\alpha}{B(\alpha)\Gamma(\alpha)} \odot \int_{t_0}^{t} \tau \odot x(\tau)$$
$$\odot (t-\tau)^{\alpha-1} d\tau$$

In the level-wise form we have,

$$\boldsymbol{D}^{\alpha}_{\boldsymbol{ABC}} x_l(t,r) = t x_l(t,r), \quad \boldsymbol{D}^{\alpha}_{\boldsymbol{ABC}} x_u(t,r) = t x_u(t,r)$$

The numerical scheme is as,

$$x_l(t_n, r) = r^2 + 4 + \frac{\alpha h^\alpha}{B(\alpha)}\left(\xi_n(r^2+4) + \sum_{i=1}^{n} \mu_{n-i} t_i x_l(t_i, r)\right)$$

$$x_u(t_n, r) = 5.5 - 0.5r + \frac{\alpha h^\alpha}{B(\alpha)}\left(\xi_n(5.5 - 0.5r) + \sum_{i=1}^{n} \mu_{n-i} t_i x_u(t_i, r)\right)$$

In this example the point $t = 0$ is a switching point of the fuzzy solution. The next example has a non-linear function in the left side (Figs. 5.10 and 5.11).

Example Consider the following non-linear fuzzy fractional differential equation,

$$\boldsymbol{D}^{\alpha}_{ABC_{gH}} x(t) = \sin t \odot x(t) \oplus t^2, \quad x(-1, r) = ((r-1)e^t, (1-r)e^t), \quad -1 \le t \le 1$$

- For $-1 \le t \le 0$, $\sin t < 0$,

$$\boldsymbol{D}^{\alpha}_{ABC} x_l(t, r) = \sin t \odot x_u(t, r) + t^2, \quad x_l(-1, r) = (r-1)e^t, \quad 0 \le r \le 1$$

$$\boldsymbol{D}^{\alpha}_{ABC} x_u(t, r) = \sin t \odot x_u(t, r) + t^2, \quad x_u(-1, r) = (1-r)e^t, \quad 0 \le r \le 1$$

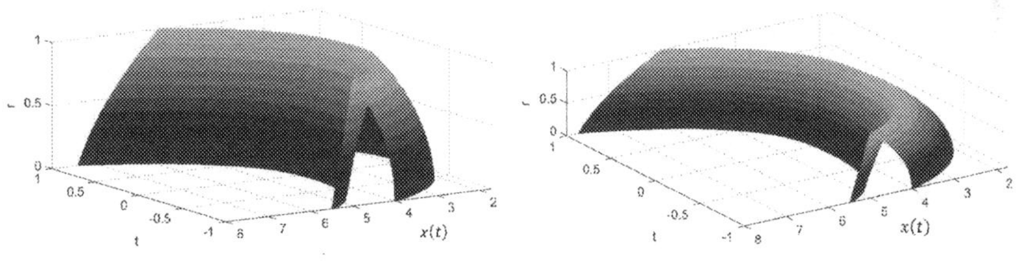

Fig. 5.10 The solution on $[-1, 1]$ for $\alpha = 0.9$ (left) and $\alpha = 0.85$ (right)

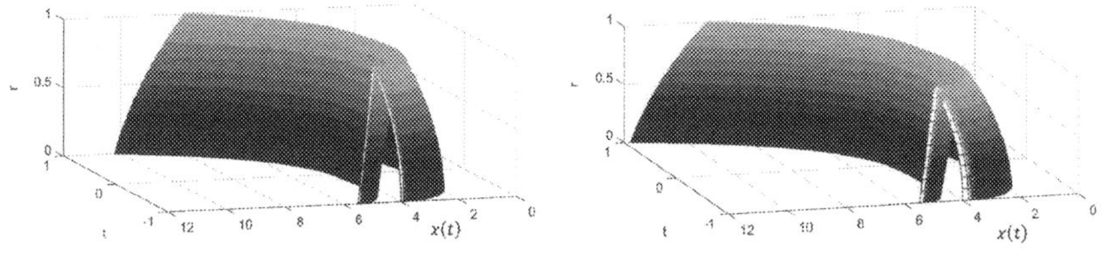

Fig. 5.11 The solution on $[-1, 1]$ for $\alpha = 0.6$ (left) and $\alpha = 0.4$ (right)

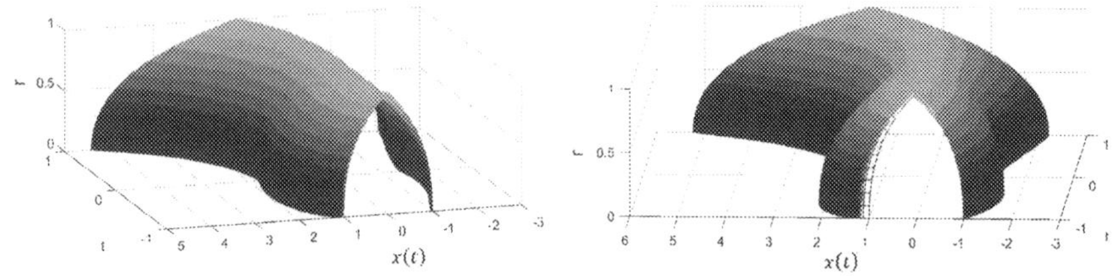

Fig. 5.13 The solution on $[-1, 1]$ for $\alpha = 0.5$ (left) and $\alpha = 0.4$ (right)

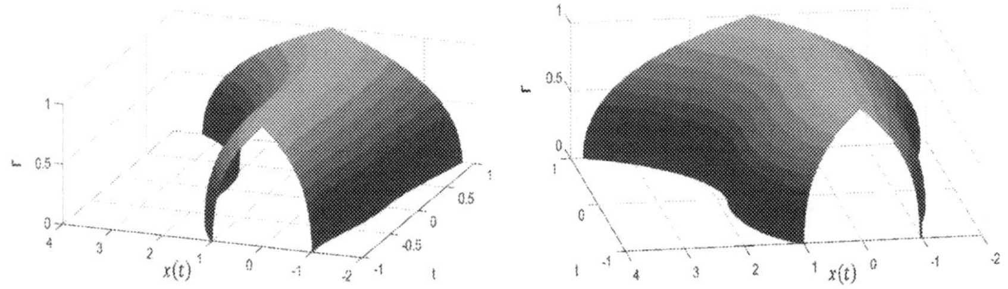

Fig. 5.12 The solution on $[-1, 1]$ for $\alpha = 0.95$ (left) and $\alpha = 0.7$ (right)

- For $0 \leq t \leq 1$, $\sin t > 0$, (Figs. 5.12 and 5.13)

$$\boldsymbol{D}^{\alpha}_{\boldsymbol{ABC}} x_l(t, r) = \sin t \odot x_l(t, r) + t^2, \quad x_l(-1, r) = (r - 1)e^t, \quad 0 \leq r \leq 1$$

$$\boldsymbol{D}^{\alpha}_{\boldsymbol{ABC}} x_u(t, r) = \sin t \odot x_u(t, r) + t^2, \quad x_u(-1, r) = (1 - r)e^t, \quad 0 \leq r \leq 1$$

5.6 Numerical Method for Fuzzy Fractional Impulsive Differential Equations

In this section, the combination of reproducing kernel Hilbert space method (RKHSM) and fractional differential transformation method (FDTM) is used to solve the fuzzy impulsive fractional differential equations with the help of the concept of generalized Hukuhara differentiability. In Sect. 4.5 of Chap. 4, we discussed the fuzzy impulsive differential equations with Caputo-Katugampola generalized fractional differentiability, now it is explained in case Caputo differentiability with $p = 1$ in the following form,

$$\begin{cases} {}_cD^{\frac{1}{\alpha}}y(t) = f(t, y(t)), & t \in [0,T],\ t \neq t_k, \quad m-1 < \frac{1}{\alpha} < m,\ m \in \mathbb{N} \\ \Delta y(t)|_{t=t_k} = I_k\left(y\left(t_k^-(t)\right)\right) \\ y(0) = y_0 \end{cases}$$

where $k = 1, 2, \ldots, m$, and ${}_cD^{\frac{1}{\alpha}}$ denotes the Caputo fractional generalized derivative of order $\frac{1}{\alpha}$, and $y(t)$ is an unknown fuzzy function of real variable t. Also suppose that f: $[0,T] \times \mathbb{F}_R \to \mathbb{F}_R$, is continuous fuzzy function, and $I_k : \mathbb{F}_R \to \mathbb{F}_R$, is a fuzzy difference transform with fuzzy number initial value $y_0 \in \mathbb{F}_R$. As it was explained in chapter four, the fuzzy solution is a peace-wise continuous fuzzy function and is defined on the subintervals, $0 = t_0 < t_1 < \ldots < t_m < t_{m+1} = T$. The fuzzy difference map is defined as,

$$\Delta|_{t=t_k} = y\left(t_k^+\right) \ominus_{gH} y\left(t_k^-\right),\ y\left(t_k^+\right) = lim_{h\to 0^+} y(t_k + h),\ y\left(t_k^-\right) = lim_{h\to 0^-} y(t_k + h)$$

The values $y\left(t_k^+\right)$ and $y\left(t_k^-\right)$ represent the right and left limits of $y(t)$ at $t = t_k$.

In Sect. 4.5, it was shown that the solution of fuzzy impulsive differential equation satisfies the following fuzzy integral equations,

$$y(t) = y_0 \oplus \frac{1}{\Gamma\left(\frac{1}{\alpha}\right)} \int_0^t (t-s)^{\frac{1}{\alpha}-1} f(s, y(s))ds, \quad t \in [0, t_1]$$

whenever $y(t)$ is $i - gH$ differentiable, and

$$y(t) = y_0 \ominus_H (-1) \frac{1}{\Gamma\left(\frac{1}{\alpha}\right)} \int_0^t (t-s)^{\frac{1}{\alpha}-1} f(s, y(s))ds, \quad t \in [0, t_1]$$

whenever $y(t)$ is $ii - gH$ differentiable, and finally if there is a switching point like t_1,

$$y(t) = \begin{cases} y_0 \oplus \frac{1}{\Gamma\left(\frac{1}{\alpha}\right)} \int_0^t (t-s)^{\frac{1}{\alpha}-1} f(s, y(s))ds, \quad t \in [0, t_1] \\ y_0 \ominus_H (-1) \frac{1}{\Gamma\left(\frac{1}{\alpha}\right)} \sum_{k=1}^{m} \int_{t_{k-1}}^{t_k} (t_k - s)^{\frac{1}{\alpha}-1} f(s, y(s))ds \ominus (-1) \\ \frac{1}{\Gamma\left(\frac{1}{\alpha}\right)} \int_{t_k}^{t} (t-s)^{\frac{1}{\alpha}-1} f(s, y(s))ds \ominus_H (-1) \sum_{k=1}^{m} I_k\left(y\left(t_k^-\right)\right), \quad t \in (t_1, t_{k+1}] \end{cases}$$

if there exists a point $t_1 \in (0, t_{k+1})$ such that $y(t)$ is $i - gH$ differentiable on $[0, t_1]$ and $ii - gH$ differentiable on (t_1, t_{k+1}).

5.6.1 Remark—Uniqueness

Let us assume, the fuzzy functions f and I_k satisfy the Lipschitz condition,

H1. There exists a constant $0 \le L_1$ such that

$$D_H(f(t, y_1(t)), f(t, y_2(t))) \le L_1 D_H(y_1(t), y_2(t)),$$

for each $t \in [0, T]$, and any $y_1, y_2 \in \mathbb{F}_R$.

H2. There exists a constant $0 \le L_2$ such that

$$D_H(I_k(y_1(t)), I_k(y_2(t))) \le L_2 D_H(y_1(t), y_2(t)),$$

for each $y_1, y_2 \in \mathbb{F}_R$, and $k = 1, 2, \ldots, m$. If

$$\left[\frac{L_1}{\Gamma\left(\frac{1}{\alpha}\right)}(m+1)T^{\frac{1}{\alpha}} + mL_2\right] < 1$$

Such that T is very small numbers the fuzzy fractional impulsive differential equation has a unique solution on $[0, T]$.

Proof By using the distance, it is easy to show, to this end, suppose the function F is defined as,

$$F(y(t)) = y_0 \ominus_H (-1)\frac{1}{\Gamma\left(\frac{1}{\alpha}\right)}\sum_{k=1}^{m}\int_{t_{k-1}}^{t_k}(t_k - s)^{\frac{1}{\alpha}-1} f(s, y(s))ds$$

$$\ominus_H (-1)\frac{1}{\Gamma\left(\frac{1}{\alpha}\right)}\int_{t_k}^{t}(t - s)^{\frac{1}{\alpha}-1} f(s, y(s))ds \ominus_H (-1)\sum_{k=1}^{m} I_k\left(y\left(t_k^-\right)\right)$$

$$D_H(F(y_1(t)), F(y_2(t))) \le \frac{1}{\Gamma\left(\frac{1}{\alpha}\right)}\sum_{k=1}^{m}\int_{t_{k-1}}^{t_k}(t_k - s)^{\frac{1}{\alpha}-1} D_H(f(s, y_1(s)), f(s, y_2(s)))ds$$

$$+\frac{1}{\Gamma\left(\frac{1}{\alpha}\right)}\int_{t_k}^{t}(t - s)^{\frac{1}{\alpha}-1} D_H(f(s, y_1(s)), f(s, y_2(s)))ds + \sum_{k=1}^{m} D_H\left(I_k\left(y_1\left(t_k^-\right)\right), I_k\left(y_2\left(t_k^-\right)\right)\right)$$

$$D_H(F(y_1(t)),F(y_2(t))) \le \frac{1}{\Gamma\left(\frac{1}{\alpha}\right)} \sum_{k=1}^{m} \int_{t_{k-1}}^{t_k} (t_k - s)^{\frac{1}{\alpha}-1} L_1 D_H(y_1(t),y_2(t))ds$$

$$+ \frac{1}{\Gamma\left(\frac{1}{\alpha}\right)} \int_{t_k}^{t} (t-s)^{\frac{1}{\alpha}-1} L_1 D_H(y_1(t),y_2(t))ds + mL_2 D_H(y_1(t),y_2(t))$$

$$D_H(F(y_1(t)),F(y_2(t))) \le \frac{1}{\Gamma\left(\frac{1}{\alpha}\right)} L_1 D_H(y_1(t),y_2(t)) \sum_{k=1}^{m} \int_{t_{k-1}}^{t_k} (t_k - s)^{\frac{1}{\alpha}-1} ds$$

$$+ \frac{1}{\Gamma\left(\frac{1}{\alpha}\right)} L_1 D_H(y_1(t),y_2(t)) \int_{t_k}^{t} (t-s)^{\frac{1}{\alpha}-1} ds + mL_2 D_H(y_1(t),y_2(t))$$

Since $\int_{t_k}^{t} (t-s)^{\frac{1}{\alpha}-1} ds \le T^{\frac{1}{\alpha}}$,

$$D_H(F(y_1(t)),F(y_2(t))) \le \frac{1}{\Gamma\left(\frac{1}{\alpha}\right)} L_1 D_H(y_1(t),y_2(t)) m T^{\frac{1}{\alpha}}$$

$$+ \frac{1}{\Gamma\left(\frac{1}{\alpha}\right)} L_1 D_H(y_1(t),y_2(t)) T^{\frac{1}{\alpha}} + mL_2 D_H(y_1(t),y_2(t))$$

$$D_H(F(y_1(t)),F(y_2(t))) \le D_H(y_1(t),y_2(t)) \left[\frac{L_1}{\Gamma\left(\frac{1}{\alpha}\right)} (m+1) T^{\frac{1}{\alpha}} + mL_2 \right]$$

The proof is completed if the map F is a fixed point Banach contraction map and this means,

$$D_H(F(y_1(t)),F(y_2(t))) \le D_H(y_1(t),y_2(t))$$

For this purpose, we should have,

$$\left[\frac{L_1}{\Gamma\left(\frac{1}{\alpha}\right)} (m+1) T^{\frac{1}{\alpha}} + mL_2 \right] < 1$$

5.6.2 *Reproducing Kernel Hilbert Space Method (RKHSM)*

This method is a semi analytical method to find the functional approximation of our problem. Here the RKHSM is explained very shortly,

Definition—Hilbert space A Hilbert space is a complete infinite-dimensional inner-product space. The elements of this space can be functions defined on a set T. In particular, the abstract (RKHS), H, is a Hilbert space of functions defined on a set T such that there exists a unique function, $R(t,y)$, defined on $T \times T$ with the following properties:

(I). $R_y(t) = R(t,y) \in H \quad \forall t \in T$
(II). $f(t), R_y(t) = f(y) \quad \forall t \in T \quad \forall f \in H$

The function $R(t,y)$ is called the reproducing kernel of the abstract RKHSM.

Definition—Kernel Let ϕ be a mapping from T into the space H such that $\phi_i = R(t,t_i)$. A function $R: T \times T \to \mathbb{R}$ such that $R_y(t) = R(t,y) = \langle \phi(t), \phi(y) \rangle$, for all $t, y \in T$ is called a kernel.

Definition—Inner product and norm
For an absolutely continuous real valued function $x^{(m-1)}(t)$, on $[a,b]$ and

$$W_2^m[a,b] = \left\{ x(t) | x^{(m)}(t) \in L^2[0,1], \quad x(a) = x(b) = 0 \right\}$$

where $L^2[a,b] = \left\{ x | \int_a^b x^2 dt < \infty \right\}$.

The inner product and norm of the functions $x, y \in$ in $W_2^m[a,b]$ are given respectively by

$$\langle x(t), y(t) \rangle = \sum_{i=0}^{m-1} x^{(i)}(a) y^{(i)}(a) + \int_a^b x^{(m)}(t) x^{(m)}(t) dt$$

and

$$\| x(t) \|_{W_2^m[a,b]} = \sqrt{\langle x, x \rangle}_{W_2^m[a,b]}$$

Note In the special case of definition, in case $m = 1, [a,b] = [0,1]$, $x(t)$ is an absolutely continuous fuzzy valued function on $[0,1]$, and

$$W_2^1[0,1] = \left\{ x(t) | x'_{gH} \in L^2[0,1], \quad x(0) = x(1) = 0 \right\}$$

The inner product and norm in $W_2^1[0,1]$ are given respectively by

$$\langle x(t), y(t) \rangle = x(0) \odot y(0) \oplus \int_0^1 x'_{gH}(t) \odot y'_{gH}(t) dx$$

and

$$\| x \|_W = \sqrt{\langle x, x \rangle_{W_2^1}}$$

$W_2^1[0,1]$ is a reproducing kernel space with reproducing kernel $R(t,y)$ that is defined as

$$R(t,y) = \begin{cases} 1+y, & y \le t \\ 1+t, & y > t \end{cases}$$

Also another special case for $m = 2$,

$$W_2^2[0,1] = \left\{ x(t) | x''_{gH} \in L^2[0,1], \quad x(0) = x(1) = 0 \right\}$$

where $x'_{gH}(t)$ are absolutely continuous real valued functions on $[0,1]$.

The inner product and norm in $W_2^2[0,1]$ are given respectively by

$$\langle x(t), y(t) \rangle = x(0) \odot y(0) \oplus x'_{gH}(0) \odot y'_{gH}(0) \oplus \int_0^1 x''_{gH}(t) \odot y''_{gH}(t) dt$$

Note In all above mentioned relations all H-differences and gH-derivatives should be defined properly (exist) and the integration can be defined in level-wise form.

The representation of the reproducing kernel function $R_t(y)$ is provided by

$$R_t(y) = \begin{cases} \frac{1}{6}(y-a)(2a^2 - y^2 + 3t(2+y) - a(6+3t+y)), & y \le t, \\ \frac{1}{6}(t-a)(2a^2 - t^2 + 3y(2+t) - a(6+3y+t)), & y > t. \end{cases}$$

5.6.3 *Numerical Solving Fuzzy Fractional Impulsive Differential Equation in* $W_2^2[0,1]$

We show how RKHSM is applied to solve integral equation. Thus

$$y(t) = \begin{cases} y_0 \oplus \frac{1}{\Gamma(\frac{1}{\alpha})} \int_0^t (t-s)^{\frac{1}{\alpha}-1} \odot f(s, y(s)) ds, \quad t \in [0, t_1] \\ y_0 \ominus_H (-1) \frac{1}{\Gamma(\frac{1}{\alpha})} \sum_{k=1}^{m} \int_{t_{k-1}}^{t_k} (t_k - s)^{\frac{1}{\alpha}-1} \odot f(s, y(s)) ds \ominus (-1) \\ \frac{1}{\Gamma(\frac{1}{\alpha})} \int_{t_k}^{t} (t-s)^{\frac{1}{\alpha}-1} \odot f(s, y(s)) ds \ominus_H (-1) \sum_{k=1}^{m} I_k(y(t_k^-)), \quad t \in (t_1, t_{k+1}] \end{cases}$$

if there exists a point $t_1 \in (0, t_{k+1})$ such that $y(t)$ is $i - gH$ differentiable on $[0, t_1]$ and $ii - gH$ differentiable on (t_1, t_{k+1}).

By impulsive effect, we have:

$$\Delta y(t) = y(0^+) \ominus_{gH} y(0^-) = I_0(y(0^-))$$

and based on the definition of gH-difference,

$$\begin{cases} y(0^+) = I_0(y(0^-)) \oplus y(0^-), & t \in [0, t_1], \\ y(0^-) = y(0^+) \oplus (-1) I_0(y(0^-)), & t \in (t_1, t_{k+1}]. \end{cases}$$

By substituting we have,

$$y(t) = \begin{cases} y(0^-) \oplus I_0(y(0^-)) \oplus \frac{1}{\Gamma(\frac{1}{\alpha})} \int\limits_0^t (t-s)^{\frac{1}{\alpha}-1} \odot f(s, y(s)) ds, t \in [0, t_1], \\ y(0^-) \ominus_H (-1) I_0(y(0^-)) \ominus_H (-1) \frac{1}{\Gamma(\frac{1}{\alpha})} \sum\limits_{k=1}^{m} \int\limits_{t_{k-1}}^{t_k} (t_k - s)^{\frac{1}{\alpha}-1} \odot f(s, y(s)) ds \\ \ominus_H (-1) \frac{1}{\Gamma(\frac{1}{\alpha})} \int\limits_{t_k}^{t} (t-s)^{\frac{1}{\alpha}-1} \odot f(s, y(s)) ds \ominus_H (-1) \sum\limits_{k=1}^{m} I_k(y(t_k)), \quad t \in (t_1, t_{k+1}], \end{cases}$$

Based on the assumptions,

$$y(t_k^+) = lim_{h \to 0^+} y(t_k + h), \quad y(t_k^-) = lim_{h \to 0^-} y(t_k + h)$$

we define $y(t^-) = y(t)$. Thus

$$y(t) = \begin{cases} I_0(y(0)) \oplus y(0) \oplus \frac{1}{\Gamma(\frac{1}{\alpha})} \int\limits_0^t (t-s)^{\frac{k}{\alpha}-1} \odot f(s, y(s)) ds, \quad t \in [0, t_1], \\ y(0) \ominus (-1) I_0(y(0)) \ominus_H (-1) \frac{1}{\Gamma(\frac{1}{\alpha})} \sum\limits_{k=1}^{m} \int\limits_{t_{k-1}}^{t_k} (t_k - s)^{\frac{1}{\alpha}-1} \odot f(s, y(s)) ds \\ \ominus_H (-1) \frac{1}{\Gamma(\frac{1}{\alpha})} \int\limits_{t_k}^{t} (t-s)^{\frac{1}{\alpha}-1} \odot f(s, y(s)) ds \ominus (-1) \sum\limits_{k=1}^{m} I_k(y(t_k)), \quad t \in (t_1, t_{k+1}], \end{cases}$$

To use RKHSM we define linear operator,

$$L = W_2^2[a, b] \to W_2^1[a, b], \quad t \in [0, t_{k+1}]$$

as follows:

$$Ly(t) = \begin{cases} y(t) \ominus_H I_0(y(0)) \ominus_H y(0) \ominus_H \frac{1}{\Gamma(\frac{1}{\alpha})} \int\limits_0^t (t-s)^{\frac{k}{\alpha}-1} \odot f(s, y(s))ds, & t \in [0, t_1], \\ y(t) \oplus (-1)I_0(y(0)) \ominus_H y(0) \oplus (-1)\frac{1}{\Gamma(\frac{1}{\alpha})} \sum\limits_{k=1}^{m} \int\limits_{t_{k-1}}^{t_k} (t_k - s)^{\frac{1}{\alpha}-1} \odot f(s, y(s))ds \oplus \\ (-1)\frac{1}{\Gamma(\frac{1}{\alpha})} \int\limits_{t_k}^{t} (t-s)^{\frac{1}{\alpha}-1} \odot f(s, y(s))ds \oplus (-1) \sum\limits_{k=1}^{m} I_k(y(t_k)), & t \in (t_1, t_{k+1}], \end{cases}$$

It is clear that L is a bounded linear operator so the model changes to the following problems:

$$Ly(t) = \begin{cases} F(t, y(t), T_1(y(t))), & t \in [0, t_1] \\ F(t, y(t), T(y(t)), S(y(t))), & t \in (t_1, t_{k+1}] \end{cases}$$

Such that

$$F(t, y(t), T_1(y(t))) = y(t) \ominus_H I_0(y(0)) \ominus_H y(0) \ominus_H T_1(y(t))$$

$$T_1(y(t)) = \frac{1}{\Gamma(\frac{k}{\alpha})} \int\limits_0^t (t-s)^{\frac{k}{\alpha}-1} \odot f(s, y(s))ds$$

Subject to all H-differences exist and

$$F(t, y(t), T(y(t)), S(y(t)))$$
$$= y(t) \oplus (-1)I_0(y(0)) \ominus_{\mathrm{H}} y(0) \oplus (-1) \sum_{k=1}^{k} I_k(y(t_k)) \oplus (-1)T(y(t)) \oplus (-1)S(y(t))$$

where

$$T(y(t)) = \frac{1}{\Gamma(\frac{k}{\alpha})} \sum_{k=1}^{m} \int\limits_{t_{k-1}}^{t_k} (t_k - s)^{\frac{k}{\alpha}-1} \odot f(s, y(s))ds$$

$$S(y(t)) = \frac{1}{\Gamma(\frac{k}{\alpha})} \int\limits_{t_k}^{t} (t-s)^{\frac{k}{\alpha}-1} \odot f(s, y(s))ds$$

where

$$y(t), L(y(t)) \in W_2^1[a, b],$$

$$F(t, y(t), T_1(y(t))), F(t, y(t), T(y(t)), S(y(t))) \in W_2^1[a, b]$$

For all $t \in [a, b]$.
Using the following reproducing kernel function $R_t(y)$,

$$R_t(y) = \begin{cases} \frac{1}{6}(y-a)(2a^2 - y^2 + 3t(2+y) - a(6+3t+y)), & y \leq t, \\ \frac{1}{6}(t-a)(2a^2 - t^2 + 3y(2+t) - a(6+3y+t)), & y > t. \end{cases}$$

Put $\phi_i(t) = R(t, t_i)$ and $\psi_i(t) = L^* \phi_i(t)$ where $\{t_i\}_{i=1}^{\infty}$ is dense in $[a, b]$ and L^* is the adjoint operator of L. It is easy to see that

$$\psi_i(t) = \left[L_y R_t(y)\right]_{y=t_i} = \left[L_y R(t, t_i)\right]_{y=t_i}$$

$$\begin{cases} R(t, t_i) \ominus I_0(R(0, t_i))) \ominus_H R(0, t_i) \ominus_H \frac{1}{\Gamma(\frac{1}{\alpha})} \displaystyle\int_0^t (t-s)^{\frac{1}{\alpha}-1} \odot f(s, R(s, t_i))ds, & t \in [0, t_1] \\ R(t, t_i) \oplus (-1) I_0(R(0, t_i)) \oplus R(0, t_i) \oplus (-1) \displaystyle\sum_{k=1}^{m} I_k(R(t_k, t_i)) \oplus (-1) \\ \frac{1}{\Gamma(\frac{1}{\alpha})} \displaystyle\sum_{k=1}^{m} \int_{t_{k-1}}^{t_k} (t_k - s)^{\frac{1}{\alpha}-1} \odot f(s, R(s, t_i))ds & t \in (t_1, t_{k+1}] \\ \oplus (-1) \frac{1}{\Gamma(\frac{1}{\alpha})} \displaystyle\int_{t_k}^{t} (t-s)^{\frac{1}{\alpha}-1} \odot f(s, R(s, t_i))ds, \end{cases}$$

Note If $\{t_i\}_{i=1}^{\infty}$ is dense on $[a, b]$ then $\{\psi_i\}_{i=1}^{\infty}$ is the complete function system of the space $W_2^2[a, b]$ and $\psi_i(t) = [L_t R_t(y)]_{y=t_i}$ where the subscript t in the operator L indicates that the operator L applies to the function of t.

Note

$$\psi_i(t) = \langle (L^* \phi_i)), R_t(y) \rangle_{W_2^2} = \langle \phi_i, L_t R_t(y) \rangle_{W_2^2} = L_t R_t(y)|_{y=t_i}, \quad t \in [0, t_{k+1}]$$

Clearly $\psi_i \in W_2^2[a, b]$.

Note For each fixed $y(t) \in W_2^2[a,b]$, let

$$\langle y(t), \psi_i(t) \rangle_{W_2^2[a,b]} = 0, \quad t \in [0, t_{k+1}]$$

$i = 1, 2, \ldots$, it means

$$\langle y(t), L^* \phi_i(t) \rangle_{W_2^2[a,b]} = \langle L_t y(t), \phi_i(t) \rangle_{W_2^2[a,b]} = L_t y(t_i) = 0, \quad t \in [0, t_{k+1}]$$

Assume that $\{t_i\}_{i=1}^{\infty}$ is dense on $[a,b]$ and so $L_t y(t) = 0$. It follows that $y = 0$ from the existence of L_t^{-1}.

Definition—Orthonormal system The orthonormal system $\{\widehat{\psi}_i(t)\}_{i=1}^{\infty}$ of $W^2[a,b]$ can be derived from the Gram-Schmidt orthogonalization process of $\{\psi_i(t)_{i=1}^{\infty}\}$,

$$\widehat{\psi}_i(t) = \sum_{k=1}^{i} \beta_{ik} \psi_k(t), \quad t \in [0, t_{k+1}]$$

where $\beta_{ik}, \quad i = 1, 2, \ldots, k = 1, 2, \ldots$ are coefficients of Gram-Schmidt orthonormalizarion and $\{\widehat{\psi}_i(t)\}_{i=1}^{\infty}$ is an orthonormal system, could be determined by solving the following equations.

$$B_{ik} = \psi_i, \widehat{\psi}_i = \psi_i(a)\widehat{\psi}_i(a) + \psi_i'(a)\widehat{\psi}_i'(a) + \int_a^b \psi_i''(t)\widehat{\psi}_i''(t)\mathrm{d}t, \quad t \in [0, t_{k+1}]$$

where

$$\beta_{ii} = \frac{1}{\sqrt{\left[(\psi_i(a))^2 + (\psi_i'(a))^2 + \int_a^b \left(\psi_i''(t)\right)^2 \mathrm{d}t - \sum_{k=1}^{i-1} B_{ik}^2\right]}}, \quad t \in [0, t_{k+1}]$$

$$\beta_{ij} = \beta_{ii} * \left(-\sum_{k=j}^{i-1} B_{ik} * \beta_{kj}\right) \quad i = 1, 2, \ldots, \quad j = 1, 2, \ldots, i-1,$$
$$k = 1, 2, \ldots, i-1$$

Remark If $\{t_i\}_{i=1}^{\infty}$ is dense on $[a,b]$ and the solution of

$$Ly(t) = \begin{cases} F(t, y(t), T_1(y(t))), & t \in [0, t_1] \\ F(t, y(t), T(y(t)), S(y(t))), & t \in (t_1, t_{k+1}] \end{cases}$$

Is unique the it is in the form of,

$$u(t) = \sum_{i=1}^{\infty}\sum_{k=1}^{i} \beta_{ik} \odot y(t_k) \odot \widehat{\psi}_i(t), \quad t \in [0, t_{k+1}]$$

Using

$$\widehat{\psi}_i(t) = \sum_{k=1}^{i} \beta_{ik}\psi_k(t), \quad u(t) = \sum_{i=1}^{\infty} \langle y(t), \widehat{\psi}_i(t)\rangle_{W_2^1} \widehat{\psi}_i(t)$$

we have

$$\begin{aligned} u(t) &= \sum_{i=1}^{\infty} \langle y(t), \widehat{\psi}_i(t)\rangle_{W_2^1} \widehat{\psi}_i(t) = \sum_{i=1}^{\infty} \langle y(t), \sum_{k=1}^{i} \beta_{ik}\psi_k(t)\rangle_{W_2^1} \widehat{\psi}_i(t) \\ &= \sum_{i=1}^{\infty}\sum_{k=1}^{i} \beta_{ik} \odot \langle y(t), \psi_k(t)\rangle_{W_2^1} \widehat{\psi}_i(t) = \sum_{i=1}^{\infty}\sum_{k=1}^{i} \beta_{ik} \odot \langle y(t), L^*\phi_k(t)\rangle_{W_2^1} \widehat{\psi}_i(t) \\ &= \sum_{i=1}^{\infty}\sum_{k=1}^{i} \beta_{ik} \odot \langle Ly(t), \phi_k(t)\rangle_{W_2^1} \widehat{\psi}_i(t) = \sum_{i=1}^{\infty}\sum_{k=1}^{i} \beta_{ik} \odot Ly(t_k)\widehat{\psi}_i(t) \\ &= \sum_{i=1}^{\infty}\sum_{k=1}^{i} \beta_{ik} y(t_k)\widehat{\psi}_i(t), \quad t \in [0, t_{k+1}] \end{aligned}$$

On the other hand, $u(t) \in W_2^2[a,b]$ and $Lu(t) = 0, \quad t \in [0, t_{k+1}]$.

$$y(t) = \sum_{i=1}^{\infty} a_i \odot \widehat{\psi}_i(t), \quad t \in [0, t_{k+1}]$$

where $a_i \in \mathbb{F}_R$,

$$a_i = y(t), \widehat{\psi}_i(t), \quad t \in [0, t_{k+1}]$$

are the Fourier series expansion about normal orthogonal system $\{\widehat{\psi}_i(t)\}_{i=1}^{\infty}$ on the Hilbert space $W_2^2[a,b]$. Thus the series,

$$\sum_{i=1}^{\infty} a_i \odot \widehat{\psi}_i(t), \quad t \in [0, t_{k+1}]$$

is convergent in the sense of $\| \cdot \|_{W_2^2}$ and the proof would be completed.

Now the approximate solution $y_N(t)$ can be obtained by the N-term intercept of the exact solution $y(t)$ and

$$y_N(t) = \sum_{i=1}^{N}\sum_{k=1}^{i} \beta_{ik} \odot y(t_k) \odot \widehat{\psi}_i(t), \quad t \in [0, t_{k+1}]$$

In the sequel, a new iterative method to achieve the solution is also presented. If

$$A_i = \sum_{k=1}^{i} \beta_{ik} \odot y(t_k), \quad t \in [0, t_{k+1}]$$

The equation can be written as,

$$y(t) = \sum_{k=1}^{\infty} A_i \odot \widehat{\psi}_i(t), \quad t \in [0, t_{k+1}]$$

Now suppose, for some $(t_i, y(t_i))$ the value of $y(t)$ is known. There is no problem if we assume $i = 1$. We put $y_0(t_1) = y(t_1)$ and define the N-term approximation to $y(t)$ by

$$y_N(t) = \sum_{k=1}^{N} c_i \odot \widehat{\psi}_i(t), \quad t \in [0, t_{k+1}]$$

where

$$c_i = \sum_{i=1}^{k} \beta_{ik} \odot y_{k-1}(t_k), \quad t \in [0, t_{k+1}]$$

In the following, it would be proven that the approximate solution $y_N(t)$ in the iterative is convergent to the exact solution of the following equations uniformly.

$$Ly(t) = \begin{cases} F(t, y(t), T_1(y(t))), & t \in [0, t_1] \\ F(t, y(t), T(y(t)), S(y(t))), & t \in (t_1, t_{k+1}] \end{cases}$$

Remark Suppose that $\|y_N(t)\|_{W_2^2}$ is bounded. If $\{t_i\}_{i=1}^{\infty}$ is dense on $[a, b]$ then N-term approximate solution $y_N(t)$ converges to the exact solution $y(t)$ of $Ly(t)$ and

$$y(t) = \lim_{n\to\infty} \sum_{i=1}^{n} c_i \odot \psi_i(t), \quad t \in [0, t_{k+1}]$$

where c_i are give by

$$c_i = \sum_{i=1}^{k} \beta_{ik} \odot y_{k-1}(t_k), \quad t \in [0, t_{k+1}]$$

In level-wise form for all $0 \le r \le 1$ we have,

$$y_l(t,r) = \lim_{n\to\infty} \sum_{i=1}^{n} c_{i,l}(t,r)\psi_i(t), \quad t \in [0, t_{k+1}]$$

$$y_u(t,r) = \lim_{n\to\infty} \sum_{i=1}^{n} c_{i,u}(t,r)\psi_i(t), \quad t \in [0, t_{k+1}]$$

where

$$c_{i,l}(t,r) = \sum_{i=1}^{k} \beta_{ik} \odot y_{k-1,l}(t_k, r), \quad t \in [0, t_{k+1}]$$

$$c_{i,u}(t,r) = \sum_{i=1}^{k} \beta_{ik} \odot y_{k-1,u}(t_k, r), \quad t \in [0, t_{k+1}]$$

Proof. First of all, the convergence of $y_N(t)$ is going to be proved. We infer

$$y_{n+1,l}(t,r) = y_{n,l}(t,r) + c_{n+1,l}(t,r)\widehat{\psi}_{n+1}(t), \quad t \in [0, t_{k+1}]$$

$$y_{n+1,u}(t,r) = y_{n,u}(t,r) + c_{n+1,u}(t,r)\widehat{\psi}_{n+1}(t), \quad t \in [0, t_{k+1}]$$

For any arbitrary and fixed r, it is obvious that the sequences $\left\|y_{n,l}(t)\right\|_{W_2^2}$ and $\left\|y_{n,u}(t)\right\|_{W_2^2}$ are monotonically increasing. Because both are bounded and convergent. Then $\sum_{i=1}^{\infty} c_{i,l}^2$ and $\sum_{i=1}^{\infty} c_{i,u}^2$ are bounded and this implies that $\{c_{i,l}\}_{i=1}^{\infty}, \{c_{i,u}\}_{i=1}^{\infty} \in L^2$. If $m > n$ then

$$||y_{m,l} - y_{n,l}||^2_{W_2^2[a,b]} = ||\sum_{i=m}^{n+1} (y_{i,l} - y_{i-1,l})||^2_{W_2^2[a,b]} = \sum_{i=m}^{n+1} ||(y_{i,l} - y_{i-1,l})||^2_{W_2^2[a,b]}$$

$$||y_{m,u} - y_{n,u}||^2_{W_2^2[a,b]} = ||\sum_{i=m}^{n+1} (y_{i,u} - y_{i-1,u})||^2_{W_2^2[a,b]} = \sum_{i=m}^{n+1} ||(y_{i,u} - y_{i-1,u})||^2_{W_2^2[a,b]}$$

So $||(y_{i,l} - y_{i-1,l})||^2_{W_2^2[a,b]} = c_{i,l}^2, ||(y_{i,u} - y_{i-1,u})||^2_{W_2^2[a,b]} = c_{i,u}^2$.
Consequently if $n \to \infty$

$$||y_{m,l} - y_{n,l}||^2_{W_2^2[a,b]} = \sum_{i=1}^{\infty} c_{i,l}^2 \to 0, \quad ||y_{m,u} - y_{n,u}||^2_{W_2^2[a,b]} = \sum_{i=1}^{\infty} c_{i,u}^2 \to 0$$

5.6.4 Combination of RKHM and FDTM

One of the applications of fuzzy Taylor expansion is fractional differential transformation method. In fact, in the Taylor series instead of the differentials some functions in terms of time and order of diferential. In the fuzzy fractional Taylor expansion assume that the derivative is defined as,

$$y_{gH}^{(k)}(t) = \frac{d^k}{dt^k} y_{gH}(t) := \phi(t,k), \quad k = 0, 1, \ldots, m-1, \quad t \in [a,b]$$

Such that

$$Y(k) := \phi(t_i, k) = \left[\frac{d^k}{dt^k} y_{gH}(t)\right]_{t=t_i}$$

where $Y(k)$ is called the spectrum of $y(t)$ at $t = t_i$ and $y(t) \equiv D^{-1}Y(k)$ and the symbol "D" denoting the differential transformation process.

This transformation can be explained in the fractional derivative in the following form. As it is mentioned, the solution of fuzzy fractional impulsive differential equations,

$$\begin{cases} {}_cD^{\frac{1}{\alpha}}y(t) = f(t, y(t)), \quad t \in [0,T], \quad t \neq t_k, \quad m-1 < \frac{1}{\alpha} < m, \quad m \in \mathbb{N} \\ \Delta y(t)|_{t=t_k} = I_k\left(y(t_k^-(t))\right) \\ y(0) = y_0 \end{cases}$$

Do have a fuzzy peace-wise continuous solution with linear and nonlinear terms. Such that

$$Ly(t) = \begin{cases} y(t) \ominus_H I_0(y(0)) \ominus_H y(0) \ominus_H \frac{1}{\Gamma(\frac{1}{\alpha})} \int\limits_0^t (t-s)^{\frac{k}{\alpha}-1} \odot f(s, y(s))ds, \quad t \in [0, t_1], \\ y(t) \oplus (-1)I_0(y(0)) \ominus_H y(0) \oplus (-1)\frac{1}{\Gamma(\frac{1}{\alpha})} \sum\limits_{k=1}^{m} \int\limits_{t_{k-1}}^{t_k} (t_k - s)^{\frac{1}{\alpha}-1} \odot f(s, y(s))ds \oplus \\ (-1)\frac{1}{\Gamma(\frac{1}{\alpha})} \int\limits_{t_k}^{t} (t-s)^{\frac{1}{\alpha}-1} \odot f(s, y(s))ds \oplus (-1)\sum\limits_{k=1}^{m} I_k(y(t_k)), \quad t \in (t_1, t_{k+1}], \end{cases}$$

The linear term,

$$Ly(t) = \begin{cases} F(t, y(t), T_1(y(t))), & t \in [0, t_1] \\ F(t, y(t), T(y(t)), S(y(t))), & t \in (t_1, t_{k+1}] \end{cases}$$

Such that

$$F(t, y(t), T_1(y(t))) = y(t) \ominus_H I_0(y(0)) \ominus_H y(0) \ominus_H T_1(y(t))$$

where

$$T_1(y(t)) = \frac{1}{\Gamma\left(\frac{k}{\alpha}\right)} \int_0^t (t-s)^{\frac{k}{\alpha}-1} \odot f(s, y(s))ds$$

Subject to all H-differences exist and

$$F(t, y(t), T(y(t)), S(y(t)))$$
$$= y(t) \oplus (-1)I_0(y(0)) \ominus_H y(0) \oplus (-1)\sum_{k=1}^{k} I_k(y(t_k)) \oplus (-1)T(y(t)) \oplus (-1)S(y(t))$$

where

$$T(y(t)) = \frac{1}{\Gamma\left(\frac{k}{\alpha}\right)} \sum_{k=1}^{m} \int_{t_{k-1}}^{t_k} (t_k - s)^{\frac{k}{\alpha}-1} \odot f(s, y(s))ds$$

$$S(y(t)) = \frac{1}{\Gamma\left(\frac{k}{\alpha}\right)} \int_{t_k}^{t} (t-s)^{\frac{k}{\alpha}-1} \odot f(s, y(s))ds$$

By applying L_t^{-1} to both sides of

$$L_t(y(t)) = \begin{cases} F(t, y(t), T_1(y(t))), & t \in [0, t_1] \\ F(t, y(t), T(y(t)), S(y(t))), & t \in (t_1, t_{k+1}] \end{cases}$$

We have

$$y(t) = \begin{cases} L_t^{-1}(F(t, y(t), T_1(y(t)))), & t \in [0, t_1] \\ L_t^{-1}(F(t, y(t), T(y(t)), S(y(t)))), & t \in (t_1, t_{k+1}] \end{cases}$$

The FDTM introduces the exact solution $y(t)$ and the nonlinear function $N(t, y(t))$ by infinite series

$$L_t\left(y(t) \oplus \sum_{i=0}^{\infty} y_i(t)\right) = \begin{cases} F(t, y(t), T_1(y(t))) \oplus N(t, y(t)), & t \in [0, t_1] \\ F(t, y(t), T(y(t)), S(y(t))) \oplus N(t, y(t)), & t \in (t_1, t_{k+1}] \end{cases}$$

where

$$N(t, y(t)) = \sum_{k=0}^{\infty} Y_{\frac{k}{\alpha}}(k) t^{\frac{k}{\alpha}}$$

where $Y_{\frac{k}{\alpha}}(k), k = 0, 1, \ldots$ are fractional FDTM polynomials for the nonlinear term $N(t, y(t))$ and can be found from the following formula,

$$Y_{\frac{k}{\alpha}}(k) = \begin{cases} \frac{1}{\Gamma(\frac{k}{\alpha}+1)} \left[\boldsymbol{D}_{C_{gH}}^{\frac{k}{\alpha}} y(t) \right]_{t=t_0} & \frac{k}{\alpha} \in \mathbb{Z}^+ \\ 0 & \frac{k}{\alpha} \notin \mathbb{Z}^+ \end{cases}$$

Now, we can evaluate $y(t_i), i = 0, 1, \ldots$ as follows,

$$y_i(t) := y(t_i) = \begin{cases} L_t^{-1}(F(t_i, y(t_i), T_1(y(t_i)))), & t_i \in [0, t_1] \\ L_t^{-1}(F(t_i, y(t_i), T(y(t_i)), S(y(t_i)))), & t_i \in (t_1, t_{k+1}] \end{cases}$$

Also

$$y(t) = \begin{cases} L_t^{-1}(F(t, y(t), T_1(y(t)))), & t \in [0, t_1] \\ L_t^{-1}(F(t, y(t), T(y(t)), S(y(t)))), & t \in (t_1, t_{k+1}] \end{cases}$$

Substituting yields,

$$y(t) \oplus \sum_{i=0}^{\infty} y_i(t) = y(t) \oplus \begin{cases} \sum_{i=0}^{\infty} L_t^{-1}(F(t_i, y(t_i), T_1(y(t_i)))), & t \in [0, t_1] \\ \sum_{i=0}^{\infty} L_t^{-1}(F(t_i, y(t_i), T(y(t_i)), S(y(t_i)))), & t \in (t_1, t_{k+1}] \end{cases}$$

$$y(t) \oplus \sum_{i=0}^{\infty} y_i(t)$$
$$= \begin{cases} L_t^{-1}(F(t, y(t), T_1(y(t)))) \oplus L_t^{-1}\left(\sum_{k=0}^{\infty} Y_{\frac{k}{\alpha}}(k) t^{\frac{k}{\alpha}} \right), & t \in [0, t_1] \\ L_t^{-1}(F(t, y(t), T(y(t)), S(y(t)))) \oplus L_t^{-1}\left(\sum_{k=0}^{\infty} Y_{\frac{k}{\alpha}}(k) t^{\frac{k}{\alpha}} \right), & t \in (t_1, t_{k+1}] \end{cases}$$

According to the FFDTM, the components $y_k(t)$ can be determined as,

$$Y_0(0)t^0 = \begin{cases} L_t^{-1}(F(t_0, y(t_0), T_1(y(t_0)))), & t_0 \in [0, t_1] \\ L_t^{-1}(F(t_0, y(t_0), T(y(t_0)), S(y(t_0)))), & t_0 \in (t_1, t_{k+1}] \end{cases}$$

$$Y_{\frac{1}{\alpha}}(1)t^{\frac{1}{\alpha}} = \begin{cases} L_t^{-1}(F(t_1, y(t_1), T_1(y(t_1)))), & t_1 \in [0, t_1] \\ L_t^{-1}(F(t_1, y(t_1), T(y(t_1)), S(y(t_1)))), & t_1 \in (t_1, t_{k+1}] \end{cases}$$

$$Y_{\frac{2}{\alpha}}(2)t^{\frac{2}{\alpha}} = \begin{cases} L_t^{-1}(F(t_2, y(t_2), T_1(y(t_2)))), & t_2 \in [0, t_1] \\ L_t^{-1}(F(t_2, y(t_2), T(y(t_2)), S(y(t_2)))), & t_2 \in (t_1, t_{k+1}] \end{cases}$$

$$\vdots$$

$$Y_{\frac{m}{\alpha}}(m)t^{\frac{m}{\alpha}} = \begin{cases} L_t^{-1}(F(t_m, y(t_m), T_1(y(t_m)))), & t_m \in [0, t_1] \\ L_t^{-1}(F(t_m, y(t_m), T(y(t_m)), S(y(t_m)))), & t_m \in (t_1, t_{k+1}] \end{cases}$$

Therefore, it can be approximated by $\{\widehat{\psi}_i(t)\}_{i=1}^{\infty}$ of $W^2[a, b]$,

$$\sum_{k=1}^{m} \frac{t^{\frac{k}{\alpha}}}{\Gamma\left(\frac{k}{\alpha}+1\right)} \left[\boldsymbol{D}^{\frac{k}{\alpha}}_{C_{gH}} y(t) \right]_{t=t_0} = \sum_{i=0}^{\infty} \sum_{k=1}^{i} \beta_{ik} \odot y(t_k) \widehat{\psi}_i(t), \quad t \in [0, t_{k+1}]$$

5.6.5 Algorithm

For finding the approximate and exact solutions $y_N(t, r)$ and $y(t, r)$ for fuzzy impulsive fractional differential equations respectively, we do the following main steps:

Step 1. Fix $t, y \in [a, b]$,

1.1 If $y \le t$ set $R_t^2(y) = \frac{1}{6}(y-a)(2a^2 - y^2 + 3t(2+y) - a(6+3t+y))$
1.2 Else set $R_t^2(y) = \frac{1}{6}(t-a)(2a^2 - t^2 + 3y(2+t) - a(6+3y+t))$
1.3 For $i = 1, 2, \ldots, n$, $\quad h = 1, 2, \ldots, m$ and $= 1, 2$, do the following:
1.4 Set $t_i = \frac{i-1}{n-1}$,
1.5 Set $r_h = \frac{h-1}{m-1}$,
1.6 Set

$$\psi_i(t_i) = L^{-1} R_t^2(y)\big|_{y=t_i}, \quad t \in [0, t_{k+1}]$$

Output: The orthogonal function system $\psi_i(t_i)$.
Step 2.

$$B_{ik} = \psi_i, \widehat{\psi}_i = \psi_i(a)\widehat{\psi}_i(a) + \psi_i'(a)\widehat{\psi}_i'(a) + \int_a^b \psi_i''(t)\widehat{\psi}_i''(t)\mathrm{d}t, \quad t \in [0, t_{k+1}]$$

$$\beta_{ii} = \frac{1}{\sqrt{\left[(\psi_i(a))^2 + (\psi_i'(a))^2 + \int_a^b \left(\psi_i''(t)\right)^2 \mathrm{d}t - \sum_{k=1}^{i-1} B_{ik}^2\right]}}, \quad t \in [0, t_{k+1}]$$

$$\beta_{ij} = \beta_{ii} * \left(-\sum_{k=j}^{i-1} B_{ik} * \beta_{kj}\right) \quad i = 1, 2, \ldots, \quad j = 1, 2, \ldots, i-1,$$
$$k = 1, 2, \ldots, i-1$$

Output: The orthogonalization coefficients β_{ik}
Step 3.
Set $\widehat{\psi}_i(t) = \sum_{k=1}^{i} \beta_{ik}\psi_k(t), \quad (\beta_{ii} > 0, i = 1, 2, \ldots)$.
Output: The orthogonal function system $\widehat{\psi}_i(t)$
Step 4. Set $y_0(t_1) = y(t_1)$
Step 5. Set $n = 1$
Step 6. Set

$$c_n = \sum_{k=1}^{n} \beta_{nk} y_{k-1}(t_k), \quad t \in (0, t_{k+1}].$$

Step 7. Set

Example Let us consider the fuzzy impulsive fractional differential equation,

$$D_{C_{gH}}^{\frac{1}{\alpha}} y(t) = \frac{t}{10(1+t)}, \quad t \in [0, 1], \quad t \neq \frac{1}{2}, \quad m - 1 < \frac{1}{\alpha} < m, \quad m \in N,$$

Set

$$I_k(t) = \left[\frac{(3r-1)t}{t+3}, \frac{(3-r)t}{t+3}\right], \quad f(t, y(t)) = \left[\frac{(r-1)t}{10(1+t)}, \frac{(1-r)t}{10(1+t)}\right]$$

The singleton zero number is considered as the initial value.

Choosing $k = 2$ and $\frac{1}{\alpha} = \frac{1}{2}$ the results for $y_l(t, r)$ are shown in Tables 5.6 and 5.7 for sevel r-levels and t.

Table 5.7 shows the results for the upper function of solution $y_u(t, r)$ for several values of levels and time.

Example Let us consider the fuzzy impulsive fractional equation,

$$D_{C_{gH}}^{\frac{k}{\alpha}} y(t) = \frac{ty^2(t)}{(3+t)(1+y^2(t))}, \quad t \in J := [0, 1], \quad t \neq \frac{1}{2}, \quad m - 1 < \frac{1}{\alpha} < m,$$
$$m \in N$$

Table 5.6 The results for $y_l(t, r)$ with $k = 2$ and $\frac{1}{\alpha} = \frac{1}{2}$

r/t	0.1	0.2	0.3	0.4	0.5	0.6	0.7	0.8	0.9	1
0.55	−0.00540	−0.00480	−0.00420	−0.00360	−0.0030	−0.00243	−0.00180	−0.00120	−0.00060	0
0.60	−0.00662	−0.00589	−0.00515	−0.00441	−0.00368	−0.00294	−0.00221	−0.00147	−0.00074	0
0.65	−0.00802	−0.00713	−0.00748	−0.00535	−0.00445	−0.00365	−0.00267	−0.00178	−0.00089	0
0.70	−0.00961	−0.00854	−0 .00888	−0.00761	−0.00534	−0.00427	−0.00320	−0.00213	−0.00107	0
0.75	−0.01142	−0.01015	−0.01047	−0.00897	−0.00634	−0.00507	−0.00380	−0.00254	−0.00127	0
0.80	−0.01346	−0.01196	−0.01225	−0.01012	−0.00747	−0.00598	−0.00448	−0.00350	−0.00149	0
0.85	−0.01576	−0.01401	−0.01426	−0.01050	−0.00875	−0.00700	−0.00525	−0.00407	−0.00175	0
0.90	−0.01834	−0.01630	−0.01426	−0.01212	−0.01018	−0.00814	−0.00610	−0.22024	−0.00203	0
0.95	−0.02122	−0.01886	−0.01650	−0.01414	−0.01178	−0.00942	−0.00706	−0.00471	−0.00235	0
1	−0.02443	−0.02171	−0.01899	−0.01628	−0.01365	−0.01084	−0.00813	−0.00542	−0.00271	0

Table 5.7 The results for $y_u(t, r)$ with $k = 2$ and $\frac{1}{\alpha} = \frac{1}{2}$

r/t	0.1	0.2	0.3	0.4	0.5	0.6	0.7	0.8	0.9	1
0.55	0.00539	0.004792	0 .004193	0.003595	0.002886	0.002397	0.001798	0.001199	0.000599	0
0.60	0.006608	0.005874	0.005141	0.004407	0.003673	0.002939	0.002204	0.001470	0.000735	0
0.65	0.008001	0.007113	0.006225	0.005336	0.004447	0.003558	0.002669	0.003422	0.00089	0
0.70	0.009588	0.008524	0 .007459	0.006395	0.005330	0.004264	0.003199	0.001780	0.001076	0
0.75	0.011386	0.010123	0 .008859	0.007595	0.00633 0	0.005065	0.003799	0.002533	0.001267	0
0.80	0.013417	0.011929	0 .010439	0.00895	0.00746 0	0.005969	0.004477	0.002986	0.001493	0
0.85	0.015702	0.013960	0 .012218	0.010475	0.000873	0.006986	0.005241	0.003494	0.001748	0
0.905	0.018264	0.016239	0 .014212	0.012184	0.010156	0.008126	0.006096	0.004065	0.002033	0
0.95	0.021128	0.018785	0 .016441	0.014095	0.011749	0.009401	0.007035	0.004703	0.002352	0
1	0.024319	0.021623	0 .018925	0.016225	0.013524	0.018220	0.008119	0.005414	0.002708	0

Table 5.8 The results for $y_l(t, r)$ with $k = 2$ and $\frac{1}{\alpha} = \frac{1}{2}$

r/t	0.1	0.2	0.3	0.4	0.5	0.6	0.7	0.8	0.9	1
0.55	0.013806	−0.01804	−0.04988	−0.00817	−0.11385	−0.14543	−0.17729	−0.20915	−0.27288	0
0.60	0.027612	−0.005760	−0.03913	−0.07251	−0.10589	0.139280	−0.17267	−0.20607	−0.23947	0
0.65	0.044673	−0.009418	−0.02584	−0.06111	−0.09639	−0.13167	−0.16696	−0.20226	−0.23757	0
0.70	0.065519	0.027961	−0.009610	−0.04719	−0.08477	−0.12237	−0.15998	−0.19760	−0.23523	0
0.75	0.090734	0.050394	0 .010039	−0.03033	−0.07072	−0.11112	−0.15153	−0.19176	−0.23241	0
0.80	0.120964	0.077291	0 .033597	−0.01012	−0.05386	−0.09762	−0.14140	−0.18520	−0.22903	0
0.85	0.156914	0.109282	0 .061620	0.013927	−0.03380	−0.08155	−0.12933	−0.17715	−0.22500	0
0.90	0.199357	0.147954	0 .094710	0.042324	−0.01010	−0.06257	−0.11509	−0.22024	−0.27288	0
0.95	0.249129	0.191353	0 .133522	0.075635	0.01769	−0.04031	−0.09836	−0.15648	−0.21465	0
1	0.307135	0.242986	0 .178764	0.114469	0.05010	−0.01434	−0.07886	−0.14346	−0.20813	0

Table 5.9 The results for $y_u(t, r)$ with $k = 2$ and $\frac{1}{\alpha} = \frac{1}{2}$

r/t	0.1	0.2	0.3	0.4	0.5	0.6	0.7	0.8	0.9	1
0.55	0.028928	0.025721	0 .022512	0.019302	0.016089	0.012875	0.009659	0.006441	0.003221	0
0.60	0.020779	0.018476	0.016171	0.013865	0.011557	0.009249	0.006938	0.004627	0.002314	0
0.65	0.015366	0.013663	0.011958	0.011025	0.008547	0.006839	0.005131	0.003422	0.001711	0
0.70	0.011649	0.010357	0 .009065	0.007773	0.006479	0.005185	0.003890	0.002594	0.001297	0
0.75	0.009022	0.008022	0 .007022	0.006020	0.005019	0.004016	0.003013	0.002009	0.001005	0
0.80	0.007121	0.006331	0 .005542	0.007452	0.003961	0.003170	0.002378	0.001586	0.000793	0
0.85	0.005713	0.00508	0 .004447	0.003813	0.003178	0.002543	0.001908	0.001273	0.000636	0
0.90	0.004652	0.004137	0 .003621	0.003105	0.002588	0.002071	0.001554	0.001036	0.000518	0
0.95	0.003838	0.003413	0 .002987	0.002561	0.002135	0.001709	0.001282	0.000855	0.000428	0
1	0.003204	0.002849	0 .002494	0.002138	0.001783	0.001427	0.001070	0.000714	0.000357	0

The singltone zero number is considered as the initial value. Set

$$f(t, y(t)) = \left[\frac{t^3(r-1)}{(3+t)(1+t^2)}, \frac{t^3(1-r)}{(3+t)(1+t^2)}\right], \quad I_k(t) = \frac{t}{t+2}, \quad t \in [0, \infty)$$

Again $k = 2$ and $\frac{1}{\alpha} = \frac{1}{2}$, the results are shown in Tables 5.8 and 5.9.

Chapter 6
Applications of Fuzzy Fractional Differential Equations

6.1 Introduction

In general, fractional calculus deals with the generalization of differentiation and integration of non-integer orders. In recent years, fractional calculus has played a significant role in several sciences such as physics, chemistry, biology, electronics, and control theory. In this chapter, first the fuzzy fractional optimal control problem is investigated and then the fuzzy fractional diffusion with applications in drug release are explained.

6.2 Fuzzy Fractional Calculus—Preliminaries for Control Problem

Fractional optimal control problems are optimal control problems associated with fractional dynamic systems. As defined by Agrawal, a fractional dynamic system is a system whose dynamics is described by fractional differential equations.

It has been demonstrated that fractional order differential equations model dynamic systems and processes more accurately than integer order differential equations. Therefore, the solution of fractional differential equations and the problem containing them with analytical and numerical schemes are of growing interest. The fractional optimal control theory is a novel topic in mathematics. The fractional optimal control problems may be defined in terms of different types of fractional derivatives. But the most important types of fractional derivatives are the Riemann-Liouville and the Caputo fractional derivatives.

It is notable to mention that the uncertainty is inherent in most real-world systems. Fuzzy set theory is a powerful tool for modeling uncertainty and for processing vague or subjective information in mathematical models, which has been applied to a wide variety of real problems. Fuzzy fractional optimal control problems are

T. Allahviranloo, *Fuzzy Fractional Differential Operators and Equations*, Studies in Fuzziness and Soft Computing 397, https://doi.org/10.1007/978-3-030-51272-9_6

fractional optimal control problems with ambiguity, which could appear, for example, in parameter values, functional relationships, or initial conditions.

Like integer-order optimal control problems, the formulations stem from Fractional variational calculus (which is extended for fuzzy case). However, to our best knowledge, there are few reports on the necessary optimality conditions for fuzzy fractional optimal control problems. Only limited work has been done in the area of fuzzy variational problems, or more specifically, in the area of fuzzy fractional optimal control.

It seems that, it is a new idea to derive the necessary optimality conditions for fuzzy fractional optimal control problems by using the fuzzy generalized Hakahara differentiability concept. To this end, some basic concepts such as generalized Hakahara differentiability (gH-differentiability) and integration for fully fuzzy functions that have been introduced in Chap. 2, are used. Furthermore, some fuzzy fractional theorems such as a formula for fuzzy fractional integration by parts are proved. Then, Euler-Lagrange equations (necessary optimality conditions) for them are derived. Finally, the sufficient conditions are discussed. Here, we reconsider some required and considerable fractional definitions and concepts of fuzzy valued functions. The fuzzy fractional Riemann-Liouville derivative operators (RL) and Caputo are brought (all have been fully discussed in Chap. 2).

To remind, the derivatives and integral operators are in the following form.

RL—Fuzzy fractional integral

$$\boldsymbol{I}^{\alpha}_{\boldsymbol{RL}}f(x) = \frac{1}{\Gamma(\alpha)} \odot \int_{x}^{b} (t-x)^{\alpha-1} \odot f(t)dt, \quad 0<\alpha<1$$

Level-wise form

$$\boldsymbol{I}^{\alpha}_{\boldsymbol{RL}}f(x,r) = \left[\boldsymbol{I}^{\alpha}_{\boldsymbol{RL}}f_l(x,r), \boldsymbol{I}^{\alpha}_{\boldsymbol{RL}}f_u(x,r)\right], \quad x \in [a,b]$$

where

$$\boldsymbol{I}^{\alpha}_{\boldsymbol{RL}}f_l(x,r) = \frac{1}{\Gamma(\alpha)} \int_{a}^{x} (x-t)^{\alpha-1} f_l(t,r)dt, \quad \boldsymbol{I}^{\alpha}_{\boldsymbol{RL}}f_u(x,r) = \frac{1}{\Gamma(\alpha)} \int_{a}^{x} (x-t)^{\alpha-1} f_u(t,r)dt$$

RL—Fuzzy fractional derivative

$$\boldsymbol{D}^{\alpha}_{\boldsymbol{RL}_{gH}}f(x) = \frac{1}{\Gamma(1-\alpha)} \odot \frac{d}{dx} \int_{a}^{x} (x-t)^{-\alpha} \odot f(t)dt, \quad 0<\alpha<1$$

or

$$\boldsymbol{D}^{\alpha}_{\boldsymbol{RL}_{gH}}f(x) = \frac{1}{\Gamma(1-\alpha)} \odot \left(\int_a^x (x-t)^{-\alpha} \odot f(t)dt \right)'_{gH}, \quad 0<\alpha<1$$

In the definition of RL-derivative if the right area is used then

$$\boldsymbol{D}^{\alpha}_{\boldsymbol{RL}_{gH}}f(x) = \frac{1}{\Gamma(1-\alpha)} \odot \left(\int_x^b (t-x)^{-\alpha} \odot f(t)dt \right)'_{gH}, \quad 0<\alpha<1$$

Level-wise form

$$\boldsymbol{D}^{\alpha}_{\boldsymbol{RL}_{gH}}f(x,r) = \left[\min\left\{\boldsymbol{D}^{\alpha}_{\boldsymbol{RL}}f_l(x,r), \boldsymbol{D}^{\alpha}_{\boldsymbol{RL}}f_u(x,r)\right\}, \max\left\{\boldsymbol{D}^{\alpha}_{\boldsymbol{RL}}f_l(x,r), \boldsymbol{D}^{\alpha}_{\boldsymbol{RL}}f_u(x,r)\right\}\right]$$

where

$$\boldsymbol{D}^{\alpha}_{\boldsymbol{RL}}f_l(x,r) = \frac{1}{\Gamma(1-\alpha)} \frac{d}{dx} \int_a^x (x-t)^{-\alpha} f_l(t,r)dt,$$

$$\boldsymbol{D}^{\alpha}_{\boldsymbol{RL}}f_u(t,r) = \frac{1}{\Gamma(1-\alpha)} \frac{d}{dx} \int_a^x (x-t)^{-\alpha} f_u(t,r)dt$$

Caputo—Fuzzy fractional gH-derivative

$$\boldsymbol{D}^{\alpha}_{\boldsymbol{C}_{gH}}f(x) = \frac{1}{\Gamma(1-\alpha)} \odot \int_a^x (x-t)^{-\alpha} \odot f'_{gH}(t)dt, \quad 0<\alpha<1$$

Level-wise form

$$\boldsymbol{D}^{\alpha}_{\boldsymbol{C}_{gH}}f(x,r) = \left[\min\left\{\boldsymbol{D}^{\alpha}_{\boldsymbol{C}}f_l(x,r), \boldsymbol{D}^{\alpha}_{\boldsymbol{C}}f_u(x,r)\right\}, \max\left\{\boldsymbol{D}^{\alpha}_{\boldsymbol{C}}f_l(x,r), \boldsymbol{D}^{\alpha}_{\boldsymbol{C}}f_u(x,r)\right\}\right]$$

where

$$\boldsymbol{D}^{\alpha}_{\boldsymbol{C}}f_l(x,r) = \frac{1}{\Gamma(1-\alpha)} \int_a^x (x-t)^{-\alpha} f'_l(t,r)dt,$$

$$\boldsymbol{D}^{\alpha}_{\boldsymbol{C}}f_u(x,r) = \frac{1}{\Gamma(1-\alpha)} \int_a^x (x-t)^{-\alpha} f'_u(t,r)dt$$

Relation between fuzzy RL and CK derivatives

If the RL derivative $\boldsymbol{D}^{\alpha}_{RL_{gH}}f(x)$ exists in $[a,b]$ and $0<\alpha<1$,

$$\boldsymbol{D}^{\alpha}_{CK_{gH}}f(x)=\boldsymbol{D}^{\alpha}_{RL_{gH}}\left(f(x)\ominus_{gH}f(a)\right)$$

We know that,

$$\boldsymbol{D}^{\alpha}_{RL_{gH}}f(x)=\left(\boldsymbol{I}^{(1-\alpha)}_{RL}f\right)'_{gH}(x).$$

where

$$\boldsymbol{I}^{(1-\alpha)}_{RL}f(x)=\frac{1}{\Gamma(1-\alpha)}\odot\int_{a}^{x}(x-s)^{-\alpha}\odot f(s)ds$$

And we have,

$$\boldsymbol{I}^{(1-\alpha)}_{RL}f(a)=\frac{(x-a)^{1-\alpha}}{\Gamma(2-\alpha)}\odot f(a)$$

If $\alpha:\rightarrow 1+\alpha$,

$$\boldsymbol{D}^{\alpha}_{RL_{gH}}f(a)=\boldsymbol{I}^{-\alpha}_{RL}f(a)=\frac{(x-a)^{-\alpha}}{\Gamma(1-\alpha)}\odot f(a)$$

Then

$$\begin{aligned}\boldsymbol{D}^{\alpha}_{CK_{gH}}f(x)&=\boldsymbol{D}^{\alpha}_{RL_{gH}}\left(f(x)\ominus_{gH}f(a)\right)\\&=\boldsymbol{D}^{\alpha}_{RL_{gH}}f(x)\ominus_{gH}\boldsymbol{D}^{\alpha}_{RL_{gH}}f(a)=\boldsymbol{D}^{\alpha}_{RL_{gH}}f(x)\ominus_{gH}\frac{(x-a)^{-\alpha}}{\Gamma(1-\alpha)}\odot f(a)\end{aligned}$$

So, the relation is,

$$\boldsymbol{D}^{\alpha}_{CK_{gH}}f(x)=\boldsymbol{D}^{\alpha}_{RL_{gH}}f(x)\ominus_{gH}\frac{(x-a)^{-\alpha}}{\Gamma(1-\alpha)}\odot f(a)$$

Two sides in the level-form and based on the first case of gH difference, we have,

Case $i-gH$ difference,

$$\boldsymbol{D}^{\alpha}_{RL_{gH}}f(x)=\boldsymbol{D}^{\alpha}_{CK_{gH}}f(x)\oplus\frac{(x-a)^{-\alpha}}{\Gamma(1-\alpha)}\odot f(a)$$

Case $ii - gH$ difference,

$$\boldsymbol{D}^{\alpha}_{\boldsymbol{CK}_{gH}} f(x) = \boldsymbol{D}^{\alpha}_{\boldsymbol{RL}_{gH}} f(x) \oplus (-1)\frac{(x-a)^{-\alpha}}{\Gamma(1-\alpha)} \odot f(a)$$

Note Please note that this relation can also be obtained by using the integration by part that is explained in Chap. 2. To prove we need the following relation,

$$\left(\int_{u(x)}^{v(x)} g(t,x)dt\right)'_{gH} = \int_{u(x)}^{v(x)} g'_{gH}(t,x)dt \oplus v'(x) \odot g(v(x),x) \ominus_{gH} u'(x) \odot g(v(x),x)$$

where $u(x), v(x)$ are real valued functions. This can be proved very easily by using the level-wise form.

Proof (*Relation of Caputo and RL derivative*) By integration part and assuming, $u = f(t), (x-t)^{-\alpha}dt = dv$ we have,

$$\begin{aligned}
\boldsymbol{D}^{\alpha}_{\boldsymbol{RL}_{gH}} f(x) &= \frac{1}{\Gamma(1-\alpha)} \odot \left(\int_a^x (x-t)^{-\alpha} \odot f(t)dt\right)'_{gH} \\
&= \frac{1}{\Gamma(1-\alpha)} \odot \left(\frac{-(x-t)^{-\alpha+1}}{-\alpha+1} \odot f(t)\Big|_a^x \ominus_{gH} \int_a^x \frac{-(x-t)^{-\alpha+1}}{1-\alpha} \odot f'_{gH}(t)dt\right)'_{gH} \\
&= \frac{1}{\Gamma(1-\alpha)} \odot \left(0 \ominus_{gH} \frac{-(x-a)^{-\alpha+1}}{1-\alpha} \odot f(a) \ominus_{gH} \int_a^x \frac{-(x-t)^{-\alpha+1}}{1-\alpha} \odot f'_{gH}(t)dt\right)'_{gH} \\
&= \frac{1}{\Gamma(1-\alpha)} \odot \left(\frac{(x-a)^{-\alpha+1}}{1-\alpha} \odot f(a) \ominus_{gH} \int_a^x \frac{-(x-t)^{-\alpha+1}}{1-\alpha} \odot f'_{gH}(t)dt\right)'_{gH}
\end{aligned}$$

Since we have,

$$\begin{aligned}
&\left(\frac{(x-a)^{-\alpha+1}}{1-\alpha} \odot f(a) \ominus_{gH} \int_a^x \frac{-(x-t)^{-\alpha+1}}{1-\alpha} \odot f'_{gH}(t)dt\right)'_{gH} \\
&= \frac{1}{1-\alpha}\left[\left((x-a)^{-\alpha+1} \odot f(a)\right)'_{gH} \ominus_{gH} \left(\int_a^x -(x-t)^{-\alpha+1} \odot f'_{gH}(t)dt\right)'_{gH}\right]
\end{aligned}$$

And also, using $\left(\int_{u(x)}^{v(x)} g(t,x)dt\right)'_{gH}$ the second derivative,

$$\left(\int_a^x -(x-t)^{-\alpha+1} \odot f'_{gH}(t)dt\right)'_{xgH} = \int_a^x -(x-t)^{-\alpha} \odot f'_{gH}(t)dt \oplus 0 \ominus_{gH} 0$$

By substituting,

$$\left(\frac{(x-a)^{-\alpha+1}}{1-\alpha} \odot f(a) \ominus_{gH} \int_a^x \frac{-(x-t)^{-\alpha+1}}{1-\alpha} \odot f'_{gH}(t)dt\right)'_{gH}$$
$$= \frac{1}{1-\alpha}\left[(x-a)^{-\alpha} \odot f(a) \ominus_{gH} \int_a^x -(x-t)^{-\alpha} \odot f'_{gH}(t)dt\right]$$

The proof van be completed.

6.2.1 Theorem—Interchanging Operators

Let us continuous $f, g : [a, b] \to \mathbb{F}_R$ are two continuous fuzzy functions on $[a, b]$ for $0 < \alpha < 1$ then

$$\int_a^b g(x) \odot \boldsymbol{I}^{\alpha}_{\boldsymbol{RL}} f(x)dx = \int_a^b f(x) \odot \boldsymbol{I}^{\alpha}_{\boldsymbol{RL}} g(x)dx$$

It seems that looks like the Fubini theorem for two regular integrals. The left side is,

$$\int_a^b g(x) \odot \boldsymbol{I}^{\alpha}_{\boldsymbol{RL}} f(x)dx = \int_a^b g(x) \odot \left(\frac{1}{\Gamma(\alpha)} \odot \int_x^b (t-x)^{\alpha-1} \odot f(t)dt\right) dx$$

where

$$\boldsymbol{I}^{\alpha}_{\boldsymbol{RL}} f(x) = \frac{1}{\Gamma(\alpha)} \odot \int_x^b (t-x)^{\alpha-1} \odot f(t)dt$$

$$\int_a^b g(x) \odot \boldsymbol{I}_{RL}^{\alpha} f(x)dx = \frac{1}{\Gamma(\alpha)} \odot \int_a^b g(x) \odot \left(\int_x^b (t-x)^{\alpha-1} \odot f(t)dt \right) dx$$

$$= \frac{1}{\Gamma(\alpha)} \odot \int_a^b \int_t^b (t-x)^{\alpha-1} \odot g(x) \odot f(t)dxdt$$

$$= \frac{1}{\Gamma(\alpha)} \odot \int_a^b f(t) \odot \left(\int_t^b (t-x)^{\alpha-1} \odot g(x)dx \right) dt$$

$$= \int_a^b f(t) \odot \left(\frac{1}{\Gamma(\alpha)} \odot \int_t^b (t-x)^{\alpha-1} \odot g(x)dx \right) dt$$

$$= \int_a^b f(t) \odot \boldsymbol{I}_{RL}^{\alpha} g(t)dt := \int_a^b f(x) \odot \boldsymbol{I}_{RL}^{\alpha} g(x)dx$$

6.2.2 Theorem—Fuzzy Fractional Integration by Part I

Let us continuous $f, g : [a, b] \to \mathbb{F}_R$ are two continuous fuzzy functions on $[a, b]$ for $0 < \alpha < 1$ then

$$\int_a^b g(x) \odot \boldsymbol{D}_{C_{gH}}^{\alpha} f(x)dx = f(x) \odot \boldsymbol{I}_{RL}^{(1-\alpha)} g(x)\Big|_a^b \ominus_{gH} (-1) \int_a^b f(x) \odot \boldsymbol{D}_{RL_{gH}}^{\alpha} g(x)dx$$

By using the definition of fuzzy fractional derivative and mentioned above interchanging theorem, the left side is as,

$$\int_a^b g(x) \odot \boldsymbol{D}_{C_{gH}}^{\alpha} f(x)dx = \int_a^b g(x) \odot \left(\frac{1}{\Gamma(1-\alpha)} \odot \int_a^x (x-t)^{-\alpha} \odot f'_{gH}(t)dt \right) dx$$

Since we have,

$$\boldsymbol{D}_{RL_{gH}}^{\alpha} f(x) = \left(\boldsymbol{I}_{RL}^{(1-\alpha)} f \right)'_{gH} (x)$$

Substituting,

$$\int_a^b g(x) \odot \boldsymbol{D}^{\alpha}_{C_{gH}} f(x)dx = \int_a^b g(x) \odot \boldsymbol{I}^{(1-\alpha)}_{\boldsymbol{RL}} f'_{gH}(x)dx = \int_a^b f'_{gH}(x) \odot \boldsymbol{I}^{(1-\alpha)}_{\boldsymbol{RL}} g(x)dx$$

Now using integral by part, we set

$$u = \boldsymbol{I}^{(1-\alpha)}_{\boldsymbol{RL}} g(x), dv = f'_{gH}(x)dx, du = \left(\boldsymbol{I}^{(1-\alpha)}_{\boldsymbol{RL}} g(x)\right)'_{gH} = \boldsymbol{D}^{\alpha}_{\boldsymbol{RL}_{gH}} g(x)$$

Therefore,

$$\int_a^b f'_{gH}(x) \odot \boldsymbol{I}^{(1-\alpha)}_{\boldsymbol{RL}} g(x)dx$$

$$= f(b) \odot \boldsymbol{I}^{(1-\alpha)}_{\boldsymbol{RL}} g(b) \ominus_{gH} f(a) \odot \boldsymbol{I}^{(1-\alpha)}_{\boldsymbol{RL}} g(a) \ominus_{gH} (-1) \int_a^b f(x) \odot \boldsymbol{D}^{\alpha}_{\boldsymbol{RL}_{gH}} g(x)dx$$

$$= f(x) \odot \boldsymbol{I}^{(1-\alpha)}_{\boldsymbol{RL}} g(x)\Big|_a^b \ominus_{gH} (-1) \int_a^b f(x) \odot \boldsymbol{D}^{\alpha}_{\boldsymbol{RL}_{gH}} g(x)dx$$

The proof is completed.

If the right area or part of RL-derivative is used, then another face of integration by part is designed and denoted by integration by part II where,

$$\boldsymbol{D}^{\alpha}_{\boldsymbol{RL}_{gH}} f(x) = \frac{-1}{\Gamma(1-\alpha)} \odot \left(\int_x^b (t-x)^{-\alpha} \odot f(t)dt\right)'_{gH}$$

6.2.3 Theorem—Fuzzy Fractional Integration by Part II

Let us continuous $f, g : [a, b] \to \mathbb{F}_R$ are two continuous fuzzy functions on $[a, b]$ for $0 < \alpha < 1$ then

$$\int_a^b g(x) \odot \boldsymbol{D}^{\alpha}_{\boldsymbol{RL}_{gH}} f(x)dx$$

$$= (-1)g(x) \odot \boldsymbol{I}^{(1-\alpha)}_{\boldsymbol{RL}} f(x)\Big|_a^b \ominus_{gH} (-1) \int_a^b f(x) \odot \boldsymbol{D}^{\alpha}_{C_{gH}} g(x)dx$$

By using the definition of fuzzy fractional derivative and mentioned above interchanging theorem, the left side is as,

$$\int_a^b g(x) \odot \boldsymbol{D}^{\alpha}_{\boldsymbol{RL}_{gH}} f(x)dx = \int_a^b g(x) \odot \frac{-1}{\Gamma(1-\alpha)} \odot \left(\int_x^b (t-x)^{-\alpha} \odot f(t)dt \right)'_{gH} dx$$

We set,

$$u = g(x), dv = \frac{-1}{\Gamma(1-\alpha)} \odot \left(\int_x^b (t-x)^{-\alpha} \odot f(t)dt \right)'_{gH} dx, du = g'_{gH}(x)$$

$$v = \frac{-1}{\Gamma(1-\alpha)} \odot \int_x^b (t-x)^{-\alpha} \odot f(t)dt = (-1)\boldsymbol{I}^{(1-\alpha)}_{\boldsymbol{RL}} f(x)$$

Substituting,

$$\int_a^b g(x) \odot \boldsymbol{D}^{\alpha}_{\boldsymbol{RL}_{gH}} f(x)dx = \int_a^b g(x) \odot \left(\frac{-1}{\Gamma(1-\alpha)} \odot \int_x^b (t-x)^{-\alpha} \odot f(t)dt \right)'_{gH} dx$$

$$= (-1)g(x) \odot \boldsymbol{I}^{(1-\alpha)}_{\boldsymbol{RL}} f(x)\Big|_a^b \ominus_{gH} (-1) \int_a^b \boldsymbol{I}^{(1-\alpha)}_{\boldsymbol{RL}} f(x) \odot g'_{gH}(x)dx$$

$$= (-1)g(x) \odot \boldsymbol{I}^{(1-\alpha)}_{\boldsymbol{RL}} f(x)\Big|_a^b \ominus_{gH} (-1) \int_a^b f(x) \odot \boldsymbol{I}^{(1-\alpha)}_{\boldsymbol{RL}} g'_{gH}(x)dx$$

$$= (-1)g(x) \odot \boldsymbol{I}^{(1-\alpha)}_{\boldsymbol{RL}} f(x)\Big|_a^b \ominus_{gH} (-1) \int_a^b f(x) \odot \boldsymbol{D}^{\alpha}_{\boldsymbol{C}_{gH}} g(x)dx$$

The proof is finished.

6.3 Fuzzy Optimal Control Problem

In this section, after stating some necessary definitions and theorems, the main problem is proposed, i.e., Fuzzy fractional optimal control problem. Furthermore, the necessary conditions for this problem are obtained.

Note In this problem, all the functions are in terms of the time variable t.

Definition (*Fuzzy functional*) The fuzzy rule like J that assigns any fuzzy valued function $x \in \mathbb{F}_R$ to a unique fuzzy number is called fuzzy functional.

Definition (*Increment*) If x and $x \oplus \delta x$ are two fuzzy valued functions the increment of fuzzy functional J is defined and denoted as,

$$\Delta J(x, \delta x) = J(x \oplus \delta x) \ominus_{gH} J(x)$$

Definition (*Variation*) The first variation of a fuzzy functional $J(x)$ is denoted by $\delta J(x \oplus \delta x)$ that includes the terms of $\Delta J(x, \delta x)$ and it is linear in terms of δx.

6.3.1 Definition—Relative Extremum of Fuzzy Functional Function

A fuzzy functional function J has a relative extremum at x^* on its domain, if there is an $\exists \epsilon > 0$ such that for all fuzzy valued functions $\forall x \in \mathbb{F}_R D_H(x, x^*) < \epsilon$,

- $J(x^*)$ is relative minimum, if $\Delta J = J(x) \ominus_{gH} J(x^*) \succcurlyeq 0$
- $J(x^*)$ is relative maximum, if $\Delta J = J(x) \ominus_{gH} J(x^*) \preccurlyeq 0$.

6.3.2 Theorem—Fuzzy Fundamental Theorem of Calculus of Variation

If x^* is a fuzzy extremum then the variation of fuzzy functional $J(x)$ must vanishes on x^* indeed, $\delta J(x^* \oplus \delta x) = 0$ for all x and $x \oplus \delta x$ from the domain of J.

Proof Based on the fuzzy Taylor expansion, in each case of the expansion we have,

$$J(x^* \oplus \delta x) = J(x^*) \oplus \delta x \odot J'(x^*) \oplus R$$

Since x^* is a fuzzy extremum then $J'(x^*) = 0$ (it is shown in Chap. 2). Then we get,

$$J(x^* \oplus \delta x) = J(x^*)$$

This means that, there is no variation or the variation is zero.

$$\delta J(x^* \oplus \delta x) = 0$$

Remark If the function $f(x)$ is a continuous fuzzy function on $[a,b]$ and if $\int_a^b f(x) \odot h(x)dx = 0$ for every continuous fuzzy function $h(x)$ then $f(x) = 0$ on $[a,b]$.

Proof We should show that,

$$\forall h(x)\left(\left(\int_a^b f(x) \odot h(x)dx = 0\right) = 0 \quad \Rightarrow \quad f(x) = 0\right.$$

It looks like such a transformation in which its effect on any function is zero, thus itself is zero transformation.

$$T(\star) := \int_a^b f(x) \odot \star$$

We can suppose that $f(x), x_1, x_2 \in \mathbb{F}_R$ is non-zero such that,

$$h(x) = \begin{cases} (x \ominus_{gH} x_1) \odot (x_2 \ominus_{gH} x), & x \in [x_1, x_2] \\ 0, & \textit{otherwise} \end{cases}$$

Now,

$$\int_a^b f(x) \odot h(x)dx = \int_a^{x_1} f(x) \odot h(x)dx \oplus \int_{x_1}^{x_2} f(x) \odot h(x)dx \oplus \int_{x_2}^{b} f(x) \odot h(x)dx$$

$$\int_a^b f(x) \odot h(x)dx = \int_{x_1}^{x_2} f(x) \odot h(x)dx \succ 0$$

So this is a contradiction.

6.3.3 Necessary Optimality Conditions

Now we focus on our main problem, finding the necessary conditions for fuzzy fractional optimal control problem. The fuzzy fractional optimal control problem can be defined as follows:

Find the fuzzy optimal control $u(t)$ that minimizes the fuzzy performance index

$$J(u) = \int_0^1 F(x, u, t)dt$$

Subject to,

$$\boldsymbol{D}^{\alpha}_{C_{gH}} x(t) = G(x, u, t), \quad x(0) = x_0$$

where $x(t)$ is a fuzzy state variable and t represents the time and F, G are two arbitrary fuzzy functions in which $0 < \alpha < 1$.

In summarize,

Problem Find $u(t)$ as a fuzzy optimal control,

$$\begin{cases} \min J(u) = \int_0^1 F(x, u, t)dt \\ s.t. \\ \boldsymbol{D}^{\alpha}_{C_{gH}} x(t) = G(x, u, t) \\ x(0) = x_0 \end{cases}$$

To find the fuzzy optimal control we follow the traditional approach and define a fuzzy modified performance index as

$$\widetilde{J}(u) = \int_0^1 \left(F(x, u, t) \oplus \lambda \odot \left(\boldsymbol{D}^{\alpha}_{C_{gH}} x(t) \ominus_{gH} G(x, u, t) \right) \right) dt$$

where λ is Lagrange multiplier. Now by using the increment, the variation of $\widetilde{J}(u)$ is determined.

$$\begin{aligned} \Delta \widetilde{J} &= \widetilde{J}(u \oplus \delta u) \ominus_{gH} \widetilde{J}(u) \\ &= \int_0^1 \left(F(x \oplus \delta x, u \oplus \delta u, t) \oplus (\lambda \oplus \delta \lambda) \odot \left(\boldsymbol{D}^{\alpha}_{C_{gH}} (x \oplus \delta x)(t) \ominus_{gH} G(x \oplus \delta x, u \oplus \delta u, t) \right) \right) dt \end{aligned}$$

$$\ominus_{gH} \int_0^1 \left(F(x,u,t) \oplus \lambda \odot \left(\boldsymbol{D}^{\alpha}_{C_{gH}} x(t) \ominus_{gH} G(x,u,t)\right)\right) dt$$

$$= \int_0^1 \left[\left(F(x \oplus \delta x, u \oplus \delta u, t) \ominus_{gH} F(x,u,t)\right) \oplus\right.$$

$$(\lambda \oplus \delta\lambda) \odot \left(\boldsymbol{D}^{\alpha}_{C_{gH}}(x \oplus \delta x)(t) \ominus_{gH} G(x \oplus \delta x, u \oplus \delta u, t)\right) \ominus_{gH} \lambda$$

$$\left.\odot \boldsymbol{D}^{\alpha}_{C_{gH}}(x)(t) \ominus_{gH} G(x,u,t)\right] dt$$

$$= \int_0^1 \left[\left(F(x \oplus \delta x, u \oplus \delta u, t) \ominus_{gH} F(x,u,t)\right) \oplus \left((\lambda \oplus \delta\lambda) \odot \boldsymbol{D}^{\alpha}_{C_{gH}}(x \oplus \delta x)(t) \ominus_{gH} \lambda\right.\right.$$

$$\left.\odot \boldsymbol{D}^{\alpha}_{C_{gH}}(x)(t)\right) \ominus_{gH} (\lambda \oplus \delta\lambda) \odot G(x \oplus \delta x, u \oplus \delta u, t) \ominus_{gH} \lambda$$

$$\odot G(x,u,t)$$

Now using the fuzzy two dimensional first order Taylor expansion for the functions F and G (please see the Chap. 2). We have,

$$\delta \widetilde{J} = \int_0^1 \left(\left[\partial_{xgH} F \odot \delta x \oplus \partial_{ugH} F \odot \delta u\right]\right.$$

$$\oplus \left[\lambda \odot \delta\left(\boldsymbol{D}^{\alpha}_{C_{gH}} x\right) \oplus \delta\lambda \odot \boldsymbol{D}^{\alpha}_{C_{gH}} x\right] \ominus_{gH} \left[\lambda \odot \partial G_{xgH} \odot \delta x \oplus \lambda \odot \partial G_{ugH}\right.$$

$$\left.\left.\odot \delta u \oplus \delta\lambda \odot G\right]\right) dt$$

where $\delta\widetilde{J}$ is the linear part of $\Delta\widetilde{J}$ in terms of δx, δu and $\delta\lambda$. Moreover δx, δu and $\delta\lambda$ are the variations of x, u and λ respectively. Now by using the integration by part I, we get that,

$$\int_0^1 \lambda \odot \delta\left(\boldsymbol{D}^{\alpha}_{C_{gH}} x\right) dt = \int_0^1 \lambda \odot \left(\boldsymbol{D}^{\alpha}_{C_{gH}} \delta x\right) dt$$

$$= \delta x \odot \boldsymbol{I}^{(1-\alpha)}_{\boldsymbol{RL}} \lambda(t)\Big|_0^1 \ominus_{gH} (-1) \int_0^1 \delta x \odot \boldsymbol{D}^{\alpha}_{\boldsymbol{RL}_{gH}} \lambda(t) dt$$

$$= 0 \ominus_{GH} (-1) \int_0^1 \delta x \odot \boldsymbol{D}^{\alpha}_{\boldsymbol{RL}_{gH}} \lambda(t) dt = \int_0^1 \delta x \odot \boldsymbol{D}^{\alpha}_{\boldsymbol{RL}_{gH}} \lambda(t) dt$$

Provided that $\delta x(0) \approx (0,0,0)$ or $\lambda(0) \approx (0,0,0)$ and $\delta x(1) \approx (0,0,0)$ or $\lambda(1) \approx (0,0,0)$. Because $x(0)$ is specified and we have $\delta x(0) \approx (0,0,0)$ and $x(1)$ is not specified, we require that $\lambda(1) = 0$. Now by substituting in $\delta\widetilde{J}$ we have,

$$\delta\widetilde{J} = \int_0^1 \left(\left[\partial F_{xgH} \oplus \left(\boldsymbol{D}^{\alpha}_{\boldsymbol{RL}_{gH}}\lambda\right)\ominus_{gH}\lambda \odot \partial G_{xgH}\right] \odot \delta x \oplus \left[\partial F_{ugH}\ominus_{gH}\lambda \odot \partial G_{ugH}\right] \odot \delta u \right.$$
$$\left. \oplus\left[\boldsymbol{D}^{\alpha}_{\boldsymbol{C}_{gH}}x\ominus_{gH}G\right] \odot \delta\lambda\right)dt$$

In this equation, if we apply the fundamental theorem of calculus of variation then $\delta\widetilde{J} = 0$. Also minimization of $\widetilde{J}(u)$ and hence minimization of $J(u)$ requires that the coefficients of $\delta x, \delta u$ and $\delta\lambda$ be zero. This leads that,

$$\begin{cases} \partial F_{xgH} \oplus \left(\boldsymbol{D}^{\alpha}_{\boldsymbol{RL}_{gH}}\hat{\lambda}\right)\ominus_{gH}\hat{\lambda} \odot \partial G_{xgH} = 0 \\ \partial F_{ugH}\ominus_{gH}\hat{\lambda} \odot \partial G_{ugH} = 0 \\ \boldsymbol{D}^{\alpha}_{\boldsymbol{C}_{gH}}x(t)\ominus_{gH}G(x,u,t) = 0 \\ x(0) = x_0,\, \hat{\lambda}(1) \approx (0,0,0) \end{cases}$$

These equations represent the Euler-Lagrange equations for the fuzzy fractional optimal control problem. These equations give the necessary conditions for the optimality of the fuzzy fractional optimal control problem that is considered here.

$$\begin{cases} \min J(u) = \int_0^1 F(x,u,t)dt \\ s.t. \\ \boldsymbol{D}^{\alpha}_{\boldsymbol{C}_{gH}}x(t) = G(x,u,t) \\ x(0) = x_0 \end{cases}$$

They are very similar to the Euler Lagrange equations for crisp fractional optimal control problem. Determination of the optimal control for the fuzzy fractional system requires solution of Euler-Lagrange equations. In the end, we obtain the necessary conditions for a special case of this problems.

Now consider the following problem: Find the control $u(t)$ that minimizes the quadratic performance index

$$J(u) = \frac{1}{2}\int_0^1 \left[q(t) \odot x^2(t) \oplus r(t) \odot u^2\right]dt$$

where $q(t) \geq 0$ and $u(t) > 0$, and the system whose dynamic is described by the following linear fractional differential equation.

$$\boldsymbol{D}^{\alpha}_{\boldsymbol{C}_{gH}}x(t) = a(t) \odot x(t) \oplus b(t) \odot u, \quad x(0) = 0$$

Comparing with the previous one,

$$F(x,u,t) := \frac{1}{2}q(t) \odot x^2(t) \oplus \frac{1}{2}r(t) \odot u^2$$

And

$$G(x,u,t) = a(t) \odot x(t) \oplus b(t) \odot u(t)$$

Now to have the new conditions for our new problem, we need for satisfaction of all the compared conditions. So,

$$\partial F_{xgH} = q(t) \odot x(t), \partial F_{ugH} = r(t) \odot u(t), \partial G_{xgH} = a(t), \partial G_{ugH} = b(t)$$

Therefor the conditions are considered as,

$$\begin{cases} q(t) \odot x(t) \oplus \left(\boldsymbol{D}^{\alpha}_{\boldsymbol{RL}_{gH}}\lambda\right) \ominus_{gH} \lambda \odot a(t) = 0 \\ r(t) \odot u(t) \ominus_{gH} \lambda \odot b(t) = 0 \\ \boldsymbol{D}^{\alpha}_{\boldsymbol{C}_{gH}} x(t) = a(t) \odot x(t) \oplus b(t) \odot u(t) \\ x(0) = x_0, \lambda(1) \approx (0,0,0) \end{cases}$$

So far, we have provided a theoretical approach to fuzzy fractional optimal control problems, which involves solving fuzzy fractional differential equations. As it is known, solving such equations is in most cases impossible to do, and numerical methods are used to find approximated solutions for the problem.

6.3.4 *Sufficient Optimality Conditions*

Here, it has been proved that the obtained fuzzy necessary optimality conditions are sufficient by considering some appropriate assumptions for the fuzzy fractional optimal control problem.

Theorem (Solution of fuzzy fractional optimal control) *Let us consider* (x^*, u^*, λ^*) *satisfying the following fuzzy necessary optimality conditions,*

$$\begin{cases} \partial F_{xgH} \oplus \left(\boldsymbol{D}^{\alpha}_{\boldsymbol{RL}_{gH}}\lambda\right) \ominus_{gH} \lambda \odot \partial G_{xgH} = 0 \\ \partial F_{ugH} \ominus_{gH} \lambda \odot \partial G_{ugH} = 0 \\ \boldsymbol{D}^{\alpha}_{\boldsymbol{C}_{gH}} x(t) \ominus_{gH} G(x,u,t) = 0 \\ x(0) = x_0, \lambda(1) \approx (0,0,0) \end{cases}$$

And assume that $\delta x = x(t) \ominus_{gH} x^*(t)$ and $\delta u = u(t) \ominus_{gH} u^*(t)$, moreover suppose that,

1. $F(x(t), u(t), t) \ominus_{gH} F(x^*(t), u^*(t), t) \succcurlyeq \partial_{xgH} F(x^*(t), u^*(t), t) \odot \delta x \ominus_{gH} \partial_{ugH} F(x^*(t), u^*(t), t) \odot \delta u$
2. $G(x(t), u(t), t) \ominus_{gH} G(x^*(t), u^*(t), t) \succcurlyeq \partial_{xgH} G(x^*(t), u^*(t), t) \odot \delta x \ominus_{gH} \partial_{ugH} G(x^*(t), u^*(t), t) \odot \delta u$
3. For all $\hat{\lambda} \in [0, 1]$ or G is linear in terms of x and u.

Then $(x^*(t), u^*(t)) := (x^*(t), u^*(t), t)$ is a fuzzy optimal solution to the fuzzy fractional optimal control problem.

Proof For the equation,

$$\partial F_{xgH} \oplus \left(\boldsymbol{D}^{\alpha}_{\boldsymbol{RL}_{gH}} \hat{\lambda} \right) \ominus_{gH} \hat{\lambda} \odot \partial G_{xgH} = 0$$

We deduce that

$$\partial F_{xgH}(x^*(t), u^*(t), t) \oplus \left(\boldsymbol{D}^{\alpha}_{\boldsymbol{RL}_{gH}} \hat{\lambda} \right) \ominus_{gH} \hat{\lambda} \odot \partial G_{xgH}(x^*(t), u^*(t), t) = 0$$

Or from the definition of gH-difference,

$$\partial F_{xgH}(x^*(t), u^*(t), t) = \hat{\lambda} \odot \partial G_{xgH}(x^*(t), u^*(t), t) \ominus_{gH} \left(\boldsymbol{D}^{\alpha}_{\boldsymbol{RL}_{gH}} \hat{\lambda} \right)$$

Using the second condition of necessity,

$$\partial F_{ugH} \ominus_{gH} \hat{\lambda} \odot \partial G_{ugH} = 0$$

We have,

$$\partial F_{ugH}(x^*(t), u^*(t), t) = \hat{\lambda} \odot \partial G_{ugH}(x^*(t), u^*(t), t)$$

From the assumptions,

$$\delta x = x(t) \ominus_{gH} x^*(t), \quad \delta u = u(t) \ominus_{gH} u^*(t)$$

Means that,

$$x(t) = x^*(t) \oplus \delta x, \quad u(t) = u^*(t) \oplus \delta u$$

Also this means the solution (x, u) should be considered as admissible solution in which satisfies the

$$\boldsymbol{D}^{\alpha}_{\boldsymbol{C}_{gH}} x(t) = G(x, u, t), \quad x(0) = x_0$$

In this case we are going to show that $J(u) \ominus_{gH} J(u^*) \succcurlyeq 0$ then u^* will be the minimum solution of the problem. So,

$$\begin{aligned} J(u) \ominus_{gH} J(u^*) &= \int_0^1 \left(F(x(t), u(t), t) \ominus_{gH} F(x^*(t), u^*(t), t)\right) dt \\ &\succcurlyeq \int_0^1 \left(\partial_{xgH} F(x^*(t), u^*(t), t) \odot \delta x \ominus_{gH} \partial_{ugH} F(x^*(t), u^*(t), t) \odot \delta u\right) dt \\ &= \int_0^1 [\left(\lambda \odot \partial_{xgH} G(x^*(t), u^*(t), t) \ominus_{gH} \boldsymbol{D}^{\alpha}_{\boldsymbol{RL}_{gH}} \lambda\right) \odot \left(x(t) \ominus_{gH} x^*(t)\right) \ominus_{gH} \lambda \\ &\quad \odot \partial_{ugH} G(x^*(t), u^*(t), t) \odot \left(u(t) \ominus_{gH} u^*(t)\right)] dt \end{aligned}$$

Using the integrating part II rule and $x(0) = 0$ we get,

$$\begin{aligned} J(u) \ominus_{gH} J(u^*) \succcurlyeq & \int_0^1 \left[\lambda \odot \partial_{xgH} G(x^*(t), u^*(t), t) \odot \delta x\right] dt \\ & \ominus_{gH} \int_0^1 \lambda \odot \left(\boldsymbol{D}^{\alpha}_{\boldsymbol{C}_{gH}} x(t) \ominus_{gH} \boldsymbol{D}^{\alpha}_{\boldsymbol{C}_{gH}} x^*(t)\right) \ominus_{gH} \int_0^1 \left[\lambda \odot \partial_{ugH} G(x^*(t), u^*(t), t) \odot \delta u\right] dt \\ = & \int_0^1 \left[\lambda \odot \partial_{xgH} G(x^*(t), u^*(t), t) \odot \delta x\right] dt \\ & \ominus_{gH} \int_0^1 \lambda \odot \left(G(x(t), u(t), t) \ominus_{gH} G(x^*(t), u^*(t), t)\right) dt \\ & \ominus_{gH} \int_0^1 \left[\lambda \odot \partial_{ugH} G(x^*(t), u^*(t), t) \odot \delta u\right] dt \end{aligned}$$

Now using the condition 2 in the hypotheses,

$$J(u)\ominus_{gH}J(u^*)\succcurlyeq\int_0^1\left[\lambda\odot\partial_{xgH}G(x^*(t),u^*(t),t)\odot\delta x\right]dt$$
$$\ominus_{gH}\int_0^1\left(\lambda\odot\partial_{xgH}G(x(t),u(t),t)\odot\delta x\right)dt$$
$$\oplus\int_0^1\left(\lambda\odot\partial_{ugH}G(x^*(t),u^*(t),t)\delta u\right)dt$$
$$\ominus_{gH}\int_0^1\left[\lambda\odot\partial_{ugH}G(x^*(t),u^*(t),t)\odot\delta u\right]dt=0$$

Now the increment of $J(u)$ is,

$$J(u)\ominus_{gH}J(u^*)\succcurlyeq 0$$

The J has the minimum at $(x^*(t),u^*(t),t)$. In other word, $(x^*(t),u^*(t),t)$ is the optimal solution of fuzzy fractional optimal control problem.

Now some numerical examples for more illustration and effectiveness of our subjects are solved. According to the method, we require to solve a system of fuzzy fractional differential equations to find a solution for the fuzzy fractional optimal control problem. But solving such a system, analytically, is usually impossible except for simple cases. Therefore, we need convenient numerical and computational methods to reach the solution of the fuzzy fractional problem. Here, we only obtain Euler-Lagrange equations for our fuzzy fractional optimal control problem.

Numerical example Consider the problem as follows,

$$\begin{cases}\min J(u)=\frac{1}{2}\int_0^1\left[u\ominus_{gH}\left(\frac{\Gamma(1+\beta)}{\Gamma(1+\beta-\alpha)}t^{\beta-\alpha}\right)\right]^2dt\\ s\cdot t.\\ \boldsymbol{D}^{\alpha}_{C_{gH}}x(t)=u(t)\\ x(0)=(-0.5,0,0.5)\end{cases}$$

The necessary Euler-Lagrange equations are,

$$\begin{cases}\partial F_{xgH}\oplus\left(\boldsymbol{D}^{\alpha}_{RL_{gH}}\lambda\right)\ominus_{gH}\lambda\odot\partial G_{xgH}=0\\ \partial F_{ugH}\Theta_{gH}\lambda\odot\partial G_{ugH}=0\\ \boldsymbol{D}^{\alpha}_{C_{gH}}x(t)\ominus_{gH}G(x,u,t)=0\\ x(0)=x_0,\quad \lambda(1)\approx(0,0,0)\end{cases}$$

where

$$\partial F_{xgH} = 0, \quad G(x,u,t) = u, \quad \partial G_{ugH} = 1, \quad \partial G_{xgH} = 0,$$
$$\partial F_{ugH} = u \ominus_{gH} \left(\frac{\Gamma(1+\beta)}{\Gamma(1+\beta-\alpha)} t^{\beta-\alpha} \right)$$

So,

$$\begin{cases} \boldsymbol{D}^{\alpha}_{RL_{gH}} \lambda = 0 \\ \lambda = u \ominus_{gH} \left(\frac{\Gamma(1+\beta)}{\Gamma(1+\beta-\alpha)} t^{\beta-\alpha} \right) \\ \boldsymbol{D}^{\alpha}_{C_{gH}} x(t) = u \\ x(0) = (-0.5, 0, 0.5), \quad \lambda(1) \approx (0,0,0) \end{cases}$$

At this point, we encounter following fractional boundary value problem that needs to be solved in order to reach the optimal solution of the example,

$$\boldsymbol{D}^{\alpha}_{RL_{gH}} \left(u \ominus_{gH} \left(\frac{\Gamma(1+\beta)}{\Gamma(1+\beta-\alpha)} t^{\beta-\alpha} \right) \right) = 0, \quad x(0) = (-0.5, 0, 0.5),$$
$$\lambda(1) \approx (0,0,0)$$

If $x(t)$ is found out the control variable $u(t)$ can be obtained using equation,

$$\boldsymbol{D}^{\alpha}_{C_{gH}} x(t) = u \quad .$$

Example Consider the following fuzzy problem: find the fuzzy control $u(t)$ which minimizes the fuzzy quadratic performance index,

$$\begin{cases} \min J(u) = \frac{1}{2} \int_0^1 \left[1 \odot x^2(t) \oplus u^2(t) \right]^2 dt \\ s.t. \\ \boldsymbol{D}^{\alpha}_{C_{gH}} x(t) = -1 \odot x(t) \oplus 2 \odot u(t) \\ x(0) = (0,1,2) \end{cases}$$

where $1 = (0,1,2), -1 = (-2,-1,0), 2 = (1,2,3)$. Now the fuzzy necessary conditions implies that,

$$1 \odot x(t) \oplus \boldsymbol{D}^{\alpha}_{RL_{gH}} \lambda \ominus_{gH} \lambda \odot (-1) = 0, u \ominus_{gH} \lambda \odot 2 = 0$$
$$\boldsymbol{D}^{\alpha}_{C_{gH}} x(t) = (-1) \odot x \oplus 2 \odot u, \quad x(0) = (0,1,2), \quad \lambda(1) \approx (0,0,0)$$

The state function $x(t)$, the control $u(t)$ and the Lagrange multiplier $\lambda(t)$ are obtained by solving the fractional differential equations subject to the given conditions. As we mentioned before, we need numerical scheme to obtain approximate solution.

6.4 Fuzzy Fractional Diffusion Equations

In this section the fuzzy solution of the following fuzzy fractional diffusion equation is investigated,

$$\partial^{\alpha}_{tgH}u(x,t) = a \odot \partial^{2}_{xgH}u(x,t)\,, \qquad t > 0, \qquad -\infty < x < \infty$$

where $0 \neq a \in R, 0<\alpha \leq 2$ and

$$\partial^{\alpha}_{tgH}u(x,t) := \boldsymbol{D}^{\alpha}_{C_{gH}}u(x,t) = \frac{1}{\Gamma(2-\alpha)} \odot \int_0^t (t-\tau)^{\alpha} \odot \partial_{\tau gH}u(x,\tau)d\tau$$

where $p - gH$ differentiability of the solution is defined as,

$$\partial_{\tau gH}u(x,\tau) = \lim_{h\to 0}\frac{u(x,\tau+h) \ominus_{gH} u(x,\tau)}{h}$$

To discuss the concept, we need some results about the differentiability of the production of functions in different cases. For instance consider two functions in which the one is real valued function and the second is fuzzy valued function. For the real one, the options are: positive, negative, increasing and decreasing. For the second fuzzy function the options are: $i - gH$ and $ii - gH$ differentiability. In the following remark they are brought briefly. All have been fully discussed in Chap. 4.

Remark Let suppose that the function $f \in \mathbb{F}_R$ is a fuzzy valued function on $[a, b]$ and $h \in R$ is a real valued function. Then the following cases are stablished.

1. f is $i - gH$ differentiable and

   1.1 If h is a positive and increasing function then $h \odot f$ is $i - gH$ differentiable and

$$\left(h(t) \odot f(t)\right)'_{i-gH} = h'(t) \odot f(t) \oplus h(t) \odot f'_{i-gH}(t)$$

$$\int_a^b h(t) \odot f'_{i-gH}(t)\,dt = h(b) \odot f(b) \ominus_H h(a) \odot f(a) \ominus_H \int_a^b h'(t) \odot f(t)\,dt$$

   1.2 If h is a positive and decreasing function then $h \odot f$ is $i - gH$ differentiable and

$$\left(\hbar(t) \odot f(t)\right)'_{i-gH} = \ominus_H (-1)\hbar'(t) \odot f(t) \oplus \hbar(t) \odot f'_{i-gH}(t)$$

$$\int_a^b \hbar(t) \odot f'_{i-gH}(t)\,dt = \hbar(b) \odot f(b) \ominus_H \hbar(a) \odot f(a) \oplus (-1) \int_a^b \hbar'(t) \odot f(t)\,dt$$

1.3 If $\hbar$ is a negative and increasing function then $\hbar \odot f$ is $ii - gH$ differentiable and

$$\left(\hbar(t) \odot f(t)\right)'_{ii-gH} = \ominus_H (-1)\hbar'(t) \odot f(t) \oplus \hbar(t) \odot f'_{i-gH}(t)$$

$$\int_a^b \hbar(t) \odot f'_{i-gH}(t)\,dt =$$

$$\ominus_H (-1)\hbar(b) \odot f(b) \oplus (-1)\hbar(a) \odot f(a) \oplus (-1) \int_a^b \hbar'(t) \odot f(t)\,dt$$

1.4 If $\hbar$ is a negative and decreasing function then $\hbar \odot f$ is $ii - gH$ differentiable and

$$\left(\hbar(t) \odot f(t)\right)'_{ii-gH} = \hbar'(t) \odot f(t) \oplus \hbar(t) \odot f'_{i-gH}(t)$$

$$\int_a^b \hbar(t) \odot f'_{i-gH}(t)\,dt =$$

$$\ominus_H (-1)\hbar(b) \odot f(b) \oplus (-1)\hbar(a) \odot f(a) \ominus_H \int_a^b \hbar'(t) \odot f(t)\,dt$$

2. f is $ii - gH$ differentiable and

2.1 If $\hbar$ is a positive and increasing function then $\hbar \odot f$ is $ii - gH$ differentiable and

$$\left(\hbar(t) \odot f(t)\right)'_{ii-gH} = \ominus_H (-1)\hbar'(t) \odot f(t) \oplus \hbar(t) \odot f'_{ii-gH}(t)$$

$$\int_a^b \hbar(t) \odot f'_{ii-gH}(t)\,dt =$$

$$\ominus_H (-1)\hbar(b) \odot f(b) \oplus (-1)\hbar(a) \odot f(a) \ominus_H \int_a^b \hbar'(t) \odot f(t)\,dt$$

2.2 If $\hbar$ is a positive and decreasing function then $\hbar \odot f$ is $ii - gH$ differentiable and

$$\left(\hbar(t) \odot f(t)\right)'_{ii-gH} = \hbar'(t) \odot f(t) \oplus \hbar(t) \odot f'_{ii-gH}(t)$$
$$\int_a^b \hbar(t) \odot f'_{ii-gH}(t)\, dt =$$
$$\ominus_H (-1)\hbar(b) \odot f(b) \oplus (-1)\hbar(a) \odot f(a) \ominus_H \int_a^b \hbar'(t) \odot f(t)\, dt$$

2.3 If $\hbar$ is a negative and increasing function then $\hbar \odot f$ is $i - gH$ differentiable and

$$\left(\hbar(t) \odot f(t)\right)'_{i-gH} = \hbar'(t) \odot f(t) \oplus \hbar(t) \odot f'_{ii-gH}(t)$$
$$\int_a^b \hbar(t) \odot f'_{ii-gH}(t)\, dt =$$
$$\hbar(b) \odot f(b) \ominus_H \hbar(a) \odot f(a) \ominus_H \int_a^b \hbar'(t) \odot f(t)\, dt$$

2.4 If $\hbar$ is a negative and decreasing function then $\hbar \odot f$ is $i - gH$ differentiable and

$$\left(\hbar(t) \odot f(t)\right)'_{i-gH} = \ominus_H (-1)\hbar'(t) \odot f(t) \oplus \hbar(t) \odot f'_{ii-gH}(t)$$
$$\int_a^b \hbar(t) \odot f'_{ii-gH}(t)\, dt =$$
$$\hbar(b) \odot f(b) \ominus_H \hbar(a) \odot f(a) \oplus (-1) \int_a^b \hbar'(t) \odot f(t)\, dt$$

Remark Let suppose that the function $f, f'_{gH} \in \mathbb{F}_R$ are gH-differentiable fuzzy valued function on $[a, b]$ and $\hbar \in R$ is a monotonic and continuous real valued function. Then the following cases are stablished.

1. If $f(t)$ and $f'_{gH}(t)$ are $i - gH$ differentiable then $\hbar(t) \odot f'_{gH}(t)$ is $i - gH$ differentiable and

$$\left(\hbar(t) \odot f'_{i-gH}(t)\right)'_{i-gH} = \hbar(t) \odot f''_{i-gH}(t) \ominus_H (-1)\hbar'(t) \odot f'_{i-gH}(t)$$

And

$$\int_a^b \hbar(t) \odot \mathfrak{f}''_{i-gH}(t)\,dt =$$

$$\hbar(b) \odot \mathfrak{f}'_{i-gH}(b) \ominus_H \hbar(a) \odot \mathfrak{f}'_{i-gH}(a) \oplus (-1) \int_a^b \hbar'(t) \odot \mathfrak{f}'_{i-gH}(t)\,dt$$

2. If $\mathfrak{f}(t)$ and $\mathfrak{f}'_{gH}(t)$ are $ii-gH$ differentiable then $\hbar(t) \odot \mathfrak{f}'_{gH}(t)$ is $i-gH$ differentiable and

$$\left(\hbar(t) \odot \mathfrak{f}'_{ii-gH}(t)\right)'_{i-gH} = \hbar(t) \odot \mathfrak{f}''_{i-gH}(t) \oplus \hbar'(t) \odot \mathfrak{f}'_{ii-gH}(t)$$

And

$$\int_a^b \hbar(t) \odot \mathfrak{f}''_{i-gH}(t)\,dt =$$

$$\hbar(b) \odot \mathfrak{f}'_{ii-gH}(b) \ominus_H \hbar(a) \odot \mathfrak{f}'_{ii-gH}(a) \ominus_H \int_a^b \hbar'(t) \odot \mathfrak{f}'_{ii-gH}(t)\,dt$$

3. If $\mathfrak{f}(t)$ is $i-gH$ differentiable and $\mathfrak{f}'_{gH}(t)$ are $ii-gH$ differentiable then $\hbar(t) \odot \mathfrak{f}'_{gH}(t)$ is $ii-gH$ differentiable and

$$\left(\hbar(t) \odot \mathfrak{f}'_{i-gH}(t)\right)'_{ii-gH} = \hbar(t) \odot \mathfrak{f}''_{ii-gH}(t) \oplus \hbar'(t) \odot \mathfrak{f}'_{i-gH}(t)$$

And

$$\int_a^b \hbar(t) \odot \mathfrak{f}''_{ii-gH}(t)\,dt =$$

$$\ominus_H (-1)\hbar(b) \odot \mathfrak{f}'_{i-gH}(b) \oplus (-1)\hbar(a) \odot \mathfrak{f}'_{i-gH}(a) \ominus_H \int_a^b \hbar'(t) \odot \mathfrak{f}'_{i-gH}(t)\,dt$$

4. If $\mathfrak{f}(t)$ is $ii-gH$ differentiable and $\mathfrak{f}'_{gH}(t)$ are $i-gH$ differentiable then $\hbar(t) \odot \mathfrak{f}'_{gH}(t)$ is $ii-gH$ differentiable and

$$\left(\hbar(t) \odot \mathfrak{f}'_{ii-gH}(t)\right)'_{ii-gH} = \hbar(t) \odot \mathfrak{f}''_{ii-gH}(t) \ominus_H (-1)\hbar'(t) \odot \mathfrak{f}'_{ii-gH}(t)$$

And

$$\int_a^b \hbar(t) \odot f''_{ii-gH}(t)\, dt =$$

$$\ominus_H (-1)\hbar(b) \odot f'_{ii-gH}(b) \oplus (-1)\hbar(a) \odot f'_{ii-gH}(a)$$

$$\oplus (-1) \int_a^b \hbar'(t) \odot f'_{ii-gH}(t)\, dt$$

Remark Consider $f(t)$ is a continuous and gH-differentiable fuzzy function of exponential $\exp \beta t$ without switching point function for $t \geq 0$. Also suppose $f'_{gH}(t)$ is a fuzzy peace-wise continuous and gH-differentiable on $[a, b]$. Then the Laplace transforms of $f(t)$ and $f'_{gH}(t)$ are listed in the following form for $Re(s) > \beta$.

1. $f(t)$ and $f'_{gH}(t)$ are $i - gH$ differentiable

$$\begin{cases} \boldsymbol{L}\left(f'_{gH}(t)\right) = s \odot F(s) \ominus_H f(0) \\ \boldsymbol{L}\left(f''_{gH}(t)\right) = s^2 \odot F(s) \ominus_H s \odot f(0) \ominus_H f'_{i-gH}(0) \end{cases}$$

2. $f(t)$ and $f'_{gH}(t)$ are $ii - gH$ differentiable

$$\begin{cases} \boldsymbol{L}\left(f'_{gH}(t)\right) = \ominus_H (-1)s \odot F(s) \oplus (-1)f(0) \\ \boldsymbol{L}\left(f''_{gH}(t)\right) = s^2 \odot F(s) \ominus_H s \odot f(0) \oplus (-1)f'_{ii-gH}(0) \end{cases}$$

3. $f(t)$ is $i - gH$ differentiable and $f'_{gH}(t)$ are $ii - gH$ differentiable

$$\begin{cases} \boldsymbol{L}\left(f'_{gH}(t)\right) = s \odot F(s) \ominus_H f(0) \\ \boldsymbol{L}\left(f''_{gH}(t)\right) = \ominus_H (-1)s^2 \odot F(s) \oplus (-1)s \odot f(0) \oplus (-1)f'_{i-gH}(0) \end{cases}$$

4. $f(t)$ is $ii - gH$ differentiable and $f'_{gH}(t)$ are $i - gH$ differentiable

$$\begin{cases} \boldsymbol{L}\left(f'_{gH}(t)\right) = \ominus_H (-1)s \odot F(s) \oplus (-1)f(0) \\ \boldsymbol{L}\left(f''_{gH}(t)\right) = \ominus_H (-1)s^2 \odot F(s) \oplus (-1)s \odot f(0) \oplus (-1)f'_{ii-gH}(0) \end{cases}$$

6.4.1 Remark—Laplace Transform of Caputo Derivative with Order $0<\alpha<2$

If $f(t), f'_{gH}(t)$ are continuous functions on $[0,\infty)$ and $D^{\alpha}_{C_{gH}} f(t)$ is a fuzzy peace-wise continuous Caputo derivative then

1. $f(t)$ and $f'_{gH}(t)$ are $i-gH$ differentiable

$$L\left(D^{\alpha}_{C_{i-gH}} f(t)\right) = s^{\alpha} \odot F(s) \ominus_H s^{\alpha-1} \odot f(0) \ominus_H s^{\alpha-2} f'_{i-gH}(0)$$

2. $f(t)$ and $f'_{gH}(t)$ are $ii-gH$ differentiable

$$L\left(D^{\alpha}_{C_{i-gH}} f(t)\right) = s^{\alpha} \odot F(s) \ominus_H s^{\alpha-1} \odot f(0) \oplus (-1)s^{\alpha-2} f'_{ii-gH}(0)$$

3. $f(t)$ is $i-gH$ differentiable and $f'_{gH}(t)$ are $ii-gH$ differentiable

$$L\left(D^{\alpha}_{C_{ii-gH}} f(t)\right) = \ominus_H (-1)s^{\alpha} \odot F(s) \oplus (-1)s^{\alpha-1} \odot f(0) \oplus (-1)s^{\alpha-2} f'_{i-gH}(0)$$

4. $f(t)$ is $ii-gH$ differentiable and $f'_{gH}(t)$ are $i-gH$ differentiable

$$L\left(D^{\alpha}_{C_{ii-gH}} f(t)\right) = \ominus_H (-1)s^{\alpha} \odot F(s) \oplus (-1)s^{\alpha-1} \odot f(0) \oplus (-1)s^{\alpha-2} f'_{ii-gH}(0)$$

6.4.2 Fundamental Solution of Fuzzy Fractional Diffusion Equation

In this section the fuzzy solution of the following equations is going to be considered. In this equations the order is $0<\alpha\leq 2$ generaly. But in each initial condition the order can be different.

$$\begin{cases} \partial^{\alpha}_{tgH} u(x,t) = a \odot \partial^{2}_{xgH} u(x,t), & t>0, \quad -\infty<x<\infty \\ u(x,0) = p_0 \odot \delta(x), & 0<\alpha\leq 2 \\ u'_{gH}(x,0) = 0, & 1<\alpha\leq 2 \end{cases}$$

where $0 \neq a \in R$, p_0 is a triangular fuzzy number and $\delta(x)$ is the Dirac delta function. The partial fractional derivative with respect to t is also defined as follow such that the Dirac delta function is defined in the measure sense and

$$\delta(x) = \begin{cases} 1, & x \geq 0 \\ 0, & x < 0 \end{cases}$$

And

$$\partial_{tgH}^{\alpha} u(x,t) := \mathbf{D}_{C_{gH}}^{\alpha} u(x,t) = \frac{1}{\Gamma(2-\alpha)} \odot \int_{0}^{t} (t-\tau)^{\alpha} \odot \partial_{\tau gH} u(x,\tau) d\tau$$

where $p - gH$ differentiability of the solution is defined as,

$$\partial_{tgH} u(x,t) = \lim_{h \to 0} \frac{u(x,t+h) \ominus_{gH} u(x,t)}{h}$$

The fuzzy solution for these equations with fuzzy initial conditions is known as the fundamental solution. This fuzzy solution is derived by the fuzzy Laplace transform of a fuzzy function $u(t,x)$ with respect to t,

$$\mathbf{L}_t(u(x,t)) = \int_{0}^{\infty} e^{-st} \odot u(x,t) dt = U_t(x,s)$$

Also consider fuzzy Fourier transform with respect to x for fixed $t > 0$,

$$\mathcal{F}_x\{u(x,t)\} = \int_{-\infty}^{\infty} u(x,t) \odot e^{-iwx} dx = \mathcal{U}_x(w,t)$$

Let $\mathcal{LF}$ denotes the space of all fuzzy number valued functions $u(x,t)$ such that the fuzzy Laplace transform and the fuzzy Fourier transform exist with the following notation,

$$\mathcal{F}_x \mathbf{L}_t(u(x,t)) = \mathcal{F}_x(\mathbf{L}_t(u(x,t))) = \int_{-\infty}^{\infty} \mathbf{L}_t(u(x,t)) \odot e^{-iwx} dx$$

$$= \int_{-\infty}^{\infty} \left(\int_{0}^{\infty} e^{-st} \odot u(x,t) dt \right) \odot e^{-iwx} dx$$

$$= \int_{-\infty}^{\infty} \int_{0}^{\infty} e^{-(s+w)t} \odot u(x,t)\, dt dx := \mathcal{V}(w,s)$$

6.4.3 Theorem—Fundamental Solution

Let $u(x,t)$ and $\partial_{tgH}u(x,t)$ both are $i-p-gH$ or $ii-p-gH$ differentiable with respect to t. Then the fuzzy solution $u(x,t) \in \mathcal{LF}$ is defined as,

$$u(x,t) = \frac{p_0}{2\sqrt{a}t^{\frac{\alpha}{2}}} M\left(\frac{\alpha}{2}; \frac{|x|}{\sqrt{a}t^{\frac{\alpha}{2}}}\right)$$

where

$$M(\alpha; z) := W(-\alpha, 1-\alpha; z) = \sum_{k=0}^{\infty} \frac{(-1)^z z^k}{k!\Gamma[-\alpha k + (1-\alpha)]}$$

The functions $M(\alpha; z)$ and $W(\alpha, \beta; z)$ are Mainardi and Wright functions respectively.

Proof In case $i-p-gH$ differentiability, by applying the Laplace transform for two sides of

$$\begin{cases} \partial_{tgH}^{\alpha}u(x,t) = a \odot \partial_{xgH}^{2}u(x,t) \\ u(x,0) = p_0 \odot \delta(x), & 0 < \alpha \leq 2 \\ u'_{gH}(x,0) = 0\,, & 1 < \alpha \leq 2 \end{cases}$$

We have,

$$\mathbf{L}_t\left(\partial_{tgH}^{\alpha}u(x,t)\right) = \mathbf{L}_t\left(a \odot \partial_{xgH}^{2}u(x,t)\right) = a \odot \partial_{xgH}^{2}\mathbf{L}_t(u(x,t))$$

Since

$$\mathbf{L}_t\left(\partial_{tgH}^{\alpha}u(x,t)\right) = s^{\alpha} \odot \mathbf{L}_t(u(x,t)) \ominus_H s^{\alpha-1} \odot p_0 \odot \delta(x)$$

Thus,

$$s^{\alpha} \odot \mathbf{L}_t(u(x,t)) \ominus_H s^{\alpha-1} \odot p_0 \odot \delta(x) = a \odot \partial_{xgH}^{2}\mathbf{L}_t(u(x,t))$$

Now we can apply the fuzzy Fourier transform,

$$s^{\alpha} \odot \mathcal{F}_x\mathbf{L}_t(u(x,t)) \ominus_H s^{\alpha-1} \odot p_0 \odot \mathcal{F}_x(\delta(x)) = a \odot \mathcal{F}_x\left(\partial_{xgH}^{2}\mathbf{L}_t(u(x,t))\right)$$

where

$$\mathcal{F}_x\mathbf{L}_t(u(x,t)) = \mathcal{V}(w,s), \quad \mathcal{F}_x(\delta(x)) = 1$$

And $\mathcal{F}_x\left(\partial_{xgH}^2 \boldsymbol{L}_t(u(x,t))\right)$ is discovered using the following procedure. We know,

$$\mathcal{F}_x\left(\partial_{xgH}^2 \boldsymbol{L}_t(u(x,t))\right) = \mathcal{F}_x\left(\partial_{xgH}^2 U_t(x,s)\right)$$

Suppose $U_t(x,s) = g(x)$

$$\mathcal{F}_x\left(g''_{gH}(x)\right) = \lim_{\tau\to\infty} \int_{-\tau}^{\tau} e^{-iwx} \odot g''_{gH}(x)\, dx$$

On the other hand,

$$\lim_{\tau\to\infty} \int_{-\tau}^{\tau} e^{-iwx} \odot g''_{gH}(x)\, dx =$$

$$\lim_{\tau\to\infty}\left(\left(e^{-iw\tau} \odot g'_{gH}(\tau) \ominus_H e^{iw\tau} \odot g'_{gH}(\tau)\right) \oplus iw \int_{-\tau}^{\tau} e^{-iwx} \odot g'_{gH}(x)\, dx\right)$$

$$= iw\mathcal{F}_x\left(g'_{gH}(x)\right)$$

Because, $\lim_{|\tau|\to\infty} g'_{gH}(\tau) = 0$. Finally,

$$\mathcal{F}_x\left(g''_{gH}(x)\right) = iw\mathcal{F}_x\left(g'_{gH}(x)\right)$$

Since,

$$\mathcal{F}_x\left(g'_{gH}(x)\right) = \lim_{\tau\to\infty} \int_{-\tau}^{\tau} e^{-iwx} \odot g'_{gH}(x)\, dx$$

$$= \lim_{\tau\to\infty} \int_{-\tau}^{\tau} e^{-iwx} \odot g'_{gH}(x)\, dx =$$

$$= \lim_{\tau\to\infty}\left(\left(e^{-iw\tau} \odot g(\tau) \ominus_H e^{iw\tau} \odot g(\tau)\right) \oplus iw \int_{-\tau}^{\tau} e^{-iwx} \odot g(x)\, dx\right)$$

$$= iw\mathcal{F}_x(g(x))$$

$$\mathcal{F}_x\left(\partial_{xgH}^2 U_t(x,s)\right) = \mathcal{F}_x\left(g''_{gH}(x)\right) = (-1)w^2\mathcal{F}_x(g(x)))$$

Now by substituting,

$$s^{\alpha} \odot \mathcal{F}_x \mathbf{L}_t(u(x,t)) \ominus_H s^{\alpha-1} \odot p_0 \odot \mathcal{F}_x(\delta(x)) = a \odot (-1) w^2 \mathcal{F}_x(g(x)))$$

$$s^{\alpha} \odot \mathcal{F}_x \mathbf{L}_t(u(x,t)) \ominus_H s^{\alpha-1} \odot p_0 = (-1) a w^2 \mathcal{F}_x(g(x)))$$

So we have

$$\begin{aligned} & s^{\alpha} \odot \mathcal{V}(w,s) \ominus_H s^{\alpha-1} \odot p_0 = (-1) a w^2 \odot \mathcal{V}(w,s) \\ & s^{\alpha-1} \odot p_0 = s^{\alpha} \odot \mathcal{V}(w,s) \oplus a w^2 \odot \mathcal{V}(w,s) = \left(s^{\alpha} + a w^2\right) \odot \mathcal{V}(w,s) \\ & \mathcal{V}(w,s) = \left(\frac{s^{\alpha-1}}{s^{\alpha} + a w^2}\right) \odot p_0 \end{aligned}$$

Now by using the inverse Laplace we get,

$$\mathcal{F}_x(u(x,t)) = \frac{p_0}{2\pi i} \int_{\gamma - i\infty}^{\gamma + i\infty} e^{st} \odot \left(\frac{s^{\alpha-1}}{s^{\alpha} + a w^2}\right) ds$$

According to Laplace transform of Mittag-Leffler function, for $|\mu s^{-\alpha}| < 1$,

$$\mathbf{L}(E_{\alpha}(\mu t^{\alpha})) = \frac{s^{\alpha-1}}{s^{\alpha} - \mu}$$

Assuming $\mu = -a w^2$ it is concluded that,

$$\mathcal{F}_x(u(x,t)) = p_0 E_{\alpha}(-a w^2 t^{\alpha})$$

Finally by using inverse Fourier, the solution of our problem

$$\begin{cases} \partial^{\alpha}_{tgH} u(x,t) = a \odot \partial^{2}_{xgH} u(x,t) \\ u(x,0) = p_0 \odot \delta(x), & 0 < \alpha \leq 2 \\ \partial_{tgH} u(x,0) = 0\,, & 1 < \alpha \leq 2 \end{cases}$$

Is found as follows,

$$u(x,t) = \frac{p_0}{2\pi} \int_{\gamma - i\infty}^{\gamma + i\infty} E_{\alpha}(-a w^2 t^{\alpha}) \cos xw \, dw$$

Now, if the order of the inverse fuzzy integral transforms is changed and the inverse Fourier transform is applied first, we have,

$$\mathbf{L}(u(x,t)) = \frac{p_0}{2\pi}\int_{-\infty}^{\infty} \cos xw\left(\frac{s^{\alpha-1}}{s^\alpha + aw^2}\right)dw = \frac{p_0}{2\sqrt{a}} \odot s^{\frac{\alpha}{2}-1}\exp\left(-\frac{|x|}{\sqrt{a}}s^{\frac{\alpha}{2}}\right)$$

By inverse Laplace it is resulted as,

$$u(x,t) = \frac{p_0}{2\sqrt{a}} \odot \mathbf{L}^{-1}\left(s^{\frac{\alpha}{2}-1}\exp\left(-\frac{|x|}{\sqrt{a}}s^{\frac{\alpha}{2}}\right)\right)$$

We have,

$$\mathbf{L}^{-1}\left(s^{\alpha-1}\exp(-as^\alpha)\right) = \frac{1}{t^\alpha}M\left(\alpha;\frac{a}{t^\alpha}\right)$$

So

$$u(x,t) = \frac{p_0}{2\sqrt{a}t^{\frac{\alpha}{2}}}M\left(\frac{\alpha}{2};\frac{|x|}{\sqrt{a}t^{\frac{\alpha}{2}}}\right)$$

Finally we get the final form of the solution in the compact form. This is also the fuzzy fundamental solution in case two derivatives $u(x,t)$ and $\partial_{tgH}u(x,t)$ both are $i-p-gH$ or $ii-p-gH$ differentiable.

Note In case the type of differentials are different the fuzzy fundamental solution can be also obtained as,

$$u(x,t) = \frac{p_0}{2\sqrt{a}t^{\frac{\alpha}{2}}}M\left(\frac{\alpha}{2};\frac{|x|}{\sqrt{a}t^{\frac{\alpha}{2}}}\right)$$

6.4.4 Application of Fuzzy Fractional Diffusion in Drug Release

In this section, the diffusion of an anti-cancer drug in the tumor is considered. The drug is delivered to the patient's body through intravenous injection. If this diffusion process studies for fractals, we have the following fractional diffusion equation,

$$\partial_{tgH}^{2\alpha}u(x,t) = C^{2(2\alpha-1)}D^{2(1-\alpha)}\partial_{xgH}^{2}u(x,t), \qquad (t>0), \qquad 0 \le \alpha \le 1$$

where D is the diffusion coefficient, define the diffusivity of the drug in the tumor. C is the speed of drug delivery to the tumor. The term D is the diffusion coefficient as the property of the tumor is related to a tumor's resistance to the anti-cancer drug. The real value $0 \le \alpha \le 1$ demonstrates the Hurst exponent. Measuring the initial amount of the anti-cancer drug is an uncertain problem and this vagueness may be appearing in the initial conditions. Therefore the fuzzy initial conditions for equation also are considered as follows

$$u(x,0) = p_0 \odot \delta(x), \qquad u'_{gH}(x,0) = 0$$

Here p_0 is a triangular fuzzy number and $\delta(x)$ is Dirac delta function in the sense of measure.

Note If $\alpha = 0$ the equation is defined as non-fractional diffusion equation

$$u(x,t) = C^{-2}D^2\partial^2_{xxgH}u(x,t)$$

If $\alpha = \frac{1}{2}$ then

$$u'_{gH}(x,t) = D\partial^2_{xgH}u(x,t)$$

If $\alpha = 1$ then we have a Poisson equation

$$\partial^2_{tgH}u(x,t) = C^2\partial^2_{xxgH}u(x,t)$$

Based on the mentioned fundamental theorem the fuzzy solution can be found as follows,

$$u(x,t) = \frac{p_0}{2C^{2\alpha-1}D^{1-\alpha}t^{\alpha}} M\left(\alpha; \frac{|x|}{C^{2\alpha-1}D^{1-\alpha}t^{\alpha}}\right)$$

Example Let $u(x,t)$ is the diffusion of Temozolomide in the tumor. Temozolomide is an anti-cancer chemotherapy drug. In adults with Anaplastic Astrocytoma, the dosage is based on medical condition, height, weight, and response to treatment. The fuzzy initial dosage of Temozolomide for a patient with 160 cm height and 87 kg weight is $(284.23, 284.93, 285.02)\mathrm{mg/m^2}$ and it is $(293.2, 293.41, 293.95)\mathrm{mg/m^2}$ for patient with 175 cm height and 80 kg weight which is prescribed once daily for 5 days.

Now, suppose that $D = C = 1$ and $\alpha = \frac{1}{2}$, then by using the diffusion equation,

- For patient with 160 cm height and 87 kg weight we obtain

$$u'_{gH}(x,t) = \partial^2_{xgH}u(x,t), \qquad (t > 0), \qquad 0 \leq \alpha \leq 1$$

the solution is as,

$$u(x,t) = \frac{(284.23, 284.93, 285.02)}{2\sqrt{t}} M\left(\frac{1}{2}; \frac{|x|}{\sqrt{t}}\right)$$

- For patient with 175 cm height and 80 kg weight we obtain the solution is as,

$$u(x,t) = \frac{(293.2, 293.41, 293.95)}{2\sqrt{t}} M\left(\frac{1}{2}; \frac{|x|}{\sqrt{t}}\right)$$

Example Vincristine (VCR) is important for the treatment of acute lymphoblastic leukemia (ALL). ALL is a disease that accounts for approximately one-third of all childhood cancer diagnoses. Let $u(x,t)$ is the diffusion of VCR in the tumor. The fuzzy initial dosage of VCR for patient with 160 cm height and 87 kg weight is $(2.64, 2.66, 2.69)\text{mg/m}^2$ and it is $(2.70, 2.74, 2.82)\text{mg/m}^2$ for patient with 175 cm height and 80 kg weight which is prescribed once daily.

Now, suppose that $D = C = 1$ and $\alpha = \frac{1}{3}$, then by using the diffusion equation,

- For patient with 160 cm height and 87 kg weight we obtain

$$\partial_{tgH}^{\frac{2}{3}} u(x,t) = C^{-\frac{2}{3}} D^{\frac{4}{3}} \partial_{xgH}^{2} u(x,t)$$

the solution is as,

$$u(x,t) = \frac{(2.64, 2.66, 2.69)}{2\sqrt{t}} M\left(\frac{1}{2}; \frac{|x|}{t^{\frac{1}{3}}}\right)$$

- For patient with 175 cm height and 80 kg weight we obtain the solution is as,

$$u(x,t) = \frac{(2.70, 2.74, 2.82)}{2\sqrt{t}} M\left(\frac{1}{2}; \frac{|x|}{t^{\frac{1}{3}}}\right)$$

Bibliography

1. Abdi, M., Allahviranloo, T.: Fuzzy finite difference method for solving fuzzy poisson's equation. J. Intell. Fuzzy Syst 5281–5296 (2019)
2. Ahmady, N., Allahviranloo, T., Ahmady, E.: A modified Euler method for solving fuzzy differential equations under generalized differentiability. Comp. Appl. Math. **39**, 104 (2020). https://doi.org/10.1007/s40314-020-1112-1
3. Ahmadian, A., Senu, F., Larki, S., Salahshour, M., Suleiman, M., Islam, M.S.: A legendre approximation for solving a fuzzy fractional drug transduction model into the bloodsatream. Recent Adv. Soft Comput. Data Mining **287**, 25–34 (2014)
4. Ahmadian, A., Suleiman, M., Salahshour, S.: An operational matrix based on Legendre polynomials for solving fuzzy fractional-order differential equations. Abstr. Appl. Anal. **2013**, 29. (Article ID 505903)
5. Ahmadian, A., Suleiman, M., Salahshour, S., Baleanu, D.: A Jacobi operational matrix for solving a fuzzy linear fractional differential equation. Adv. Differ. Equ. **2013**, 104 (2013)
6. Ahmadian, A., Chang, C.S., Salahshour, S.: Fuzzy approximate solutions to fractional differential equations under uncertainty: operational matrices approach. IEEE Tran. Fuzzy Syst. **25**, 218–236 (2017)
7. Ahmadian, A., Salahshour, S., Amirkhani, H., Baleanu, D., Yunus, R.: An efficient tau method for numerical solution of a fuzzy fractional kinetic model and its application to oil palm frond as a promising source of Xylose. J. Comput. Phys. **264**, 562–564 (2015)
8. Ahmadian, A., Salahshour, S., Ali-Akbari, M., Ismail, F., Baleanu, D.: A Novel Approach to Approximate Fractional Derivative with Uncertain Conditions. Chaos, Solitons and Fractals, Elsevier, **104**, 68–76 (2017)
9. Ahmadian, A., Ismail, F., Salahshour, S., Baleanu, D., Ghaemi, F.: Uncertain viscoelastic models with fractional order: a new spectral tau method to study the numerical simulations of the solution. Commun. Nonlinear Sci. Numer. Simul. **53**, 44–64 (2017)
10. Alinezhad, M., Allahviranloo, T.: On the solution of fuzzy fractional optimal control problems with the Caputo derivative. Inf. Sci. **421**, 218–236 (2017)
11. Allahviranloo, T.: Uncertain information and linear systems. In: Studies in Systems, Decision and Control, vol. 254. Springer (2020). ISBN 978-3-030-31323-4
12. Allahviranloo, T., Abbasbandy, S., Touhparvar, H.: The exact solutions of fuzzy wave-like equations with variable coefficients by a variational iteration method. Appl. Soft Comput. **11** (2), 2186–2192 (2011)
13. Allahviranloo, T., Salahshour, S., Abbasbandy, S.: Explicit solutions of fractional differential equations with uncertainty. Soft Comput. **16**, 297–302 (2012). https://doi.org/10.1007/s00500-011-0743-y

T. Allahviranloo, *Fuzzy Fractional Differential Operators and Equations*, Studies in Fuzziness and Soft Computing 397, https://doi.org/10.1007/978-3-030-51272-9

14. Allahviranloo, T., Abbasbandy, S., Balooch Shahryari, M.R., Salahshour, S., Baleanu, D.: On solutions of linear fractional differential equations with uncertainty. Abstr. Appl. Anal. **2013**. (Article ID 178378). https://doi.org/10.1155/2013/178378
15. Allahviranloo, T., Armand, A., Gouyandeh, Z.: Fuzzy fractional differential equations under generalized fuzzy caputo derivative. J. Intell. Fuzzy Syst. 1481–1490 (2014)
16. Allahviranloo, T., Gouyandeh, Z., Armand, A.: A Full fuzzy method for solving differential equation based on taylor expansion. J. Intell. Fuzzy Syst. 1039–1055 (2015)
17. Allahviranloo, T., Ghanbari, B.: On the Fuzzy Fractional Differential Equation with Interval Atangana–Baleanu Fractional Derivative Approach, vol. 130, pp. 109397. Chaos, Solitons & Fractals (2020)
18. Atangana, A., Baleanu, D.: New fractional derivatives with nonlocal and non-singular kernel: theory and application to heat transfer model. Therm. Sci. **20**, 763 (2016)
19. Armand, A., Allahviranloo, T., Gouyandeh, Z.: Some fundamental results on fuzzy calculus. Iran. J. Fuzzy Syst. **15**(3), 27–46 (2018). https://doi.org/10.22111/ijfs.2018.3948
20. Armand, A., Allahviranloo, T., Abbasbandy, S., Gouyandeh, Z.: Fractional relaxation-oscillation differential equations via fuzzy variational iteration method. J. Intell. Fuzzy Syst. **32**, 363–371 (2017)
21. Bede, B., Gal, S.G.: Generalizations of the differentiability of fuzzy-number-valued functions with applications to fuzzy differential equations. Fuzzy Set Syst. **151**, 581–599 (2005)
22. Chehlabi, M., Allahviranloo, T.: Concreted solutions to fuzzy linear fractional differential equations. Appl. Soft Comput. **44**, 108–116 (2016)
23. Dirbaz, M., Allahviranloo, T.: Fuzzy multiquadric radial basis functions for solving fuzzy partial differential equations. Comp. Appl. Math. **38**, 192 (2019). https://doi.org/10.1007/s40314-019-0942-1
24. Garrappa, R., Kaslik, E., Popolizio, M.: Evaluation of fractional integrals and derivatives of elementary functions: overview and tutorial. Mathematics **7**, 407 (2019)
25. Ghaemi, F., Yunus, R., Ahmadian, A., Salahshour, S., Suleiman, M.B.: Application of fuzzy fractional kinetic equations to modelling of the acid hydrolysis reaction. Abstr. Appl. Anal. **2013**, 19. (Article ID 610314)
26. Long, H., Son, N., Hoa, N.: Fuzzy fractional partial differential equations in partially ordered metric spaces. Iran. J. Fuzzy Syst. **14**(2), 107–126 (2017). https://doi.org/10.22111/ijfs.2017.3136
27. Mathal, A.M., Haubold, H.J.: An Introduction to Fractional Calculus. Nova Sciences Publishers, Inc., New York (2017). ISBN 9781536120639
28. Shabestari, R.M., Ezzati, R., Allahviranloo, T.: Solving fuzzy volterra integro-differential equations of fractional order by Bernoulli Wavelet method. Adv. Fuzzy Syst. **2018**, 11 (Article ID: 5603560). https://doi.org/10.1155/2018/5603560
29. Najafi, N., Allahviranloo, T.: Semi-analytical Methods for solving fuzzy impulsive fractional differential equations. J. Intell. Fuzzy Syst. 353–3560 (2017)
30. Noeiaghdam, Z., Allahviranloo, T., Nieto, J.J.: q-fractional differential equations with uncertainty. Soft Comput. **23**, 9507–9524 (2019). https://doi.org/10.1007/s00500-019-03830-w
31. Najafi, N., Allahviranloo, T.: Combining fractional differential transform method and reproducing kernel Hilbert space method to solve fuzzy impulsive fractional differential equations. Comp. Appl. Math. **39**, 122 (2020). https://doi.org/10.1007/s40314-020-01140-8
32. VanHoa, N., Lupulescu, V., O'Regand, D.: Solving interval-valued fractional initial value problems under Caputo gH-fractional differentiability. Fuzzy Sets Syst. **309**(15), 1–34 (2017)
33. Van Hoa, N.: Fuzzy fractional functional differential equations under Caputo gH-differentiability. Commun. Nonlinear Sci. Numer. Simul. **22**(1–3), 1134–1157 (2015)
34. Van Hoa, N., Ho, V.: Tran Minh Duc, Fuzzy fractional differential equations under Caputo-Katugampola fractional derivative approach. Fuzzy Sets Syst. **375**, 70–99 (2019)

35. Hoa, N.V.: On the initial value problem for fuzzy differential equations of non-integer order \alpha \in (1,2). Soft Comput. **24**, 935–954 (2020). https://doi.org/10.1007/s00500-019-04619-7
36. Senol, M., Atpinar, S., Zararsiz, Z., Salahshour, S., Ahmadian, A.: Approximate Solution of Time-Fractional Fuzzy Partial Differential Equations, Computational and Applied Mathematics, vol. 38, 18 p. Springer, (2019)
37. Salahshour, S., Allahviranloo, T., Abbasbandy, S.: Solving fuzzy fractional differential equations by fuzzy Laplace transforms. Commun. Nonlinear Sci. Numer. Simul. **17**(3), 1372–1381 (2012)
38. Salahshour, S., Ahmadian, A., Salimi, M., Ferarra, M., Baleanu, D.: Asymptotic solutions of fractional interval differential equations with nonsingular kernel derivative, Chaos: an interdisciplinary. J. Nonlinear Sci. AIP **29**, 083110 (2019)
39. Sabzi, K., Allahviranloo, T., Abbasbandy, S.: A fuzzy generalized power series method under generalized Hukuhara differentiability for solving fuzzy Legendre differential equation. Soft Comput. (2020). https://doi.org/10.1007/s00500-020-04913-9
40. Salahshour, S., Ahmadian, A., Ali-Akbari, M., Senu, N., Baleanu, D.: Uncertain fractional operator with application arising in the steady heat flow. Therm. Sci. **23**, 1–8 (2019)
41. Salahshour, S., Ahmadian, A., Abbasbandy, S., Baleanu, D.: M-fractional Derivative Under Interval Uncertainty: Theory, Properties and Applications, vol. 116, pp. 121–125. Chaos, Solitons and Fractals, Elsevier (2018)
42. Salahhsour, S., Ahmadian, A., Ismail, F., Baleanu, D.: A fractional derivative with non-singular kernel for interval-valued functions under uncertainty. Optik Int. J. Light Electron Opt. **130**, 273–286 (2017)
43. Salahshour, S., Ahmadian, A., Ismail, F., Baleanu, D., Senu, N.: A New fractional derivative for differential equation of fractional order under interval uncertainty. Adv. Mechan. Eng. **7** (12), 1687814015619138 (2015)
44. Salahshour, S., Ahmadian, A., Senu, N., Baleanu, D., Agarwal, P.: On analytical solutions of fractional differential equation with uncertainty: application to basset problem. Entropy **17**, 885–902 (2015)
45. Shabestari, M.R.M., Reza, E., Allahviranloo, T.: Numerical solution of fuzzy fractional integro-differential equation via two-dimensional legendre wavelet method. J. Intell. Fuzzy Syst. 2453–2465 (2018)
46. Vu, H, An, T.V., Van Hoa, N.: On the initial value problem for random fuzzy differential equations with Riemann-Liouville fractional derivative: existence theory and analytical solution. J. Intell. Fuzzy Syst. 6503–6520 (2019)
47. Vu, H., An, T.V., Van Hoa, N.: Random fractional differential equations with Riemann-Liouville-type fuzzy differentiability concept. J. Intell. Fuzzy Syst. 6467–6480 (2019)
48. An, T.V., Vu, H., Van Hoa, N.: A new technique to solve the initial value problems for fractional fuzzy delay differential equations. Adv. Differ. Equ. **2017**, 181 (2017). https://doi.org/10.1186/s13662-017-1233-z
49. Vu, H., VanHoa, N.: On impulsive fuzzy functional differential equations. Iran. J. Fuzzy Syst. **13**(4), 79–94 (2016). https://doi.org/10.22111/ijfs.2016.2597
50. Yong, Z.: Basic Theory of Fractional Differential Equations. World Scientific Publishing (1964). ISBN 978-9814579896

Printed in the United States
By Bookmasters